SOLAR AND INFRARED
RADIATION
MEASUREMENTS

Frank Vignola
Joseph Michalsky
Thomas Stoffel

CRC Press
Taylor & Francis Group
Boca Raton London New York

CRC Press is an imprint of the
Taylor & Francis Group, an **informa** business

CRC Press
Taylor & Francis Group
6000 Broken Sound Parkway NW, Suite 300
Boca Raton, FL 33487-2742

First issued in paperback 2017

© 2012 by Taylor & Francis Group, LLC
CRC Press is an imprint of Taylor & Francis Group, an Informa business

No claim to original U.S. Government works

ISBN 13: 978-1-4398-5189-0 (hbk)
ISBN 13: 978-1-138-07552-8 (pbk)

Library of Congress Cataloging-in-Publication Data

Vignola, Frank.
    Solar and infrared radiation measurements / Frank Vignola, Joseph Michalsky, Thomas Stoffel.
      p. cm.
    Summary: "Written for students and professionals, this reference explores the various types of solar radiation measurements and how they operate. The book gives the reader a step-by-step approach to the set-up, operation, and maintenance of a solar monitoring station. The book provides the background and terminology needed to understand the uses and requirements of various solar monitoring radiometers, and it gives an overview of the various solar radiation instruments. It also discusses the calibration and maintenance necessary for instruments in the lab and field, as well as covering the traceability of calibrations to international standards"-- Provided by publisher.
    Includes bibliographical references and index.
    ISBN 978-1-4398-5189-0 (hardback)
    1. Solar radiation--Measurement. 2. Radiometers. I. Michalsky, Joseph. II. Stoffel, Thomas L. III. Title.

QC912.V56 2012
523.7'2--dc23
                                                  2012014563

Visit the Taylor & Francis Web site at
http://www.taylorandfrancis.com

and the CRC Press Web site at
http://www.crcpress.com

Dedication

---

*The authors would like to dedicate this book to our wives,
Mary Lou Vignola, Randi Michalsky, and Ann Stoffel, who provided
encouragement and support while we worked on the manuscript.*

Contents

Preface

The rather specialized field of solar and infrared radiation measurements has become increasingly important due to the increased demands by the renewable energy and climate change research communities for data with higher accuracy and increased temporal and spatial resolutions. Recent advances in radiometry, measurement systems, and information dissemination have also increased the need for refreshing the literature available for this topic.

The objective of this book is to provide the reader with an up-to-date review of the important aspects of solar and infrared radiation measurements: radiometer design; equipment installation, operation, maintenance, and calibration; data quality assessment parameters; and the knowledge necessary to properly interpret and apply the measured data to a variety of topics. Each of the authors has more than 30 years of experience with this subject, primarily as the result of developing and operating multiple measurement stations, working with the industry to improve radiometry, and conducting various research projects.

The book's scope and subject matter have been designed to help a wide audience understand the general subject and to serve as a technical reference. A student new to the field will benefit from the review of terminology and the historical perspective for radiometry before addressing the other topics that we hope will be of interest to the more experienced reader.

Acknowledgments

The authors would like to thank the sponsors of their organizations for the support over the years that helped build the experience and expertise that went into this book.

Sponsors of the University of Oregon Solar Radiation Monitoring Laboratory that should be recognized for their support are the Bonneville Power Administration, Energy Trust of Oregon, Emerald People's Utility District, Oregon Built Environment & Sustainable Technologies Center, and Eugene Water and Electric Board. We would also like to thank the University of Oregon and Luminate, LLC, for encouraging our participation in this book.

In addition to the primary sponsorship of the Global Monitoring Division's Radiation Group by the National Oceanic and Atmospheric Administration, considerable support comes from the U.S. Department of Energy (DOE) Office of Science (BER) with additional contributions from the National Aeronautics and Space Administration, the U.S. Department of Agriculture through Colorado State University, and the Environmental Protection Agency.

The DOE BER Atmospheric Radiation Measurement Program and the DOE Office of Energy Efficiency and Renewable Energy provide support for the Solar Resources and Forecasting Group at the National Renewable Energy Laboratory (NREL) and its development and operation of the Solar Radiation Research Laboratory.

Individuals who have generously given advice or aided with the figures during the writing of this book include John Augustine, Victor Cassella, Chris Cornwall, Patrick Disterhoft, Robert Dolce, Ellsworth Dutton, Chris Gueymard, Peter Harlan, Bobby Hart, John Hickey, Rich Kessler, Gary Hodges, Ed Kern, Peter Kiedron, Tom Kirk, Vikki Kourkouliotis, Kathy Lantz, Fuding Lin, Tilden Meyers, Daryl Myers, Richard Perez, Ibrahim Reda, Ruud Ringoir, Angie Skovira, and Steve Wilcox.

About the Authors

Frank Vignola is the director of the University of Oregon Solar Energy Center and runs the Solar Radiation Monitoring Laboratory (SRML). He received his B.A. in physics at the University of California–Berkeley in 1967 and his Ph.D. in physics at the University of Oregon in 1975, with his thesis on elementary particle physics. He decided to apply his skills to more practical applications and started working in solar energy in 1977 at the University of Oregon. Dr. Vignola helped establish and manage the SRML solar radiation monitoring network that has created the longest-running high-quality solar radiation data set in the United States. He has organized and participated in a number of workshops on solar resource assessment and has written and contributed to approximately 100 papers on solar resource assessment. He was technical chair of the 1994 and 2004 conferences of the American Solar Energy Society (ASES) and for several years chaired the Technical Review Committee for ASES that managed the publication of a number of white papers on solar energy. He is currently associate editor for solar resource assessment for the *Solar Energy Journal*. To help facilitate the use of solar energy, he created and maintains a solar resource assessment website that is accessed by over 100,000 users a year. In addition, he has served on the boards of ASES and the Solar Energy Industries Association and is past president of the Oregon Solar Energy Industries Association (OSEIA) and a current OSEIA board member. Dr. Vignola has helped pass solar tax credit and net metering legislation in Oregon and was author of Oregon's law requiring 1.5% of the capital for solar in new public buildings.

Joseph Michalsky is a physical scientist with the Global Monitoring Division (GMD) in the Office of Oceanic and Atmospheric Research (OAR) at the National Oceanic and Atmospheric Administration (NOAA). When he first came to NOAA in 2003, he managed the Surface Radiation Research Branch within the Air Resources Laboratory of OAR before its merger with GMD. Prior to that he was senior research associate with the Atmospheric Sciences Research Center at the State University of New York–Albany. He began his career with Battelle at what is now the Department of Energy's Pacific Northwest National Laboratory. Dr. Michalsky received his B.S. in physics at Lamar University and M.S. and Ph.D. in physics at the University of Kentucky. His early career focused on astronomical research before taking on problems in solar energy. His primary focus in the last several years has been in the atmospheric sciences. Dr. Michalsky has nearly 100 refereed publications in astronomy, solar energy, and atmospheric science.

Thomas Stoffel is a principal group manager at the National Renewable Energy Laboratory (NREL) operated by the U.S. Department of Energy. He currently manages the Solar Resources and Forecasting Group at NREL. This 12-member group of scientists and engineers is responsible for the development and dissemination of solar resource information for the advancement of solar technologies and climate change research. He received his B.S. in aerospace engineering from the University of Colorado–Boulder and his M.S. in meteorology from the University of Utah. Stoffel began his professional career as an aerospace engineer at the U.S. Air Force Propulsion Laboratory simulating gas turbine engine performance and infrared radiation signatures. He returned to school to pursue his interests in radiative transfer and atmospheric science. Upon graduation, he worked at what is now the National Oceanic and Atmospheric Administration Earth System Research Laboratory to analyze urban–rural differences in solar radiation. In 1978, Stoffel began his career in renewable energy at the Solar Energy Research Institute (now NREL). There, he developed the Solar Radiation Research Laboratory (SRRL), which continues to provide research-quality solar resource measurements, radiometer calibrations, and systems for data acquisition and quality assessment (http://www.nrel.gov/solar_radiation). Stoffel has authored and contributed to more than 80 publications and continues to provide radiometry expertise for the U.S. Department of Energy Atmospheric Radiation Measurement Program Climate Research Facility installations and the World Meteorological Organization Baseline Surface Radiation Network.

1 Measuring Solar and Infrared Radiation

If the observation of the amount of heat the sun sends the earth is among the most important and difficult in astronomical physics, it may also be termed the fundamental problem of meteorology, nearly all whose phenomena would become predictable, if we knew both the original quantity and kind of this heat.

Samuel Pierpont Langley
(1834–1906)

Solar radiation measurements are the basis for understanding Earth's primary energy source. Humans have studied the sun and applied their understanding to improve their lives since the dawn of history. Buildings have been designed to take advantage of solar radiation for daylighting and heating since the time of Greek city states, if not earlier. Today, solar-generating facilities that are 200–300 MW in size are under construction; 17,000 MW of peak solar-generating capacity were installed in 2010 with the equivalent annual power output of four to five large nuclear or coal power plants. The world is fitfully preparing for the solar age when oil will be depleted and environmentally benign sources of energy will be needed. Today a wide variety of solar technologies are being used, and new and improved solar technologies are being explored in the lab and tested in the field. The sun's energy is free and available to everyone, and the technologies that turn solar irradiance into useable energy are dropping in price as the solar industry grows and production increases. Given the right financial structure, many solar technologies have found cost-effective markets. For example, a cost-effective project to replace kerosene lamps with photovoltaic (PV) panels and batteries in the Dominican Republic has been successful (Perlin, 1999). Since 1985, the Luz plants in California, with 354 MW of generating capacity, have been producing electricity from solar thermal–electric power facilities. The future for the solar industry is bright as we work to develop sustainable energy sources for the growing world population while reducing deleterious effects of greenhouse gas emissions.

Part of the fundamental infrastructure that supports the growing solar industry includes solar resource assessment and forecasting. As in hydroelectric generation where stream flow data are necessary to design and operate the facility, knowledge of the temporal and spatial behavior of the solar resource is critical to the design and operation of solar electric-generating facilities. With multimillion and even billion dollar solar projects under development, the need for high-quality solar data is imperative. Solar data are needed in the planning and design of the facility as well as for its operation after completion. Testing new technologies to characterize and validate the design improvements requires the highest-quality solar irradiance

1

measurements because a 1–2% increase in efficiency can translate into millions of dollars in cost reductions. Developers and financial institutions are interested in the economic viability of projects and the ability of the facility to service its debt in even the cloudiest years. The better the solar resource information available, the more accurately a project's performance can be estimated, reducing uncertainty and risk for the investors. Lower risks translate into lower interest rates, which result in lower costs to the ratepayers.

The solar industry is not the only group interested in the solar resource. For decades, farmers have used solar data to evaluate their irrigation requirements. Climate and atmospheric scientists use solar resource data in their studies. Solar forecasts based on weather forecasts from the National Weather Service are being used now, although there is a need for considerable improvement in this field. Architects are using sunlight to reduce energy demands on buildings, and some buildings are being constructed that have a zero net energy balance.

There is a broad community of users of solar data and groups who supply this information. Many of these groups are federal agencies tasked to provide these data. Scientists are running solar measurement networks to study the climate and weather. Industry is prospecting for sites with "bankable" solar resource data to convince the financial community of proposed project viability. Private groups and individuals are making measurements to evaluate the performance of solar systems. Some schools are integrating the solar resource data into their science and engineering curricula. To accommodate the needs of this broad group of interested parties with their diverse requirements, a wide variety of solar instruments is available. The price and quality of these instruments vary widely, with the more expensive equipment usually providing more accurate results. This book provides an extensive survey of the solar instrumentation that is available and attempts to provide general performance characteristics that can help the reader select the appropriate instrument for the task and provide information necessary to assess the accuracy and quality of the data produced.

The main body of this book describes the solar irradiance sensors and how they function while providing a description of the type of solar radiation they are designed to measure. Useful information on solar resource assessment and auxiliary information are contained in the appendices. Chapter 2 describes the nature of solar radiation and provides the terminology and basic equations used in solar resource assessment. A thorough understanding of this background information is needed to fully comprehend the information in the following chapters. Chapter 3 is a history of the development of solar instrumentation. This chapter is useful for understanding the difficulty in obtaining accurate solar radiation measurements, and it shows the considerable improvements that have been made over the years. Some of these instruments are still in use today, and understanding historic data requires knowledge of how these instruments operated and performed. Those who want to use these older data should become familiar with the instruments that provided the data and understand their limitations.

Chapters 4, 5, and 6 cover basic solar radiation measurements and instrumentation. Chapter 4 discusses instruments used to measure direct normal irradiance (DNI) or "beam" irradiance (radiation coming directly from the sun). Astronomical-based calculations tell us exactly where the sun is in the sky, and we can calculate the coordinates and angles associated with the direct solar radiation to a high degree of accuracy. This knowledge plus the fact that most of the solar radiation received comes directly from the sun enhances the need to make sound direct normal irradiance measurements. Besides describing the instruments used to make DNI measurements, Chapter 4 discusses issues concerning the calibration and accuracy of these measurements. Chapter 5 discusses the instruments used to measure the total irradiance on a horizontal surface, referred to in this book as global horizontal irradiance (GHI). Due to their relative ease of use, global irradiance measurements are by far the most common and are useful to a wide audience from farmers to scientists. There are three types of instruments used to measure GHI, and these are characterized and evaluated in Chapter 5. The measurement of diffuse irradiance is covered in Chapter 6. While diffuse measurements use the same type of instruments that are used to measure global irradiance, only a limited number of these instruments are suitable for measuring diffuse irradiance. The reason for selecting certain instruments for measuring diffuse horizontal irradiance (DHI) will be covered in more detail in Chapter 6. In all three chapters, the uncertainties associated with the measurements and the methodology used to determine the uncertainties in the measurements are discussed. Chapter 7 discusses the rotating shadowband radiometer (RSR). This single instrument measures the global and diffuse horizontal irradiance and calculates the corresponding direct normal irradiance. The accuracy, advantages, and limitations of this approach are discussed.

Measurements of solar irradiance on tilted surfaces are discussed in Chapter 8. Most solar collectors are tilted to better intercept the incident solar radiation. Some of the instruments suitable for global or diffuse horizontal measurements are unsuitable for tilted irradiance measurements. The problems associated with tilted irradiance measurements are discussed, and the performance characteristics of sensors when tilted are described. This is an important topic as solar system performance is gauged against tilted surfaces measurements. Measurements of surface albedo (ratio of reflected to incident irradiance) are discussed in Chapter 9. Albedo is an important consideration for models that utilize satellite measurements because the albedo determines the amount and nature of the surface-reflected irradiance. In addition, calculations of irradiance on tilted surfaces also depend on characterizing the surface albedo. The amount of reflected global irradiance is dependent on the albedo of the surface in front of the collector, and albedos can vary widely from 10 to 90% reflectance depending on the surface characteristics.

Measurement of infrared (IR) irradiance is covered in Chapter 10. The earth's surface and the atmosphere emit IR radiation, and the amount of IR radiation emitted is proportional to the temperature and emissivity of the emitter. Most solar instruments do not detect IR radiation; special instruments are needed to measure the incoming and outgoing IR irradiance.

Instruments that measure net radiation are discussed in Chapter 11. Net radiation is the difference between the incident solar and infrared radiation from the sun and sky minus the solar and infrared irradiance from the ground. When studying the earth's energy balance, which drives weather and climate, good measurements of the net irradiance are essential. Net radiation is measured many ways, and accurate measurements of net radiation are extremely difficult.

Spectral measurements of solar radiation are cover in Chapter 12. Incident solar radiation is composed of different wavelengths of light, and many natural processes from photosynthesis to human vision use different portions of the solar spectrum. In addition, the performance of solar (photovoltaic, or PV) cells is dependent on the spectral composition of the incident solar radiation as well as the magnitude of the incoming irradiance. Instruments that measure the spectrum of incident solar radiation are expensive and sometimes difficult to maintain. However, good spectral measurements are of great scientific value while also having practical applications. Relatively inexpensive instruments are available to measure selected portions of the spectrum, such as that to which the eye is sensitive for daylighting applications. Instruments that provide high-resolution spectral data over most of the spectrum have considerably higher costs, but deliver a much more useful and flexible product.

Nonradiation meteorological measurements are covered in Chapter 13. Instruments that measure temperature, pressure, relative humidity, wind speed, and wind direction are discussed. These meteorological measurements are important because some parameters affect the performance of certain radiometers; for example, both thermopile and photodiode radiometers are sensitive to the ambient temperature, and a temperature correction can be used to improve the accuracy of measurements by reducing the temperature bias in the measurement. Analysis of photovoltaic panel performance is another use of meteorological measurements. PV panel performance mainly depends on the incident solar radiation, but temperature affects its power production.

Important factors to consider before establishing a quality solar monitoring station are discussed in Chapter 14. Even with the best instruments, a well-designed solar monitoring station is essential to ensure the accuracy, validity, and completeness of the data. Siting to minimize blockage by nearby obstructions and to facilitate maintenance of the station are two key factors. Good record keeping and routine calibration of the station instruments are also essential actions to ensure high-quality solar irradiance data. To get full value from solar measurements, good design, adequate maintenance, and validation of the incoming data are required.

Appendices A to F cover various topics related to modeling solar irradiance, conversion factors, and manufacturers' information that is useful for solar monitoring. Models used to estimate solar radiation are discussed in Appendix A. These include correlations between various irradiance components and models that are used to estimate irradiance on tilted surfaces. The purpose of Appendix A is to present the most common models and give an idea of how they work. These models are also useful in understanding why certain measurements are made and why they are useful. Appendix B discusses the use of satellite imagery to estimate the large-scale spatial variability of solar radiation. This appendix is an overview of how satellite-based remote sensing models work and their accuracy. Appendix C presents a series of sun path charts for various latitudes. This appendix can serve as a quick reference to the sun path at a given site and illustrates where the sun will be at any given time of the day and year. Appendix D discusses the algorithms used to calculate the position of the sun. Knowing the solar position is always useful when analyzing data, designing a solar monitoring station, or applying solar radiation data. Many different units have been used over the years to describe solar radiation: Appendix E contains a table of useful conversion factors. This table provides information on how to convert from one unit to another. A list of solar instrument manufacturers is given in Appendix F. This list is fairly extensive and includes the major sources of solar instrumentation and provides a basis for searching for available instruments.

A sample calibration report from the National Renewable Energy Laboratory is given in Appendix G. This Broadband Outdoor Radiation Calibration (BORCAL) report illustrates what is included in a comprehensive calibration of a solar sensor. Sunshine duration is discussed briefly in Appendix H. This was one of the first proxies for measurements of solar irradiance, and it is still made in many parts of the world. Although it has been superseded by actual measurements of solar irradiance, it is included for completeness. Finally, Appendix I covers some helpful troubleshooting hints that identify common problems in radiometry and how they might be addressed.

The goal of this instrumentation book is to pass on the experience that the authors have acquired over the years of setting up solar monitoring sites, measuring the solar resource, analyzing the data, and working with a wide variety of users in need of accurate solar resource information. It takes a lot of effort and expense to build a quality solar radiation database. From our collective experience we have learned that proper documentation and archiving of the measured solar irradiance and relevant meteorological data are critically important to the full and appropriate use of the information. Additionally, funding for solar resource assessment comes in waves, and it takes considerable planning, ingenuity, and determination to keep a measurement record intact to produce a long-term quality product that meets the needs of generations to come.

QUESTIONS

1. Name two uses of solar radiation data.
2. Why do you want to learn about solar measurements?
3. Who would want more accurate solar data, a farmer scheduling his irrigation or a banker financing a solar concentrating power plant? Explain your answer.
4. To which chapter or appendix in this book would you go to find a way to calculate the sun's position at any time?

REFERENCE

Anon. 2011. Copper Mountain Solar, the largest photovoltaic solar plant in the U.S., *Solar Thermal Magazine*. http://www.solarthermalmagazine.com (retrieved 3/4/12).

Neville, A. 2010, December. Top Plant: DeSoto next generation solar energy center, DeSoto County, Florida. *Power Magazine*, 1. http://www.powermag.com (retrieved 3/4/12).

Perlin, J. 1999. *From space to Earth: The story of solar electricity*. Cambridge, MA: Harvard University Press.

2 Solar Resource Definitions and Terminology

Everything should be made as simple as possible, but not simpler.

Albert Einstein
(1879–1955)

2.1 INTRODUCTION

The purpose of this chapter is to acquaint the reader with basic terminology and concepts needed to understand the measurement of incoming solar and infrared radiation. A methodology to calculate the position of the sun in the sky is described along with the specialized vocabulary and parameters employed in the calculation. The spectral nature of solar radiation, useful concepts relevant to heat transfer, and basic ideas required for understanding photodiodes are briefly discussed.

The sun has fascinated humans since the dawn of recorded history. Striving to understand the heavens and the stars has led some civilizations to worship spirits that cause the rebirth of the sun every morning and the setting of the sun at night. Ancient astronomers created solar calendars to track the changing path of the sun across the sky, and primitive civilizations built monuments such as Stonehenge to mark the equinoxes and to give clues to changing seasons.

This fascination was driven not by idle curiosity, but by the real needs to know when to prepare for planting and when to hold rituals to ensure a good harvest or celebrate holy days. More modern societies, such as the Greeks, had laws protecting access to the sun. The Anasazi in the southwestern United States designed their cliff-side dwellings to make optimal use of the solar resource.

Before going into modern-day instruments used to measure incoming solar radiation, it is worthwhile to discuss a few basics concepts about the sun, how it presents itself in the sky, and the terminology commonly used to describe solar irradiance. In addition, concepts and fundamentals of heat transfer and the fundamentals of photodiodes are discussed to serve as the basis for discussions of solar radiation sensors in later chapters.

2.2 THE SUN

The sun is considered an *average*-sized star. Nuclear fusion deep within the solar interior powers the sun as hydrogen combines to form helium. Gravity keeps the sun from blowing apart, and the nuclear fusion cycle has kept the sun radiating energy to space at a fairly constant rate for billions of years. The effective temperature of the sun is 5778 K (5505°C or 9941°F). This means that the sun radiates similar to, although not exactly as, a black body heated to 5778 K.

The earth is in a slightly elliptical orbit with an average separation of 149,598,106 km (92,955,888 miles) from the sun. Since radiant energy intensity decreases in proportion to the inverse of the distance squared from the source, the earth's surface temperature is much cooler than the sun with a mean temperature of about 287 K (14°C or 57°F). Note that without the greenhouse effect of the earth's atmosphere the earth would be even cooler.

2.3 EXTRATERRESTRIAL RADIATION

Solar radiation incident just above the earth's atmosphere is called extraterrestrial radiation. On average the accepted extraterrestrial irradiance is 1366 watts/meter^2 (Wm^{-2}). The solar output, called the total solar irradiance (TSI), varies by about 0.1% in proportion to the changing number of sunspots over the 11-year solar cycle. The term TSI is now used to refer to the extraterrestrial radiation instead of the previously used term solar constant (I_0), an acknowledgment of the very slight variation in the sun's output. The absolute accuracy to which the TSI can be measured is around ±0.5% (Fröhlich, 2006).

The eccentricity of the earth's elliptical orbit is 0.0167 and results in the earth being closest to the sun (perihelion) in early January and farthest from the sun (aphelion) in early July. The earth and the sun are 1.67% closer at perihelion and 1.67% farther apart at aphelion than their mean separation. Since the solar intensity is inversely proportional to the square of the distance from the sun, the extraterrestrial irradiance impinging on the earth varies over a range of 6.7% as the earth orbits the sun. The extraterrestrial irradiance outside the earth's atmosphere on a surface normal to the sun is

$$DNI_o = I_o \cdot (R_{av}/R)^2 [Wm^{-2}] \tag{2.1}$$

where R_{av} is the mean sun–earth distance, and R is the actual sun–earth distance depending on the day of the year. An approximate equation for $(R_{av}/R)^2$ is

$$(R_{av}/R)^2 = 1.000110 + 0.034221 \cdot \cos(\beta) + 0.001280 \cdot \sin(\beta)$$
$$+ 0.000719 \cdot \cos(2\beta) + 0.000077 \cdot \sin(2\beta) \tag{2.2}$$

where $\beta = 2\pi d/365$ radians, and d is the day of the year (Spencer, 1971). For example, January 15 is the fifteenth day of the year, and February 15 is the forty-sixth day of the year. Since there are 366 days in a leap year, there is a slight difference in the calculation in leap years. To calculate the extraterrestrial global horizontal irradiance, DNI$_o$ is multiplied by the cosine of the solar zenith angle (see Figure 2.6). Information on calculating hourly, daily, or monthly average DNI$_o$ and the extraterrestrial global horizontal irradiance (GHI$_o$) are given in appendix A.

2.4 SOLAR COORDINATES

The earth's axis is tilted approximately 23.44° with respect to the plane of the earth's orbit around the sun. As the earth moves around the sun, its rotational axis is fixed if viewed from space (Figure 2.1). In June, the orientation of the axis is such that the northern hemisphere is pointed toward the sun. In December, the earth's position is at the opposite end of the orbit around the sun, and the orientation of the earth's axis is such that the northern hemisphere is pointed away from the sun. During the spring and fall equinoxes, the earth's axis is perpendicular to a line drawn between the earth and the sun. The mean geometric orbit of the earth around the sun defines the *ecliptic* plane. For an eclipse to occur, the orbit of the new (solar) or full (lunar) moon has to be very near to the ecliptic plane, hence the name.

From an Earth-centered point of view, the sun moves among the other stars in the heavens on the celestial sphere (see Figure 2.2). The observer on Earth is stationary, and the objects in the heavens look as if they are on the celestial sphere rotating at a rate of just under 360° per day or, equivalently, 15°/hr. The plane running through the earth's equator is designated the celestial equator (or equatorial plane), and the (celestial) polar axis is the axis of rotation of the celestial sphere around the earth. This plane and axis extend to infinity. The position of any star on the celestial sphere, the sun in this discussion, can be determined by specifying two terms: the *right ascension* and the *declination*. From the ground-based observer, the stars or sun look like they are moving in a plane that is parallel to the equatorial plane at a rate of about 15°/hr. Unlike distant stars, the right ascension of the sun changes from 0° to 360°, or 0^h to 24^h as a consequence of the fact that the earth orbits the sun once a year. The two points on the equatorial plane where the plane of the earth's orbit around the sun, the ecliptic, intersects the equatorial plane are called the equinoxes. At the equinoxes the declination is equal to zero. The right ascension of the sun is zero at the vernal equinox.

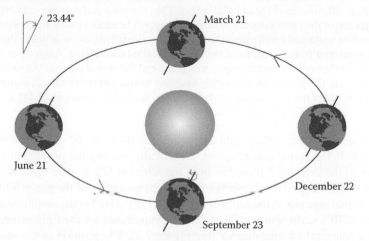

FIGURE 2.1 The earth circles the sun in an elliptical orbit. The tilt of the earth's axis with respect to the orbital plane is about 23.44°. As the earth orbits the sun, the northern hemisphere tilts toward the sun in June and away from the sun in December. On average the earth rotates just under once every 24 hours. (Courtesy of Angie Skovira of Luminate, LLC.)

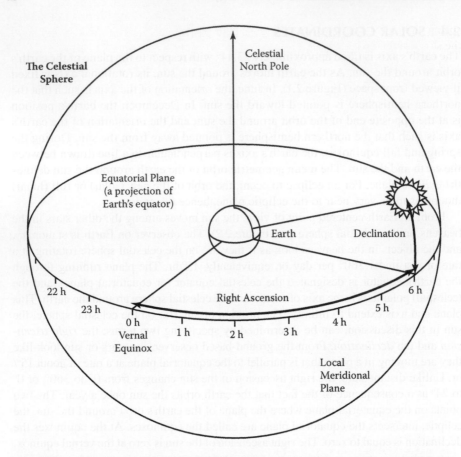

FIGURE 2.2 Illustration of the celestial sphere. The celestial equator or the equatorial plane is the projection of the earth's equator extended to space. The earth's rotation axis extended to space forms the north and south celestial poles. Angular distances above or below the equatorial plane measured from the center of the earth are called declination. Angles (measured in time units) from the point where the equatorial plane and the orbital plane of the earth around the sun cross are the equinoxes. 0 hours is assigned to the spring (or vernal) equinox. The illustration indicates the first day of summer with the sun's declination at 23.44° and a right ascension of 6 hours. (Graphics by Chris Cornwall.)

Angular measurements, called right ascension, are made in the equatorial plane as measured from the vernal equinox point. Historically, the angular position of the right ascension on the equatorial plane has been measured in hours since the earth rotates 360° in a day (24 hours) or 15° per hour. The right ascension of the sun is 6 hrs (90°) from the vernal equinox at the summer solstice, 12 hrs (180°) at the autumnal equinox and 18 hrs (270°) at the winter solstice. The measurements for the right ascension are made in a counterclockwise manner (see Figure 2.2). The position of the sun on the celestial sphere is shown at the summer solstice in Figure 2.2.

An angular measurement above or below the equatorial plane toward the celestial polar axis is a measurement of the sun's *declination*. The sun is above the equatorial plane in the summer, below it in winter, and exactly on the plane on the first day of spring and on the first day of autumn. An approximate formula for the declination of the sun in degrees is (Spencer, 1971):

$$dec = \begin{bmatrix} 0.006918 - 0.399912 \cdot \cos(\beta) + 0.070257 \cdot \sin(\beta) \\ -0.006758 \cdot \cos(2\beta) + 0.000907 \cdot \sin(2\beta) \\ -0.002697 \cdot \cos(3\beta) + 0.00148 \cdot \sin(3\beta) \end{bmatrix} \cdot (180/\pi) \qquad (2.3)$$

where β is defined as $2\pi d/365$ radians, and d is the day of the year. For example, February 20 is the fifty-first day of the year. Since there are 366 days in a leap year, there is a slight difference in the calculation for leap years.

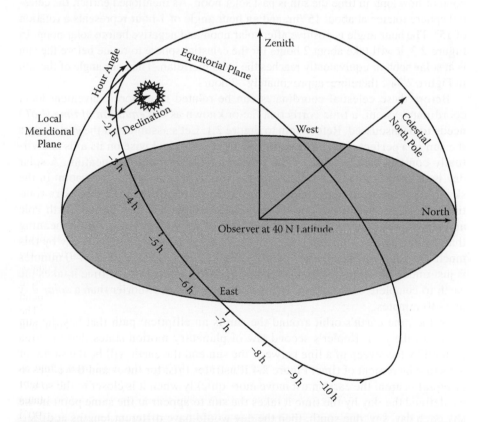

FIGURE 2.3 Relation of solar angles from the observer's point of view as seen in the northern hemisphere at 40° N. The north and south directions are in the plane of the page. The east and west directions are out of and into the page, respectively. The zenith line is perpendicular to the earth's surface. The local meridional plane is in the plane of the page and is defined by the south, north, and zenith points. The equatorial plane is perpendicular to the polar axis about which the celestial sphere rotates. The solar hour angle, ha, is the angle between the sun's current position and solar noon that is due south and measured along the equatorial plane in hours (about −2 hours in this figure). (Graphics by Chris Cornwall.)

To an observer on the earth at 40° north latitude, the celestial sphere is tilted as in Figure 2.3. Note that the closer the observer is to the equator, the closer the equatorial plane is to the zenith, and on the spring and autumn equinoxes for an observer at the equator the sun will pass directly overhead. Now the task is to relate the celestial position of the sun to a local coordinate system defined for an observer's position on the earth's surface. The local *meridional* plane is defined by three points: due south, due north, and the zenith. The meridional plane runs through the earth's axis of rotation and is perpendicular to the local East–West axis and is the plane defining the local longitude of the site. As the celestial sphere rotates around the earth, the time during the rotation when the local meridional plane bisects the sun is called *solar noon* and is the moment when the sun has an hour angle value of zero. Solar hour angle uses time units to measure how long in time the sun is from reaching solar noon or how long in time the sun is past solar noon. As mentioned earlier, the celestial sphere rotates at about 15°/hr, and an hour angle of 1 hour represents a rotation of 15°. The hour angle is positive after solar noon and negative before solar noon. In Figure 2.3, it will take about 2 hours for the celestial sphere to rotate before the sun is at solar noon or equivalently reaches the local meridian. The hour angle of the sun in Figure 2.3 is, therefore, approximately –2 hours.

Before these celestial coordinates can be related to a more convenient local coordinate system, a time correction factor known as the *equation of time (EoT)* needs to be discussed. Refer again to Figure 2.1. Let's assume that the earth orbits the sun in a perfect circle. The earth rotates counterclockwise on its axis at a virtually constant rate because of the conservation of angular momentum. A solar day is defined by the time it takes for the sun to appear at the same point in the sky. This is the time to complete one rotation plus a little more; this extra rotation is because the earth rotates counterclockwise with respect to the North Pole and also moves counterclockwise 1/365.25th of the way around its orbit meaning that the sun appears to have moved in the heavens from the previous day by this much. One day equals 24 hours or 1440 minutes, and 1/365.25th of 1440 minutes is just under 4 minutes. Note that a *sidereal day* is defined by the time it takes the earth to complete one rotation, and it is about 4 minutes shorter than a *solar day* of 1440 minutes.

In fact, the earth's orbit around the sun is an elliptical path that has the sun at one of the foci. Kepler's second law of planetary motion states that the area defined by the sweep of a line between the sun and the earth will be the same for the same increment of time. Figure 2.4 illustrates this; for the A and B wedges to be equal in area, the earth must move more quickly when it is closer to the sun. If we defined the day by the time it takes the sun to appear at the same point in the sky each day, say, due south, then the day would have different lengths at different points in its orbit around the sun. The day would be longer when the sun and earth were closest (perihelion) compared with when they are the farthest apart (aphelion). This is because the earth must turn a complete revolution and a little more as it progresses in its orbit around the sun, and this progression around the sun proceeds at a faster clip when the two bodies are closer together. Since we prefer the day length to be constant, we fix it at 24 hours for civil time. This will cause the sun to be at a different position relative to our noon clock each

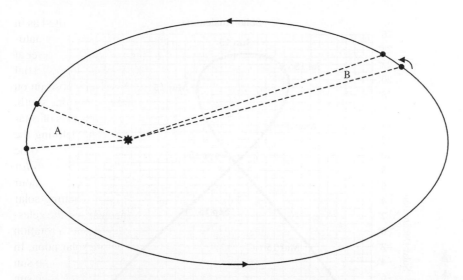

FIGURE 2.4 Illustration of the elliptical orbit of the earth around the sun with the area labeled A equal in size to the area labeled B. According to Kepler's second law, the time to traverse the arc bounding A is equal to the time it takes to traverse the arc bounding B. Therefore, the earth has to travel faster when it is closer to the sun than when it is farther from the sun.

day. This apparent repositioning of the sun with respect to local time is defined by the *equation of time*. Spencer's (1971) approximation for the equation of time in radians is

$$EoT = \begin{bmatrix} 0.000075 + 0.001868 \cdot \cos(\beta) - 0.032077 \cdot \sin(\beta) \\ -0.014615 \cdot \cos(2\beta) - 0.040849 \cdot \sin(2\beta) \end{bmatrix} \quad (2.4)$$

where β is defined as $2\pi d/365$ radians, and d is the day of the year as defined after Equation 2.3. To change to minutes from radians, multiply by $1440/2\pi = 229.183$.

The calculation provides the number of minutes that the sun is behind or ahead of the local meridian at solar noon. For example, the local meridian at Denver, Colorado, is about 105° W, which is seven hours west of Greenwich, England (0° longitude). On January 1 Equation 2.4 produces a result of –3.35 minutes, meaning that the sun will appear due south at its highest point in the sky for that date 3.35 minutes after noon local time. In Boulder, Colorado, 0.25 degrees west of Denver, it will take 1 minute more for the sun to reach Boulder's local meridian since the sun moves 1° in 4 minutes—that is, 24 hours is equal to 1440 minutes, so 1440 minutes/360 degrees is equal to 4 minutes per degree. Figure 2.5 is called the analemma that results from the equation of time and the declination changes over a year. It represents the sun's position over a year for a fixed location on Earth at the same time every day.

A more accurate (±0.01°) algorithm for calculating the sun's declination and equation of time is given in Michalsky (1988). A "C" program, solpos.c, based on this algorithm is found at the National Renewable Energy Laboratory (NREL) Renewable Resource Data Center (RReDC) website (http://rredc.nrel.gov/solar/

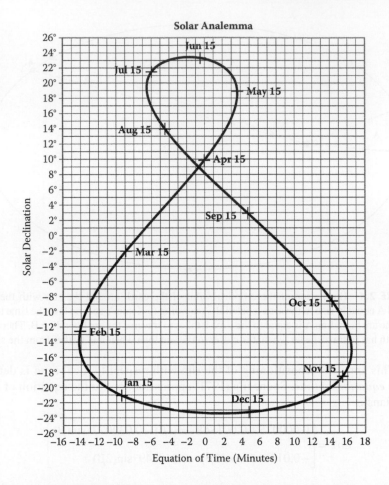

FIGURE 2.5 Solar analemma. Plot of solar declination verses equation of time. This plot has the general shape of the solar position in the sky if a photograph was taken on each day of the year at the same time of day.

codesandalgorithms/solpos/). The solpos.c code is easier to understand than the still more accurate solar position algorithm (spa.c) (Reda and Andreas, 2008) (available at http://rredc.nrel.gov/solar/codesandalgorithms/spa/). The latter contains matrix tables that make the algorithm more difficult to follow but provides a stated accuracy of +/− 0.0003° over the period −2000 to 6000. More information on the solar position algorithms is found in Appendix D.

2.5 ZENITH, AZIMUTH, AND HOUR ANGLES

A coordinate system that can be used for defining positions with respect to an observer fixed on the surface of the earth is shown in Figure 2.6. The solar zenith angle is defined as the angle between the zenith and the sun. The cosine of the solar zenith angle with respect to the observer on the surface of the earth can be obtained

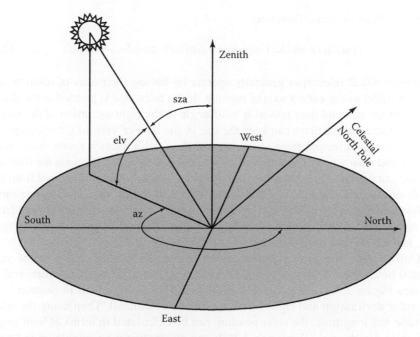

elv = elevation angle, measure up from horizon
sza = zenith angle, measured from vertical
az = azimuth angle, measured clockwise from north

FIGURE 2.6 View from the local horizontal plane. The solar zenith angle *sza* is measured from the zenith position to the sun. The solar elevation, or altitude, angle *elv* is measured from the horizon to the sun. The azimuth angle, *az*, is measured from due north going eastward from 0° to 360°. (Graphics by Chris Cornwall.)

by calculating the vector product of the normal to the surface and the unit vector coming directly from the sun. Consider a reference coordinate system with the $\hat{x}$ and $\hat{y}$ axes defined in the equatorial plane with the East–West axis for the $\hat{x}$ axis and the intersection with the local meridian plane for the $\hat{y}$ axis. The celestial polar axis is the $\hat{z}$ axis. The normal to the surface at a location of latitude *lat* degrees is then

$$\hat{n} = \cos(lat)\hat{y} + \sin(lat)\hat{z} \tag{2.5}$$

The unit vector to the sun is tilted from the equatorial plane by the declination *dec* with an hour angle of *ha*. Projecting the sun's vector onto the three reference axes yields

$$\hat{s} = \cos(dec)\sin(ha)\hat{x} + \cos(dec)\cos(ha)\hat{y} + \sin(dec)\hat{z} \tag{2.6}$$

The dot product of the unit vectors is the cosine of the angle between the two vectors. The angle between the normal to the surface and the sun's direction is the *solar zenith angle (sza).* Therefore,

$$\cos(sza) = \sin(lat) \cdot \sin(dec) + \cos(lat) \cdot \cos(dec) \cdot \cos(ha) \qquad (2.7)$$

Astronomical telescopes generally operate by having their axes of rotation oriented parallel to the earth's axis of rotation. If the telescope is pointed at the declination of the star and then moved in hour angle to the right ascension of the star, a simple sidereal clock drive can keep the star in the field of view of the telescope by rotating in one direction from east to west at the earth's rotation speed. Some telescopes and most automated solar trackers do not use equatorial mounts for tracking the sun but use alt-azimuth trackers instead. Solar *azimuth (az)* is measured from the north clockwise toward the east and varies from 0° through 360°. The solar azimuth is shown in Figure 2.6. Altitude or, more commonly, *elevation (elv)* is measured from the horizon to the solar position (0 to 90°) at a fixed azimuth. The solar zenith angle, *sza*, is the complement to elevation so $sza = 90° - elv$.

Unlike stars, the sun's right ascension is constantly changing over the year from 0 hours the first day of spring to 6 hours on the first day of summer, and so on (see Figure 2.3). Using only universal time (UT is explained in Section 2.6), the solar declination and right ascension can be calculated. Then using the site's latitude and longitude, the solar position can be calculated in terms of hour angle and solar zenith angle (Equation 2.7). It can alternatively be calculated in terms of solar zenith angle and azimuth angle. The hour angle and declination can be used to point an equatorial-mounted instrument, and the zenith angle and azimuth angle can be used to point an alt-azimuth-mounted solar device. These calculations are performed in solpos.c and spa.c, described in Section 2.4, with accuracies of 0.01° and 0.0003°, respectively.

2.6   SOLAR, UNIVERSAL, AND LOCAL STANDARD TIME

Algorithms are easier to formulate and use in solar time, but it is important to relate the solar position and descriptions to local standard time. Local standard time (LST) is different from clock time because clock time is typically shifted by 1 hour in the summer to provide daylight saving time. LST is equal to clock time in the winter and is not changed to daylight saving time in the summer. Local standard time is the same across the entire time zone, whereas solar time is related to the position of the sun with respect to the observer with solar noon the time when the sun crosses the local meridian. The relationship between local standard time and solar time is dependent on one's exact longitude and one's time zone. The earth rotates just under once every 24 hours; hence, the sun appears to travel across the sky at a rotation rate of about 15° per hour. Solar angles are calculated using Greenwich mean time (GMT). However, the earth's rotation is very gradually slowing, and atomic clocks are now used for time standards. *Coordinated universal time* would be abbreviated CUT in English and TUC in French (*temps universel coordonné*); however, neither abbreviation is used, and coordinated universal time, designated UTC, is the international time standard based on an atomic clock. UTC uses leap seconds to adjust for the ever-so-gradual slowing of Earth's rotation and to coordinate with the mean solar time or GMT or universal time (UT). UT and UTC are always within a second of each other but have the slight difference previously described. The sun moves about 0.004° in 1 second, so this difference is generally ignored.

Local standard time is related to GMT through the time zone of a person's location. In Eugene, Oregon, the local time is 8 hours earlier than GMT. In Berlin, Germany, the local time is 1 hour ahead of GMT. Every 15° of longitude roughly defines a time zone. Therefore, if one's location is within ±7.5° longitude of a longitude that divides evenly by 15°, one would guess that one was within the time zone of that longitude; however, a location's time zone is sometimes modified from this strict interpretation by other factors, usually related to commerce. In any case, correctly specifying the longitude and the UT and the latitude allows one to calculate solar position very precisely using the programs solpos.c or spa.c.

2.7 SOLAR POSITION CALCULATION

An example is provided to show how to calculate the sun's position at a given location. Assume the location is Omaha, Nebraska. It is 9:30 a.m. central daylight time (CDT) in the morning on September 13 in a non-leap year. The latitude of the city center is about 41.25°N; the longitude is 96.00°W. Omaha is in the central time zone of the United States, which is 6 hours earlier than Greenwich mean time or UT. Since it is still daylight saving time season, the local standard time is 8:30 a.m. September 13 is day 256 of the year. Using Equations 2.3 and 2.4 to calculate declination and the equation of time, the declination of the sun is found to be 3.73° and the equation of time +4.27 minutes for this date. The solar hour angle is calculated by considering that the time is 3.5 hours before solar noon; the equation of time is +4.27 minutes, meaning that the sun reaches solar noon that much before local time but that Omaha is 6° farther west than the center longitude (also called the standard meridian) 90° for the central time zone. Therefore, the hour angle of the sun is −3 hours − 30 minutes + 4.27 minutes − 6*4 minutes = −3 hours − 49.73 minutes.

The relationship between local standard time (*lst*) and solar time (*st*) is

$$st = lst + EoT + (long_{sm} - long_{ob})/15 \qquad (2.8)$$

where $long_{sm}$ is the longitude of the standard meridian from which the clock time for the time zone is initiated, and $long_{ob}$ is the longitude of the location of the observer. The units for this equation are hours. If EoT is calculated in minutes, then it needs to be divided by 60 minutes/hour to get the proper units. The hour angle in degrees is found by

$$ha = 15 \cdot (st - 12) \quad \text{or} \quad st = ha/15 + 12 \qquad (2.9)$$

When solar time is noon (12) the hour angle is zero.

Now an equatorial mount solar tracker can align with the sun using the solar coordinates 3.73° declination and an hour angle of −3.8288 hours (or −57.4325° or −1.0024 radians). For executing the sine and cosine functions, the angular arguments to these functions must be used correctly. Most often programming languages use radians for trigonometric calculations. If the solar azimuth and zenith angle are needed, it is necessary to convert these to local coordinates for an alt-azimuth tracker's use.

With the hour angle, latitude, and declination, one can now calculate the cosine of the solar zenith angle (*sza*) using Equation 2.7. The *lat* is the latitude of central Omaha 41.25°, *dec* is the declination of the sun with respect to the earth calculated from Equation 2.3, namely, 3.73° in this example, and *ha* is the hour angle at 8:30 central standard time on September 13 as determined already. The result is a solar zenith angle of 63.46°. The azimuth angle can also be determined from the zenith angle, the hour angle, and the declination of the sun.

$$\sin(azm) = \frac{-\cos(dec) \cdot \sin(ha)}{\cos(\pi/2 - sza)} \tag{2.10}$$

The arcsine has more than one solution, and therefore it is necessary to exercise the following logic to obtain the right solution between 0° and 360°:

If $[\sin(dec) - \sin(\pi/2 - sza) \cdot \sin(lat)] > 0$

then if $[\sin(azm)] < 0$

then $azm = azm + 360°$

else $azm = 180° - azm$

else $azm = azm$

For this case azimuth is found to be 109.95°. Alternately, running *solpos* as mentioned in Section 2.4 yields a solar zenith angle of 63.44° and a solar azimuth of 109.86°. These latter results are more accurate, but they are consistent with the lower accuracy approximations given by Equations 2.3 and 2.4.

To view the motion of the sun across the sky during different times of year, one can plot sun path charts. These sun path charts are two-dimensional plots of solar elevation versus solar azimuth and are used by several solar tools to evaluate how objects around a site will cast shadows on instruments or solar energy systems. A sun path chart for latitude 35°N is shown in Figure 2.7. The time lines are plotted in solar time, but it is also possible to plot them in local standard time. Sun path charts can be created in polar coordinates (Figure 2.8). If one was standing in the center of this polar coordinate plot and looking toward the equator, the sun would rise behind one's back, to the north and east in this example, and set behind one's back to the north and west for 6 months of the year. Three-dimensional sun

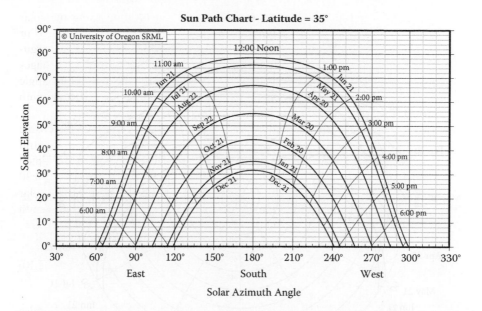

FIGURE 2.7 Sun path chart for latitude 35° N plotted in solar time. The azimuth angle varies from 60° at sunrise to 300° at sunset on June 21 and varies from 120° at sunrise to 240° at sunset on December 21. At the equinoxes, the solar zenith angle at sunrise is at 90°, due east, and 270°, due west, at sunset. At solar noon, the solar elevation angle varies from 78.44° on June 21 to 31.59° on December 21. The solar elevation is 0° at sunrise and sunset. Note that the solar zenith angle is 90° minus the solar elevation.

path charts are also useful in showing the path of the sun across the sky from an observer's perspective (see Figure 2.9). For locations above the Arctic Circle or below the Antarctic Circle, the sun will never rise or set during parts of the year. If one is slightly north of the Arctic Circle, say latitude 70°, during parts of May and June, the sun never sets, and during parts of December and January the sun never makes it above the horizon. More sun path charts for every 10° of latitude from −90° to 90° are shown in Appendix C. Site-specific sun path charts can be created for any location on earth online (http://solardata.uoregon.edu/SunChartProgram.php).

Another way to examine the path of the sun is to look at the shadow cast by a vertical pole or tree during the day. The curve traced is much like that generated by the gnomon of a sundial (see Figure 2.10). The curves trace the end of the shadow cast by the top of a 10-meter pole during different times of year. Near sunrise and sunset, the shadows can be extremely long as the sun is very low in the sky. During the middle of the day, the sun casts the shortest shadow of the day on a horizontal surface. The website that generates the sundial plots can be found online (http://solardata.uoregon.edu/SunDialProgram.html). The sun path charts and sundial plots are useful when planning and constructing a solar monitoring station as the effects of shadows cast by surrounding objects can be studied. Prudent use of the sun path programs can identify the location with minimal shading and identify times when

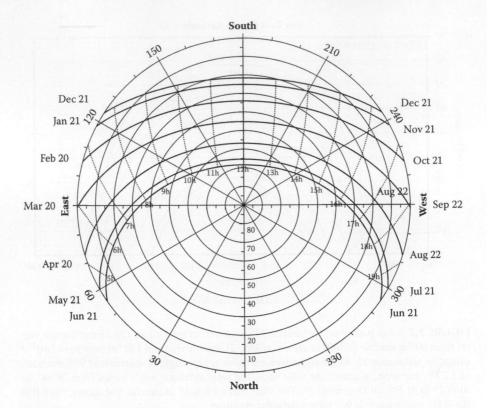

FIGURE 2.8 Sun path chart for latitude 45° N in polar coordinates. Hour lines are plotted in local solar time. As the sun moves closer to the center of the plot, it is higher in the sky. If the zenith angle was zero, the sun would be directly overhead. For the sun to be directly overhead, one would have to be at a location on Earth between 23.44° latitude and –23.44° latitude at the right time of year.

shading will affect the monitoring equipment. The importance of site surveys before establishing a solar monitoring station is discussed in chapter 14 on setting up a solar monitoring station.

2.8   SUNRISE AND SUNSET TIMES

The calculation of sunrise and sunset times provides an easy exercise to test one's understanding of the information presented so far. Sunrise and sunset occur when the sun is at the horizon, the solar zenith angle is 90°, or $\pi/2$ radians, and hence the cosine of the solar zenith angle is zero.

Setting the cosine of the solar zenith angle to zero in Equation 2.7 results in

$$ha_{sunrise,sunset} = \pm \arccos[-\tan(lat) \cdot \tan(dec)] \tag{2.11}$$

where $ha_{sunrise}$ is the *sunrise hour angle*(–), and ha_{sunset} is the *sunset hour angle*(+). For the example given for Omaha, this results in hour angles of ± 93.28°. Of course,

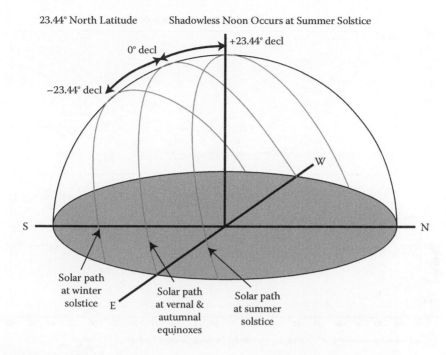

23.44° North Latitude   Shadowless Noon Occurs at Summer Solstice

+23.44° decl

0° decl

−23.44° decl

W

S

N

Solar path at winter solstice

E

Solar path at vernal & autumnal equinoxes

Solar path at summer solstice

FIGURE 2.9 Three-dimensional plot of the sun for latitude 23.44° N. At solar noon on the summer solstice, the sun is directly overhead, yielding a shadowless solar noon. The sun paths are for summer solstice, spring and fall equinoxes, and winter solstice. (Graphics by Chris Cornwall.)

the sunrise and sunset times would need to be modified for the equation of time, which is 4.27 minutes, and the fact that Omaha is 6° west of the middle of the time zone or 24 minutes later in solar time. So the 6 hour and 13 minutes before solar noon would be modified by having sunrise occur 24 minutes later because Omaha is west of the 6-hour meridian and then modifying this by the equation of time adjustment so that sunrise occurs 4.27 minutes earlier; this would result in sunrise occurring 5 hours and 53.27 minutes before noon local standard time or 6:06:44 a.m. Sunset would occur 6 hours and 32.73 minutes after noon local standard time or 6:32:44 p.m. by similar reasoning.

With all of these corrections the sunrise and sunset hour angles will still not be exactly the same value as the sunrise and sunset times that appear in the local paper. The sunrise reported in the paper may be earlier, and the sunset times may be later. The reason for this difference is primarily caused by refraction (bending of light by the atmosphere) that is, on average, approximated 0.7° for the sun on the horizon at sea level. Refraction causes the sun to appear slightly higher in the sky than simple geometrical calculations indicate. Another effect that gives an additional minute of sunlight at sunrise and sunset is the fact that sunrise and sunset refer to the rising of the leading edge of the sun and setting of the trailing edge. Since the sun is about 0.5° in diameter, the edges are about 0.25° from the center of the sun that is the reference point for the earlier calculations. The angular travel time for 0.25° is about 1 minute.

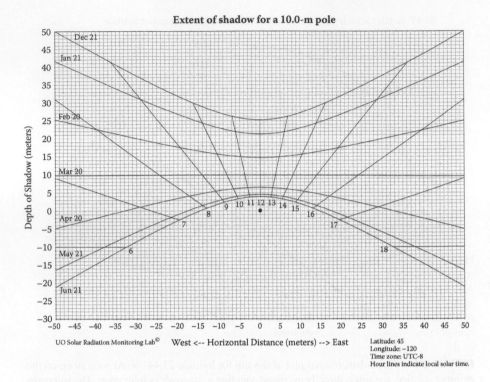

FIGURE 2.10 Sun dial plot for a latitude of 45°. The lines follow the end of the shadow from the tip of a 10-meter-tall pole located at the (0,0) coordinate. Near sunrise or sunset the distance to the end of the shadow will be extremely large. In the summer when the sun is highest in the sky, the shadow will fall closest to the pole or gnomon. In the winter, when the sun is lowest in the sky, the shadow at solar noon will have the longest noontime reach. The time marks on the chart are for reference.

2.9 GLOBAL, DIRECT NORMAL, AND DIFFUSE IRRADIANCE

The nature of solar radiation reaching the earth and the terminology commonly used to describe the incident radiation components are fundamental to choosing the instruments used to measure the solar resource. Near noon on a day without clouds, about 25% of the solar radiation incident outside the atmosphere is scattered and absorbed as it passes through the earth's atmosphere. Therefore, roughly 1000 Wm⁻² of solar irradiance reaches the earth's surface without being significantly attenuated. This "beam" radiation, coming from the direction of the sun, is called *direct normal irradiance* (DNI).

Some of the scattered sunlight is reflected back into space, and some of it reaches the surface of the earth (see Figure 2.11). The scattered radiation reaching the earth's surface is called *diffuse horizontal irradiance* (DHI). Some radiation is also scattered by the earth's surface and then rescattered by the atmosphere to the observer. This is part of the DHI an observer sees. The total amount of sunlight on a horizontal surface is called *global horizontal irradiance* (GHI). Solar radiation is a flux

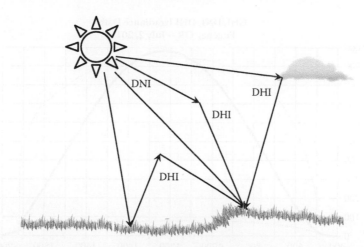

FIGURE 2.11 Components of solar radiation reaching the earth's surface. The extraterrestrial radiation is the sun's irradiance as it impinges on the top of the atmosphere of the earth. All of the radiation radiated from the sun is unscattered as it reaches the top of the earth's atmosphere in its orbit around the sun. As the sunlight passes through the atmosphere some of it is scattered and some is absorbed. The DNI measured at the surface is the portion of the sunlight that comes through the atmosphere with no scattering or absorption. Light scattered by the molecules, aerosols, and clouds that strikes the observation area is called diffuse horizontal irradiance (DHI). This also includes some irradiance that scattered from the earth's surface and then scattered again by clouds or the atmosphere to the observation site. In snow-covered areas, this rereflected radiation can be significant. The total irradiance on a horizontal surface, called GHI, is the sum of the DNI projected onto the horizontal surface plus all the diffuse radiation that makes it to the observing instrument.

of energy measured in power (usually watts) per unit area (usually square meters). When the direction of the flux is normal to the surface, the amount of flux is greatest. The amount of energy per unit area falls off as the cosine of the angle between the normal to the surface and the incident angle. In the case of DNI, the incident angle is the solar zenith angle. A plot of GHI, DNI, and DHI on a clear day is shown in Figure 2.12.

The global irradiance on a horizontal surface is equal to the direct normal irradiance times the cosine of the solar zenith angle plus the diffuse irradiance:

$$GHI = DNI * \cos(sza) + DHI \tag{2.12}$$

This is an exact formula for instantaneous irradiance and becomes an approximation if the irradiance is measured over time because the cosine of the solar zenith angle varies with time. During shorter time intervals, the GHI irradiance can be larger than the extraterrestrial irradiance reaching the earth's orbit because of clouds reflecting a good deal of incident irradiance onto the sensing instrument. This is often referred to as the *lensing effect of clouds* (see Figure 2.13).

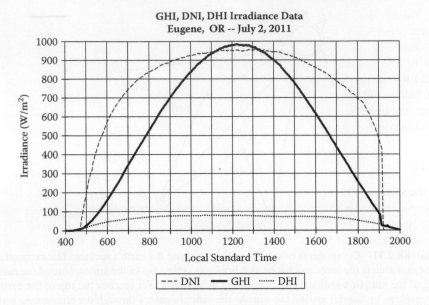

FIGURE 2.12 Plot of the solar radiation components on a clear day. The value of DNI is influenced by the amount of atmosphere through which it travels. During most of the day the air mass changes slowly but changes rapidly in the morning and evening hours. On this clear day the diffuse irradiance is only about 10% of the maximum global irradiance. The global horizontal irradiance is the direct normal irradiance projected onto a horizontal surface plus the diffuse horizontal irradiance. Hence, on a clear day the global horizontal irradiance plotted against time has a typical bell shape or cosine shape.

FIGURE 2.13 The plot illustrates the lensing effect of clouds. GHI values can momentarily exceed the extraterrestrial irradiance when additional DNI is reflected by the side of clouds toward the sensor.

The amount of sunlight scattered from the surface, called ground-reflected irradiance, depends on the albedo of the surface. The albedo is the ratio of the reflected or up-welling solar radiation that is reflected back to the sky to the incident or down-welling GHI irradiance. The albedo can vary from about 0.1 for dark surfaces to 0.2 for grasslands to greater than 0.8 for freshly fallen snow. A perfectly reflective surface would have an albedo of 1.0. Because the atmosphere will scatter some of the reflected light back to the surface, areas with high surface albedo will have a larger diffuse component due to multiple surface reflections.

2.10   SOLAR RADIATION ON TILTED SURFACES

If the surface under study is tilted with respect to the horizontal, the total irradiance received is the incident diffuse radiation on the tilted surface plus the direct normal irradiance projected onto the tilted surface plus ground-reflected irradiance that is incident on the tilted surface. The amount of direct radiation on a horizontal surface can be calculated by multiplying DNI by the cosine of the solar zenith angle. On a surface that is tilted by T degrees from the horizontal and rotated γ degrees from the north–south axis (with north facing being zero and east facing 90°), the direct component on the tilted surface is determined by multiplying the direct normal irradiance by the cosine of the incident angle to the tilted surface (θ_T).

$$\cos(\theta_T) = \sin(dec) \cdot \sin(lat) \cdot \cos(T) + \sin(dec) \cdot \cos(lat) \cdot \sin(T) \cdot \cos(\gamma)$$

$$+ \cos(dec) \cdot \cos(lat) \cdot \cos(T) \cdot \cos(ha)$$

$$- \cos(dec) \cdot \sin(lat) \cdot \sin(T) \cdot \cos(\gamma) \cdot \cos(ha)$$

$$+ \cos(dec) \cdot \sin(T) \cdot \sin(\gamma) \cdot \sin(ha) \tag{2.13}$$

where the *dec*, declination, *lat*, latitude, and *ha*, hour angle were discussed earlier. An alternate formula calculating the cosine of the incident angle to a tilted surface is

$$\cos(\theta_T) = \cos(T) \cdot \cos(sza) + \sin(T) \cdot \sin(sza) \cdot \cos(az - \gamma) \tag{2.14}$$

where *az* is the azimuthal angle. Irradiance on tilted surfaces is discussed in more detail in Chapter 8.

2.11   SPECTRAL NATURE OF SOLAR RADIATION

The ultimate source of the sun's radiative energy is the fusion of hydrogen atoms in the sun's core. This energy propagates to the surface through a series of absorption and re-emission processes finally reaching the visible part of the sun's surface, the photosphere, after tens of thousands of years. The spectral distribution of the sun's radiation is similar to that of a black body that is heated to the temperature of 5778 K. That is, the amount of radiation varies with the wavelength as shown in Figure 2.14. The majority of solar irradiance is between 300 nm and 3,000 nm. The absorption of some solar radiation by the ions in the chromosphere of the sun

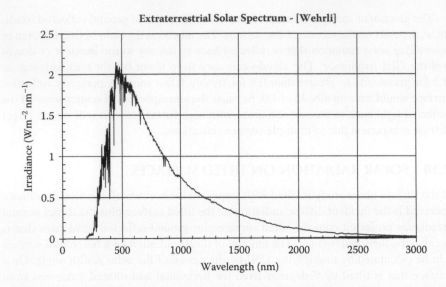

FIGURE 2.14 Spectral distribution of the extraterrestrial irradiance reaching the earth's surface. The shape of the distribution is similar to a black body radiating at 5778 K. Most of the spectral absorption lines in the plot result from the absorption of photospheric radiation in the solar chromosphere.

results in absorption lines in the extraterrestrial solar spectrum. The effect of these ions can be seen in the solar spectrum in Figure 2.14, especially for wavelengths below about 700 nm.

The spectral distribution of incident solar radiation spans a large range of wavelengths from the ultraviolet to the infrared with the peak around 500 nm. Approximately 47% of the incident extraterrestrial solar radiation is in the visible wavelengths from 380 nm to 780 nm. The infrared portion of the spectrum with wavelengths greater than 780 nm accounts for another 46% of the incident energy and the ultraviolet portion of the spectrum below 380 nm accounts for 7% of the extraterrestrial solar radiation.

As the sunlight passes through the atmosphere, a large portion of the ultraviolet (UV) radiation is absorbed and scattered. Air molecules scatter the shorter wavelengths more strongly than the longer wavelengths. On days with few clouds, the preferential scattering of blue light by air molecules is the reason the sky appears blue. Water vapor and atmospheric aerosols further reduce the amount of direct sunlight passing through the atmosphere. The information in Table 2.1 shows that on a clear day approximately 76% of the extraterrestrial direct normal irradiance passes through the atmosphere without being scattered or absorbed. The magnitude of the scattering and absorbing are about equal under these idealized conditions.

Many molecules selectively scatter or absorb radiation in specific wavelength regions, creating large variations in the solar spectrum as measured at the earth's surface (Figure 2.15). The abundance of the atmospheric constituents along the sunlight's path determines the depth of these absorption regions. Therefore, the solar spectrum depends on the length of the path through the atmosphere and the density

TABLE 2.1
Absorption and Scattering for Typical Clear-Sky Conditions

Factor	Percent Absorbed	Percent Scattered	Percent of Total Passing Through the Atmosphere
Ozone	2	0	
Water vapor	8	4	
Dry air	2	7	
Aerosol	2	3	
Total not absorbed or scattered	87	87	76

of the molecules along this path. Said another way, the atmosphere is a highly variable filter affecting the amounts of solar irradiance reaching the surface at any given time and place.

Users of solar radiation data fall mainly into two groups. The broadband group is interested in the total amount of energy available over all wavelengths, as in many solar thermal applications. The spectral group is concerned with specific regions of the solar spectrum or the spectral distribution of the irradiance.

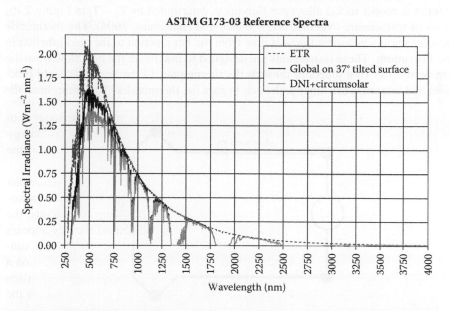

ASTM G173-03 Reference Spectra

Legend:
- - - - ETR
——— Global on 37° tilted surface
——— DNI+circumsolar

Axis labels: Spectral Irradiance (W m⁻² nm⁻¹) — $\text{(W m}^{-2}\text{ nm}^{-1})$; Wavelength (nm)

FIGURE 2.15 Spectral distribution of the incident solar radiation as it reaches the earth's surface. This plot is for a specific aerosol content of the atmosphere when the air mass is 1.5 when the instrument is tilted 37°. The dashed line is the ETR radiation, the solid black line is the total irradiance on the tilted surface, and the gray line is the DNI plus the circumsolar radiation. This plot shows, the solar spectrum for the ASTM (2008) reference solar angle and aerosol content that are used for specifying instrument performance as a function of spectral wavelength.

For example, those who evaluate the performance of photovoltaic devices use detailed spectral measurements. Others such as architects are interested in lighting for the human eye, which is sensitive only to the visible part of the spectrum. The eye's response to various wavelengths in daylight conditions is called the photopic response. More information on the photopic response is presented in Chapter 12. Daylighting illuminance is the sum, over all wavelengths, of the product of solar irradiance as a function of wavelength times the photopic response as a function of wavelength. The SI unit for illuminance is the lux (lumen m^{-2}).

2.12 FUNDAMENTALS OF THERMODYNAMICS AND HEAT TRANSFER

Thermopiles are used in a wide variety of solar sensors to measure incident solar radiation. To characterize these instruments and to understand how these instruments operate, a background on the fundamentals of thermodynamics and heat transfer is provided. Some instruments use the photoelectric effect and the basics of photodiodes and solar cells are discussed in Section 2.13.

The basis for all thermopile radiometers is the thermoelectric effect, the generation of voltage from temperature differences at the junctions of two dissimilar metals. The thermoelectric effect was discovered in 1821 by Thomas Seebeck, a German physicist. The *Seebeck Effect* predicts 41 μvolts from junctions of copper and constantan (a copper-nickel alloy) per Kelvin (as determined by $T_2 - T_1$ in Figure 2.16) at room temperature (Velmre, 2007; Bunch and Hellemans, 2004). The thermopile in solar sensors measures the heat flow from the hot junction to the cold junction in the instrument. These instruments are designed to maximize the heat flow from the sensor receiver (hot junction) through the thermopiles to the heat sink (cold junction) and minimize the other paths that bypass the thermopiles. To comprehend the

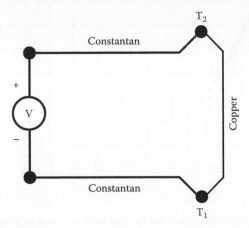

FIGURE 2.16 Circuit diagram illustrating the thermocouple where two different metals at different temperatures generate an electrical voltage when electrons flow from the hot junction (T_2) toward the cold junction (T_1).

rationale for various design features and to characterize the performance of the solar instruments it is necessary to understand how heat flows.

There are three ways that heat can flow:

- Conduction of heat through a material
- Convection of heat by liquid or gas from a surface
- Radiative transfer of heat

The nature of conduction, convection, and radiation of heat are discussed in Sections 2.12.1 to 2.12.3, and some of the basic equations governing heat transfer are given.

2.12.1 CONDUCTION

Conduction is the transfer of heat through a material as a result of molecular motion. The more heat applied to a material, the faster the molecules that make up the material will vibrate. When heat is applied to a material, the molecules where the heat is applied will vibrate faster. This vibration causes molecules nearest this heat source to vibrate faster, and slowly this vibration spreads throughout the material. For a slab, the rate of heat transfer depends on the area where the heat is applied, the type and structure of the material, the temperature difference across the material, and the thickness of the material. A general formula for heat transfer across a slab is given by

$$\frac{dQ}{dt} = -kA\frac{dT}{dx} \tag{2.15}$$

where dQ/dt is the rate of heat flow (change in heat over the time interval dt), dT/dx is the temperature gradient (the change in temperature over a change in distance traveled in the material (dx)), A is area, and k is the thermal conductivity of the material given in ($Wm^{-1}K^{-1}$). The thermal conductivities of various materials at room temperature are given in Table 2.2. The thermal conductivity varies widely, and materials can be chosen to increase or decrease the rate of heat flow. Textbooks such as the one by Yener and Kakac (2008) cover thermal conductivity.

TABLE 2.2
Conductivity at Room Temperature

Material	Conductivity ($Wm^{-1}K^{-1}$)
Brass	120.000
Copper	390.000
Steel	50.000
Styrofoam	0.033
Water	0.600

2.12.2 CONVECTION

Convection is the movement of heat by a gas or a liquid from one place to another. A more apt description of the processes of concern is advection, taking heat away from a surface and then having the gas or fluid move the heat by convection. An example of advection is wind blowing air molecules across a surface with heat transferred from the surface to the air. Thus, the air carries away the heat from the surface. Newton's law of cooling states that the rate of change in temperature of an object is proportional to the temperature difference between the object and the fluid or gas taking away, or adding, heat.

$$\frac{dQ}{dt} = hA(T_s - T_g) \tag{2.16}$$

where T_s is the temperature of the surface, T_g is the temperature of the gas or liquid, and h is the heat transfer coefficient in units of $Wm^{-2}K^{-1}$. The heat transfer coefficient varies with the type of gas or liquid, the type and geometry of the surface, and how the fluid or gas flows over the surface. The heat transfer coefficient of water is about 33 times that of air. Therefore, air with high levels of humidity has the capability of taking away more heat than dry air. Relative humidity can be a factor in the performance of solar instruments because air containing more moisture is better at transferring heat. The type of flow is also important. Laminar flow is when the gas or fluid moves in parallel layers and no currents or eddies are perpendicular to the flow. Turbulent flow is chaotic, and there is considerable mixing between layers. There is also a transitional flow that is neither laminar nor fully turbulent. Turbulent flow is better at transferring heat from the surface than laminar flow. Convection and advection can be complex, and many papers and textbooks discuss convective heat flow (see, e.g., Lienhard and Lienhard, 2002).

2.12.3 RADIATIVE HEAT TRANSFER

All bodies radiate and absorb heat. This radiation is in the form of electromagnetic waves. These waves can vary in wavelength from X-rays and ultraviolet waves to visible, infrared, and radio waves. The energy in these waves is inversely proportional to the wavelength, and the longer the wavelength, the less energy the wave carries. In 1897, Jožef Stefan (1835–1893) and Ludwig Boltzmann (1844–1906) developed the Stefan Boltzmann law, which states that the energy emitted by a body is proportional to the fourth power of the temperature of the body.

$$E = \varepsilon\sigma T^4 \tag{2.17}$$

where E is the energy emitted, ε is the emissivity of the material, σ is the Stefan Boltzmann constant equal to 5.6704×10^{-8} $Wm^{-2}K^{-4}$, and T is the temperature in K. A black body is a perfect emitter with an emissivity equal to 1. The spectral distribution of the radiation emitted is given by Planck's law

$$I(\lambda, T) = \frac{2\pi hc^2}{\lambda^5} \cdot \frac{1}{\exp^{\frac{hc}{\lambda kT}} - 1} \tag{2.18}$$

where λ is the wavelength, in micrometers for example, h is Planck's constant, c is the speed of light, k is the Boltzmann constant, and T is the temperature. Planck's constant is equal to $6.62607 \cdot 10^{-34}$ Js, the speed of light is equal to $2.99792 \cdot 10^8$ ms^{-1}, and the Boltzmann constant is 1.38065×10^{-23} m^2 kg s^{-2} K^{-1}. Equation 2.18 can be used to estimate the spectral distribution of the solar radiation incident outside the earth's atmosphere. The spectral flux density from the Planck spectral distribution has to be lowered by the square of the ratio of radius of the sun divided by the distance between the earth and the sun.

The wavelength of the maximum energy emitted by a black body, from Wien's law, is

$$\lambda = \frac{2898}{T} \tag{2.19}$$

where T is temperature in K, and λ is in micrometers.

Another important property of an object is its absorptivity, an object's ability to absorb incident solar radiation. In thermal equilibrium, Kirchhoff's law states that the absorptivity of a body is equal to its emissivity. Therefore, a good absorber is a good emitter. A perfect black body has an emissivity of 1 and an absorptivity of 1. In general, the duller and darker the material, the closer the emissivity is to 1, and the more reflective, the lower its emissivity. Highly polished silver has an emissivity of about 0.02 and is a great reflector and a poor absorber. Wavelength, angle of incidence, and shape all affect the emissivity and, hence, the absorptivity of a material. Detailed explanations of radiative heat transfer can be found in, for example, Siegel and Howell (2002).

2.13 PHOTODIODES AND SOLAR CELL PREREQUISITES

Many solar sensors are based on the silicon photodiode. A rudimentary description of how photodiodes work is provided, and this information forms a basis for characterizing solar sensors based on photodiodes.

Silicon photodiodes are made from crystalline silicon that has been transformed into a semiconductor by introducing atoms that do not have the same number of valence electrons as silicon. Silicon has four valence electrons (quadrivalent) that form tight bonds with four neighboring silicon atoms. Introduction of atoms with one fewer or one extra valence electron is called doping. A photodiode consists of n- and p-type doped silicon sections that are brought together. N-type doped silicon contains a number of atoms with one excess electron when all the electron bond pairs are filled in the crystalline structure. Typically for n-type doping, approximately every thousandth silicon atom is replaced by a phosphorous atom (pentavalent) with one extra valence electron. P-type doped silicon contains a number of atoms with one less electron than is needed to complete the electron bond pairs. Typically for p-type doping, every millionth atom is replace by a trivalent atom such as boron. This missing electron acts much like a positively charged electron and is treated as such when modeling semiconductor materials. In the manufacturing process the p-type silicon is the base material and then the cell is doped with n-type atoms. At the boundary of n-type and p-type materials, the loosely bound electrons in the n-type silicon drift over to the p-type side to fill "holes" that are created by doping. The drift of electrons from the n-type material to the p-type material results in a shift of charge because the doped atoms that were originally associated with the electrons or holes did not move. An electric field is created by the difference between the number of positively charged protons in the doping atom and the number of negatively charged electrons in the bonds. The layer of silicon where this shift of electrons and holes occurs is called the depletion layer as the number of free holes and electrons are depleted. This electric field of about 0.5 V separates the electron–hole pairs that are created when solar photons knock the electrons from their bound state in the depletion layer of the doped silicon semiconductor. This electric field allows the movement of the electron and hole to their respective collection areas. When the photodiode is connected in a circuit, a current will flow. The current created by this photodiode can then be measured.

Mathematically it is easier to treat the properties of solar cells in momentum space. For semiconductor materials, electrons can be treated as living in energy bands. Conduction bands are areas with excess electrons and valence bands are areas with excess "holes." Without light the conduction and valence bands reach equilibrium. Incident light can disturb this equilibrium and in a silicon solar cell, if the energy of the photon is greater than 1.1 eV (electron volt), electrons can jump into the conduction band. A full discussion of how semiconductors work is the subject of solid-state physics texts such as Kittel (2005).

When light is incident on a photodiode, the resulting current is the sum of the photodiode current, the shunt resistance current, and the photocurrent.

$$I_{out} = I_{ph} - I_s \cdot \left(e^{\frac{eV}{kT}} - 1\right) - I_{sh} \qquad (2.20)$$

where I_{out} is the output current, I_{ph} is the photocurrent generated by the incident light, I_{sh} is the shunt resistance current, and the middle term is the result of the diode current (I_s is the reverse diode current, e is the electron charge of 1.602 10^{-19} coulombs, V is the applied voltage in volts, k is the Boltzmann constant, and T is the absolute temperature of the photodiode in Kelvin). The shunt resistance current is an alternate path for the current to flow. Cells are designed with a large shunt resistance to minimize the loss. The diode current is the current that flows when there is no light incident on the device and is sometimes called the dark current. The circuitry for photodiodes is designed in a manner that minimizes the dark current.

2.14 MODELS

There are a wide variety of instruments for measuring solar irradiance, but historically only a limited quantity of measured solar data is available. To compensate for this limitation, many models have been created to estimate solar irradiance values from meteorological observations at the surface and from satellite-based images. Additional relationships between the various solar components have been developed to estimate and optimize the potential performance of energy conversion systems. The discussion of solar radiation models is outside the scope of this book, but an overview is given in Appendix A. These modeled estimates are often inadequate substitutes for measuring irradiance data when accurate information is required to validate a system's field performance. Today the financing of large-scale solar-systems is driving the demand for ever-increasing accuracy in solar radiation measurements.

QUESTIONS

1. If the sun was a perfect black body, what would be its temperature?
2. Calculate the solar declination for September 23.
3. What is the average extraterrestrial irradiance at the mean earth–sun distance?
4. What is the minimum solar zenith angle for your location?
5. What is your local standard time when it is solar noon on March 21 in Greenwich, England?
6. What is the local standard time at 12:00 p.m. CDT on May 21 in Chicago?
7. What is the geometric solar zenith angle at sunrise to the nearest degree on April 20?
8. What is the relationship between DNI, DHI, and GHI?
9. How would you calculate DNI from measured GHI and DHI?
10. Calculate the cosine of the incident solar angle on a surface tilted 30° from the horizontal facing due west on June 21 at 2:00 p.m. near Denver at 40° N latitude and 105° W longitude.
11. How much energy is radiated from a black body at room temperature?
12. How much energy is absorbed by a black body at room temperature?
13. Draw a plot of the extraterrestrial solar spectrum. What is the peak wavelength of extraterrestrial solar radiation?
14. Why are modeled solar values used in resource assessment instead of measured data?

REFERENCES

ASTM G173–03. 2008. *Standard tables for reference solar spectral irradiances: Direct normal and hemispherical on 37° tilted surface,* ASTM Standards (E-891) and (E-892). West Conshohocken, PA: American Society for Testing and Materials.

Bunch, B. and A. Hellemans. 2004. *The history of science and technology: A browser's guide to the great discoveries, inventions, and people who made them from the dawn of time to today.* Boston: Houghton Mifflin.

Duffie, J. A. and W. A. Beckman. 1991. *Solar engineering of thermal processes.* New York: John Wiley.

Fröhlich, C. 2006. Solar irradiance variability since 1978. *Space Science Reviews* 90:1–13.

Kittel, C. 2005. *Introduction to solid state physics.* Hoboken, NJ: John Wiley & Sons, Inc.

Lienhard, J. H. IV and J. H. Lienhard V. 2002. *A heat transfer textbook*. Cambridge, MA: Phlogiston Press.

Michalsky, J. J. 1988. The astronomical almanac's algorithm for approximate solar position (1950–2050). *Solar Energy* 40:227–235.

Paltridge, G. W. and C. M. R. Platt. 1976. *Radiative processes in meteorology and climatology*. Amsterdam: Elsevier Scientific Publishing Company.

Reda, I. and A. Andreas. 2008. Solar position algorithm for solar radiation applications (revised). NREL Report No. TP-560-34302, 56 pp.

Siegel, R. and J. R. Howell. 2002. *Thermal radiation heat transfer*. New York: Taylor & Francis.

Spencer, J. W. 1971. Fourier series representation of the position of the sun. *Search* 2:172.

Velmre, E. 2007. Thomas Johann Seebeck (1770–1831). *Proceeding of the Estonian Academy of Sciences, Engineering* 13(4):276–282.

Wehrli, C. 1985. *Extraterrestrial solar spectrum*. Publication no. 615, Physikalisch-Meteorologisches Observatorium Davos + World Radiation Center (PMOD/WRC) Davos Dorf, Switzerland.

Yener, Y. and S. Kakac. 2008. *Heat conduction*. Boca Raton, FL: CRC Press.

3 Historic Milestones in Solar and Infrared Radiation Measurement

To measure is to know.

Lord Kelvin
(1824–1907)

3.1 INTRODUCTION

Abilities to measure solar and infrared radiation have developed through centuries of scientific discovery and engineering applications. Historically, measurements have played a critical role in the development of accepted physical laws. Today, solar and infrared radiation measurements are particularly important for advancing atmospheric physics, developing renewable energy technologies, and understanding climate change. The goal of improving data accuracy is common to the development of instrumentation for measuring solar and infrared radiation. This chapter provides a selective account of important solar and infrared radiation observations and instrumentation developments.

3.2 EARLIEST OBSERVATIONS OF THE SUN AND THE NATURE OF LIGHT

Sunspots and their cyclic behavior have long been an interest of scientific observation. The invention of the telescope in the early seventeenth century provided the capability to determine that the dark spots seen on the face of the sun, noticed as early as the fourth century B.C.E., were associated with the sun itself (Eddy, 1976). In addition to viewing the number of sunspots, the early telescopes provided improvements to observations of the faculae (bright areas on the sun's surface that typically occur before a sunspot) size, shape, location, and most of the other observational detail recorded today. In 1848, a number of European observatories began regular observations using a standard scheme developed by Rudolf Wolf (Waldmeier, 1961). Combined with future efforts to standardize measurement practices, the work started by Wolf produced the most reliable historical record of sunspot numbers in the nineteenth century.

In 1666, Sir Isaac Newton discovered the spectral distribution of sunlight using glass prisms to display the visible spectrum in a darkened room. He also developed the corpuscular theory of light as a means of explaining propagation in a straight line and the property of specular reflection from mirror-like surfaces. Other discoveries in the seventeenth century include the law of refraction by Willebord Snell and the description of light diffraction, discovered independently by Francesco Grimaldi and Robert Hooke (Coulson, 1975). The electromagnetic character of visible radiation would not be discovered for another two centuries.

The measurement of direct solar radiation, the amount of "beam" radiation from the direction of the sun, is essential for understanding the total solar irradiance (TSI) output from the sun. Formerly known as the *solar constant*, determining the mean value of TSI and its slight variability continues to be the subject of research in astronomy, atmospheric physics, climate change, renewable energy technology development, and other applications. In 1837, Claude Servais Mathias Pouillet produced the first successful research on TSI measurements and introduced the term *pyrheliometer* from the Greek *fire*, *sun*, and *measure* (Crova, 1876).

In 1913, Charles Greeley Abbot and Loyal Blaine Aldrich at the Smithsonian Institution developed the first radiometer designed for measuring global horizontal irradiance (downwelling total hemispheric radiation from the sun and sky). Using the Greek words for *fire*, *above*, and *a measure*, Abbot and his colleagues named the *pyranometer* for its ability to measure solar heat from above.[*] Intended to serve as a standard instrument for solar radiation measurements, details of the design were given by Abbot and Aldrich (1916).

Other notable advances in understanding and measuring optical radiation are listed in Table 3.1, and more detailed descriptions of selected instruments are presented in this chapter.

3.3   NINETEENTH-CENTURY RADIOMETERS

3.3.1   POUILLET'S PYRHELIOMETER (1837)

As previously introduced, this instrument was the first to be called a *pyrheliometer* and was designed as a calorimeter for measuring solar energy. The radiation receiver was a water-filled cylindrical vessel with a blackened absorbing surface aligned perpendicular (normal) to the sun as shown in Figure 3.1. The other surfaces of the cylindrical vessel were silvered to reduce the heat exchange between the receiver and the atmosphere. The receiver fluid temperature was measured with a thermometer in direct contact. The cylinder was rotated about the solar alignment axis to provide mixing of the calorimeter fluid (later changed to mercury). Measurements of direct normal irradiance (DNI) were based on alternating 5-minute periods of aligning the blackened receiver surface normal to and away from the sun. The corresponding temperature rise and fall, the heat capacity of the calorimeter, the total heat received per unit of surface, and unit of time could be computed. From his clear-sky measurements, Pouillet estimated the TSI to be 1227 Wm^{-2} (1.76 cal cm^{-2} min^{-1}) (Coulson, 1975) or about 10% lower than the currently accepted value of 1366 Wm^{-2}.

---

[*] In 1961, in the second edition of the "Guide to Meteorological Instrument and Observing Practices," the World Meteorological Organization first recommended the use of the term *pyranometer* for hemispherical measurements to replace the more commonly used terminology *180° pyrheliometer*.

TABLE 3.1
Highlights of Optical Radiation Measurement

Year	Event
1611–1612	David and Johannes Fabricius, Christoph Scheiner, and Galileo Galilei independently document the existence of sunspots and the sun's rotation on its axis.
1666	Sir Isaac Newton discovers the spectral nature of visible sunlight using glass prisms.
1800	William Hershel discovers infrared radiation using thermometers and a prism to measure spectral irradiance of sunlight beyond the visible range.
1825	Herschel develops an actinometer as the first instrument to introduce cooling rate into solar radiation measurements.
1837	First use of the term *pyrheliometer*. Claude Pouillet invents his pyrheliometer and measures the *solar constant* as 1.76 cal cm^{-2} min^{-1} (1227 Wm^{-2}) within about 10% of the currently accepted value of 1366 Wm^{-2}.
1879	Campbell–Stokes burning sunshine recorder developed. The instrument remains in use to provide daily records of percent possible bright sunshine. These data are often used to estimate the equivalent global irradiation at the measurement site.
1884	Boltzmann derives the Stefan-Boltzmann law to describe the relationship between a body's temperature (T) and its radiant energy (W): $W = \varepsilon \sigma T^4$
1893	Knut Ångström invents the electrical compensation pyrheliometer.
1898	H. S. Callendar invents the Callendar pyranometer using four platinum wire grids wound on strips of mica to form electrical resistance thermometers. Two square grids were painted black, and two were naturally reflective platinum.
1900	The Smithsonian Institution in Washington, D.C., begins measuring global horizontal radiation.
1901	The U.S. Weather Bureau begins first measurements of solar radiation by deploying three Ångström's electric compensation pyrheliometers in Washington, D.C., Baltimore, Maryland, and Providence, Rhode Island, to measure direct normal irradiance.
1904	Samuel P. Langley attempts correlation between weather and solar radiation.
1905	Knut Ångström's electrical compensation pyrheliometer is adopted as a measurement standard at the Meteorological Conference, Innsbruck and by the Solar Physicist Union, Oxford.
1909	The U.S. Weather Bureau begins routine measurements of global horizontal solar irradiance using the Callendar pyranometer in Washington, D.C., with later installations in Madison, Wisconsin (1911) and Lincoln, Nebraska (1915).
1910	Smithsonian water-flow pyrheliometer is perfected into the primary standard. Abbot develops the secondary standard silver-disk pyrheliometer. The Marvin pyrheliometer is invented. Measurements of direct normal irradiance commence on Mt. Fuji, Japan, and at Madison, Wisconsin, Lincoln, Nebraska, and Santa Fe, New Mexico.
1911	The U.S. Weather Bureau begins its first *routine* measurement of direct normal irradiance with the deployment of Marvin pyrheliometers at Madison, Lincoln, and Santa Fe
1912	Measurements in Algeria show 20% reduction of sunlight due to volcanic ash from Mt. Katmai eruption in Alaska.
1913	First revision of Smithsonian Pyrheliometric Scale, based on water-flow pyrheliometer. Abbot develops balloon-borne pyranometer carried to 45,000 ft. for solar constant measurements. Mean value of the solar constant given as 1.933 cal cm^{-2} min^{-1} (1348 Wm^{-2}) within about 1% of the currently accepted value of 1366 Wm^{-2}.

(continued)

TABLE 3.1 (CONTINUED)
Highlights of Optical Radiation Measurement

Year	Event
1915	In April, the *Monthly Weather Review* begins regular publication of measured solar irradiance data from the U.S. Weather Bureau.
1916	C. G. Abbot and L. B. Aldrich at the Smithsonian Institution develop a pyranometer for measuring global horizontal irradiance.
1919	A. K. Ångström invents electrical compensation pyrgeometer for measuring infrared irradiance.
1923	The Kimball-Hobbs pyranometer (later adopted by Eppley for their "light bulb" pyranometer) is invented.
1924	Deployment of the first Moll-Gorczynski pyranometer, later known as the Kipp & Zonen solarimeter.
1927	Shulgin develops double water-flow pyrheliometer to improve measurement accuracy.
1930	Eppley Laboratory, Inc. commercializes the Kimball-Hobbs 180° pyrheliometer and it is later given the nickname, "Eppley light bulb pyranometer."
1932	The Smithsonian Pyrheliometric Scale is revised based on double water-flow pyrheliometer (adopted as the standard by Smithsonian Institution).
	Robitzsch bimetallic pyranometer (actinograph) provides self-contained recording mechanism, but not recommended by the Radiation Commission of the International Association of Meteorology for any measurements except daily total irradiation.
1950-51	The U.S. Weather Bureau increases the number of pyranometer measurement stations to 78 using the Eppley 180° pyrheliometer.
1952	World Meteorological Organization's Subcommission on Actinometry recommends the silver-disk pyrheliometer for measuring direct normal irradiance.
1954	U.S. Weather Bureau changes method of calibrating Eppley pyranometers from outdoor to an indoor integrating sphere and artificial lighting.
1956	International Radiation Conference in Davos, Switzerland, recommends new measurement scale, "International Pyrheliometric Scale (IPS)."
	Yanishevsky pyranometer is developed for measuring global and diffuse irradiance and surface albedo in the U.S.S.R.
1957	The International Pyrheliometric Scale 1956 is established. Ångström Scale is increased by 1.5%, and 1913 Smithsonian Scale is decreased by 2.0% for compliance.
1957	Eppley Laboratory, Inc. introduces Model Precision Spectral Pyranometer (PSP) and the Model Normal Incidence Pyrheliometer (NIP).
1968	First measurements of extraterrestrial irradiance (total solar irradiance) are reported from flights of the USAF/NASA X-15 by the Eppley Laboratory, Inc. and the Jet Propulsion Laboratory (TSI = 1360 Wm^{-2}).
1969	Eppley Laboratory, Inc. introduces model 8-48 "black-and-white" pyranometer. The light bulb pyranometer is discontinued.
1969	The primary absolute cavity radiometer is reported by J. M. Kendall, Sr. at the Jet Propulsion Laboratory. The design is based on electrical heating substitution equivalent to radiation heating of the cavity receiver.
1970	The World Radiation Center conducts the Third International Pyrheliometer Comparisons where the first electrically self-calibrating absolute cavity radiometer participates.

TABLE 3.1 (CONTINUED)
Highlights of Optical Radiation Measurement

Year	Event
1971	Lambda Instruments Corporation (now LI-COR) develops a pyranometer with a photodiode detector (Model LI-200).
1975	DIAL radiometer concept is developed for low-cost automated field measurements of atmospheric turbidity. The design formed the basis for future commercial systems known as *rotating shadowband pyranometer, rotating shadowband radiometer, thermopile shadowband radiometer,* and *multifilter rotating shadowband radiometer.*
1977	PACRAD III becomes the reference radiometer for the International Pyrheliometer Comparison.
	The University of Oregon in Eugene begins the measurement of direct normal and other solar irradiance elements to establish what is now the longest continuous record of solar resources available from anywhere in the United States.
1978	NIMBUS satellite radiometers begin measurement record of total solar irradiance using the Hickey–Frieden absolute cavity radiometer.
1979	World Radiometric Reference (WRR) is established by the World Radiation Center as the internationally recognized measurement standard for direct normal irradiance (above 700 Wm^{-2}).
	The U.S. National Weather Service develops and tests solar radiation forecasts for 1 to 3 days ahead using measurements from three agricultural stations in South Carolina (Jensenius and Carter, 1979).
	The National Oceanic and Atmospheric Administration (NOAA) proposes a new system to provide daily solar energy forecasts 1 and 2 days in advance for the conterminous U.S. based on data from 34 solar measurement stations in the NOAA Solar Radiation Network (Jensenius and Cotton, 1981).
1994	Multifilter rotating shadowband radiometer is commercialized.
2003	Interim World Infrared Standard Group (WISG) of Pyrgeometers is established by the World Radiation Center in Davos, Switzerland.
	Total irradiance monitor (TIM) begins measuring total solar irradiance from space as part of NASA's Earth-observing system Solar Radiation and Climate Experiment. With an estimated absolute accuracy of 350 ppm (0.035%), the TIM can detect relative changes in solar irradiance to less than 10 ppm/yr (0.001%/yr), allowing determination of possible long-term variations in solar output.
2010	Variable condition pyrheliometer comparisons completed at the National Renewable Energy Laboratory (Michalsky, Dutton, Nelson, Wendell, Wilcox, Andreas et al., 2011).

Source: After Coulson, K. L., *Solar and Terrestrial Radiation, Methods and Measurements,* Academic Press, New York, 1975.

3.3.2 CAMPBELL–STOKES SUNSHINE RECORDER (1853, 1879)

Although this instrument is not a radiometer, this device and some later variants remain in use today for measuring bright sunshine duration. These observations can then be used to estimate daily amounts of solar irradiation (Iqbal, 1983). J. F. Campbell invented his *burning* sunshine recorder in 1853 by using a glass ball filled with water.

FIGURE 3.1 Pouillet's pyrheliometer—1837. Direct normal irradiance measurements were based on the temperature changes of the water-filled blackened cylindrical receiver resulting from alternating exposure to the sun and the sky at 5-minute intervals. The end of the thermometer (T) was inserted into the receiver (A) and positioned for reading (B) ahead of the alignment disc (C). (After Fröhlich, C., *Metrologia*, 28, 1991.)

This spherical lens was placed inside a white stone bowl (later mahogany wood) with a 4-inch diameter surface painted with an oil paint or varnish and engraved with hour lines. During periods when the solar disk was free of clouds or dense haze, the concentrated solar radiation melted the paint (burned the mahogany). The total length of the bright sunshine *burns* was compared with the total day length to compute percent-possible bright sunshine. Later, a solid glass sphere replaced the water-filled ball to allow operation in freezing conditions. In 1879, Sir G. G. Stokes significantly modified the Campbell sunshine recorder to assume its present form (Figure 3.2). The solid glass sphere is supported by a metal structure that holds specially treated paper strips with precise hourly and subhourly markings. The strips are available in three different lengths: one for summer, winter, and spring/fall periods. Daily, a single card is inserted and precisely registered using a metal pin. The treated paper is stable over a range of humidity levels, including effects of precipitation, allowing the paper to burn when direct normal irradiance levels exceed 210 Wm^{-2}. In 1962, the Commission for Instruments and Methods of Observation (CIMO) of the World Meteorological Organization (WMO) adopted the Campbell–Stokes sunshine recorder as a standard of reference (WMO, 2008). In 1966, the analysis of bright sunshine data from 233 measurement stations and solar radiation measurements from an additional 668 stations were used to produce the first set of monthly world maps of the daily mean total solar radiation on a horizontal surface (Löf, Duffie, and Smith, 1966).

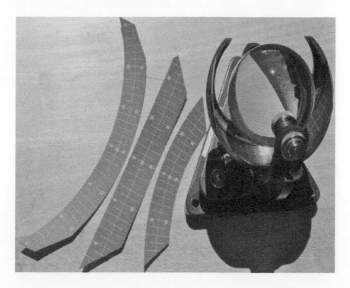

FIGURE 3.2 Campbell–Stokes sunshine recorder continues to provide measurements of bright sunshine duration at several locations around the world. Sensitized recording paper is available in three lengths corresponding to the season of measurement.

Routine measurements of bright sunshine duration using the Campbell–Stokes instrument continue to the present day. These data are included in the records maintained by the World Radiation Data Center (http://wrdc.mgo.rssi.ru/).

3.3.3 Ångström Electrical Compensation Pyrheliometer (1893)

Unlike previous designs based on calorimetry, this instrument was the first to equate electrical power with solar irradiance. Developed by Knut Ångström in the late nineteenth century, this pyrheliometer was recognized for its reliability and accuracy (Ångström, 1893, 1899). With only minor design developments based mostly on improved electrical measurement devices, the Ångström pyrheliometer remains in service today as a primary reference instrument for many solar measurement laboratories. The design uses two strips of Manganin* foil painted on one side with Parson's optical black lacquer. The strips are mounted side by side at the base of a collimator tube and aligned with two rectangular apertures at the front of the instrument (Figure 3.3). A thermojunction is attached to the back of each receiver strip. Electrical leads from the strips and the thermojunctions are connected to the instrument controls. The operator controls a reversible shutter at the front of the collimator to alternately shade and unshade each of the strips at 90-second intervals. In operation, the shaded strip is heated by a known electrical current to the same temperature as the exposed strip. The rate of energy absorption by the exposed strip (solar heating) is thermoelectrically compensated by the rate of energy supplied electrically to the shaded strip (electrical heating). The thermojunctions attached

---

* Manganin is an alloy of typically 86% copper, 12% manganese, and 2% nickel.

FIGURE 3.3 Ångström electrical compensation pyrheliometer continues to serve as a reference instrument for some countries. This unit was operated during the Eleventh International Pyrheliometer Comparisons at the Physikalisch-Meteorologisches Obseratorium Davos, World Radiation Center in 2010.

to the back of each strip are used to monitor their temperature equivalence using a galvanometer or digital null detector. Observations of solar irradiance consist of a series of shade–unshade measurements in which each strip is alternately exposed and shaded. The direct normal irradiance is determined from the equivalence of radiant and electrical power:

$$DNI \cdot \alpha \cdot b = C \cdot r \cdot i^2 \qquad (3.1)$$

where
DNI = direct normal irradiance (Wm^{-2})
α = absorptance of the receiver strip per unit length
b = width of the receiver strip [cm]
C = a constant based on measurement units (9,995.175)
r = resistance of the receiver strip per unit length (ohms/cm)
I = electric current through the shaded strip (amperes)

In practice, the instrument manufacturer supplied a constant K defined as

$$K = C \cdot r/(\alpha \cdot b) \qquad (3.2)$$

to simplify the irradiance determination

$$DNI = K \cdot i^2 \qquad (3.3)$$

The instrument functioned as the European absolute reference beginning in 1905 and ended in 1956 with the adoption of the International Pyrheliometric Scale (IPS 1956) that resolved the differences with the Smithsonian scale of 1913 (Fröhlich, 1991).

3.3.4 CALLENDAR PYRANOMETER (1898)

Invented in 1898 by H. L. Callendar, this early example of a pyranometer was at the time called a *bolometric sunshine receiver*. An improve version was available in 1905 (Callendar and Fowler, 1906). The instrument used four platinum wire grids wound on strips of mica to form an electrical resistance thermometer. As shown in Figure 3.4, one pair of grids was painted black to enhance the absorption of solar radiation, and the other pair was left as reflecting platinum metal. The checkerboard receiver had an area of about 33.6 cm² and was mounted in a protective evacuated glass bulb with an outside diameter of about 9.1 cm (Coulson, 1975). The temperature difference between the pairs of grids when exposed to sunlight was proportional to the incoming solar radiation. The popularity of this early pyranometer was due in large part to the integrated data-recording mechanism. A self-adjusting Wheatstone

FIGURE 3.4 The Callendar pyranometer was a successful all-weather design based on platinum resistance thermometry. One of the blackened receiver sections has lifted from the detector surface.

bridge with a pen and recording drum produced an ink trace of the temperature difference in the receiver pairs with time. A calibration factor was supplied with each instrument to convert the recorded data into global irradiance. The Callendar pyranometer enjoyed considerable popularity during the first two decades of the twentieth century. An example of this pyranometer was used at the University of Cambridge School of Botany for more than 50 years and agreed within 1% (more specifically, the correlation coefficient was 0.99) when compared with the British Meteorological Office pyranometer (Beaubien, Bisberg, and Beaubien, 1998). The U.S. Weather Bureau used this instrument from 1908 until the last unit was removed from service in 1941 (Hand, 1941).

3.3.5 Ångström Tulipan Pyrgeometer (1899)

Some of the first reported measurements of *nocturnal* radiation, infrared (longwave) downwelling irradiance, were collected by two radiometers designed by Knut Ångström. Measurement comparisons of the *Tulipan* and the Ångström pyrgeometer were reported at the Physical Meteorological Observatory in Davos, Switzerland, from November 1920 to October 1921. The Tulipan design was based on "the over-distillation of ether" (Dorno, 1923). The other Ångström design, which would later be called a pyrgeometer, consisted of four Manganin strips, of which two were blackened and two were polished. The blackened strips were allowed to radiate to the atmosphere, while the polished strips were shielded. The electrical power required to equalize the temperature of the four strips was assumed to equate to the longwave irradiance. Based on 16 nights of operation, the Tulipan pyrgeometer had a persistent high bias of 40% compared with the Ångström pyrgeometer.

3.4 OPERATIONAL RADIOMETERS OF THE TWENTIETH CENTURY

3.4.1 Abbot Silver-Disk Pyrheliometer (1906)

At the beginning of the twentieth century, continuing investigations to determine the *solar constant* (now referred to as the total solar irradiance, or TSI) suggested the need for establishing a measurement standard for solar irradiance. Seeking improved measurement accuracy, Charles Greeley Abbot developed his pyrheliometer based on the principles of calorimetry. Designed as a measurement reference, the instrument was disseminated in quantity by the Smithsonian Institution (more than 100 copies were constructed and distributed internationally). As shown in Figure 3.5, a blackened silver disk (38 mm in diameter and 7 mm thick) is positioned near the base of a collimating tube limiting the instrument field of view to 5.7° full angle. In the final design, a mercury-in-glass thermometer was inserted into a radial chamber bored into the edge of the disk. The disk was filled with mercury for rapid heat transfer from the blackened receiver to the thermometer. The mercury was sealed in the disk using a thin steel liner, surrounded by cord and shellacked, and wax was used to seal the thermometer stem (Coulson, 1975). Three fine steel wires held the disk inside a copper box. This assembly was then enclosed in a wooden box to provide some thermal insulation from changes in ambient temperature.

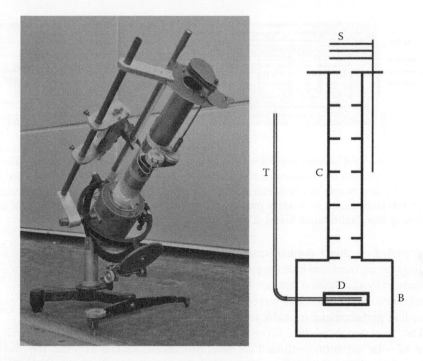

FIGURE 3.5 Design elements of the Abbot silver-disk pyrheliometer used for measuring direct normal irradiance: Thermometer (T) measures the disk (D) temperature located in the base (B) of the instrument separated by a collimating tube (C) defining the field of view that can be shuttered (S) from the direct normal irradiance.

Operation of the instrument required properly aligning the receiver disk with the sun, alternately shading and unshading the receiver, accurately keeping time to 1/5th second, and reading the thermometer to a few hundredths of a degree Celsius. Summarizing the procedure originally specified by Abbot (1911), it is apparent the instrument design was best suited for measurements under very clear and stable atmospheric conditions:

Time (seconds)	Operator Action
0	Align instrument with the sun, open instrument cover, close shutter
80	Read thermometer
180	Read thermometer, open shutter, check alignment
200	Read thermometer
300	Read thermometer, close shutter
320	Read thermometer
420	Read thermometer, open shutter

Note: Continue the readings in above order as desired.

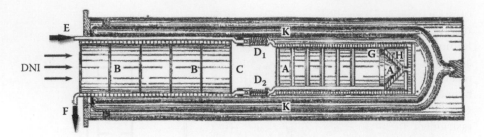

FIGURE 3.6 Smithsonian water-flow pyrheliometer measures direct normal irradiance (DNI) by monitoring the temperature rise of water between points D_1 and D_2. Abbot's design introduced the concepts of cavity enhancement for absorbing solar radiation and electrical power substitution used in modern pyrheliometers. (After Abbot, C. G., *The Sun and the Welfare of Man*, Smithsonian Series, 2, 1929.)

Determining the direct normal irradiance required the correction of all temperature readings for air, stem, and bulb temperatures and multiplying the corrected temperature differences by a calibration constant for the particular instrument used for the measurements. The calibration constant was obtained by direct comparison with the Smithsonian water-flow pyrheliometer traceable to the Smithsonian 1913 Scale of Radiation.

The importance of the *silver-disk pyrheliometer* to the accurate measurement of solar radiation continued for several decades. As late as 1952, the World Meteorological Organization recommended the instrument for measuring direct normal irradiance.[*]

3.4.2 SMITHSONIAN WATER-FLOW PYRHELIOMETER (1910)

Also developed by C. G. Abbot at the Smithsonian Institution, this absolute pyrheliometer was based on the measurement principles of calorimetry. As shown in Figure 3.6, the water-flow pyrheliometer measures DNI entering tube (BB), passing through a precision aperture (C), and absorbed by a blackened chamber (AA) with a cone at the base of the chamber to enhance absorption of the radiation. The hollow cylinder (K) insulates the receiving chamber from changes in ambient air temperature. A spiral watercourse encloses the receiving chamber. Distilled water at constant temperature is supplied to an inlet tube (E) where the water enters the chamber wall at a constant rate and passes through a spiral ivory piece (D_1) and exits by a similar ivory spiral (D_2). A measured length of "hair-like" platinum wire is threaded through each of the ivory spiral channels such that the water bathes the platinum wires, forcing them to the temperature of the water entering and exiting the instrument. When the instrument is exposed to DNI, the resulting temperature difference measured at D_1 and D_2 and the known aperture area are used to compute solar irradiance. To confirm the water-flow temperature measurements, a coil of wire (H) is used to electrically heat the receiving chamber with a measured current. These electrical substitution measurements were used to confirm the equivalence

---

[*] Meeting of the Subcommission on Actinometry, Brussels, 1952.

of solar and electrical power performance of the instrument. Abbot's design incorporated key features in use today by the pyrheliometers defining the World Radiometric Reference: (1) cavity enhancement of the receiver to increase photon absorption, (2) insulating "muffler" to limit heat exchange between the internal receiver and exterior surfaces of the instrument, and (3) electrical substitution by means of a wire-wound heater at the receiver for comparing measurements of solar-induced heating with a known amount of electrical heating. Although the instrument did not become widely used, it was the basis for the Smithsonian 1913 Scale of Radiation for solar radiation measurements (Abbot and Aldrich, 1932). The instrument's function as the American absolute reference ended in 1956 with the adoption of the International Pyrheliometric Scale (IPS 1956) that resolved the differences with the Ångström scale of 1905 (Fröhlich, 1991).

3.4.3 Marvin Pyrheliometer (1910)

Prior to serving as chief of the U.S. Weather Bureau from 1913 to 1934, C. F. Marvin was developing improved instruments for measuring bright sunshine duration and a pyrheliometer for measuring direct normal solar irradiance. Similar in many respects to the Abbot silver-disk pyrheliometer, the Marvin pyrheliometer used a resistance wire instead of a mercury-in-glass thermometer for measuring the temperature of the blackened silver-disk receiver (Foote, 1918). An electrically operated shutter mounted in front of the collimating tube aperture provided for automated shade and exposure of the receiver to solar radiation. Due to its simpler operation, the Marvin pyrheliometer replaced the Ångström pyrheliometer in 1914 as the instrument used by the U.S. Weather Bureau (Covert, 1925).

3.4.4 Ångström Pyranometer (1919)

Anders Knutsson Ångström, Swedish physicist and meteorologist and son of Knut Ångström, used his father's principle of electrical compensation developed for the Ångström pyrheliometer as the basis for his pyranometer nearly two decades later. As shown in the schematic diagram in Figure 3.7, the Ångström pyranometer detector consisted of two white strips and two black strips mounted on an insulating framework at the end of a nickel-plated cylinder. A glass hemisphere was mounted to a metal disk to protect the detector from the weather. Thermojunctions were attached to the backs of the strips in good thermal contact and electrically isolated. The operation of the pyranometer was based on maintaining the temperature of the white strips by supplying a measured electrical heating to match the amount of radiant energy absorbed by the black strips (Coulson, 1975). In spite of some attractive design attributes, the Ångström pyranometer was never put into widespread use.

3.4.5 Kipp & Zonen Solarimeter (1924)

A significant advance in the practical measurement of solar radiation is attributed to Dr. W. J. H. Moll, who, as a physics professor at the University of Utrecht, designed the Moll thermopile (Moll, 1923). His design, depicted in Figure 3.8, addressed the

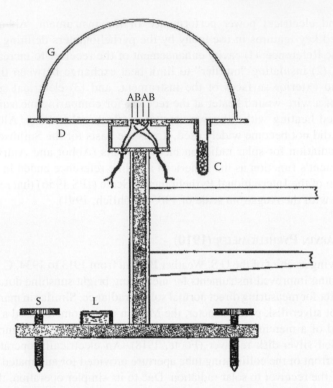

FIGURE 3.7 Schematic diagram of the Ångström electrical compensation pyranometer with white thermopile strips (A), black thermopile strips (B), glass hemisphere (G) mounted in a supporting disk (D), desiccant chamber (C), adjusting screws (S), and spirit level (L), (After Coulson, K. L., *Solar and Terrestrial Radiation, Methods and Measurements,* Academic Press, New York, 1975.)

measurement performance characteristics still sought in modern pyranometers and pyrheliometers:

- High level of responsivity (detector output signal produced per unit of incident solar irradiance)
- Rapid response to changes in solar irradiance (instrument time constant)
- Linearity over the range of expected levels of solar irradiance (consistent responsivity)
- Uniform response to radiation within the solar spectrum (generally 0.3 to 3.0 μm)
- Minimum influence of ambient temperature on the detector output

In 1924, the Moll thermopile was used by Dr. Ladislas Gorczynski, director of the Polish Meteorological Institute, as the basis for both a pyrheliometer and pyranometer. The resulting Moll–Gorczynski pyranometer was presented to Kipp & Zonen, a scientific instrument company in Delft, the Netherlands, and by 1927, the company

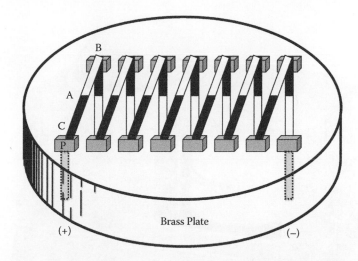

FIGURE 3.8 Concept of the Moll thermopile using thermojunctions of thin (0.005 mm) blackened strips of Manganin and constantan joined at A and soldered at points B and C to copper posts P. The mounting posts are fastened to a thick brass plate about 2 cm in diameter and coated with lacquer to provide electrical isolation and good thermal conductivity for the active junctions.

was offering "Solarimeters and Pyrheliometers" in their sales catalog.[*] The hand-made Moll thermopiles continued to serve as the basis of Kipp & Zonen radiometers until 1989 when the designs for models CM 1, CM 3, CM 5, and CM 6B were based on electroplated polyamide foil thermopiles.

3.4.6 ROBITZSCH BIMETALLIC ACTINOGRAPH (1932)

Based on a simple thermo-mechanical design, this portable and self-contained recording pyranometer was popular for monitoring solar radiation at remote locations and contributed to the measured data used for estimating the world distribution of global horizontal irradiation (Löf, Duffie, and Smith, 1966). Since its introduction in 1932, the instrument design has been refined and remains commercially available today. Designed by Max Robitzsch (1887–1952), a German meteorologist and professor at the Meteorological Observatory at Lindenberg, the instrument uses a blackened bimetallic strip about 8.5 × 1.5 cm mounted under a protective glass dome about 10 cm in diameter. The differing thermal expansion properties of the materials used to form the bimetallic strip create a simple bimetallic thermometer that responds proportionally to changes in solar irradiance. Using an ink pen and a mechanical linkage attached to the free end of the bimetallic strip, the Robitzsch design records the receiver deflections on chart paper wrapped on a clock-driven drum. The drum rotation can be made in 24 hours or in seven days to produce records of daily irradiation. The bimetallic receiver, mechanical linkage, and recording drum were housed in a

---

[*] A brief history of Kipp & Zonen is available from http://www.kippzonen.com.

FIGURE 3.9 The Robitzsch bimetallic actinograph was a popular instrument for measuring daily amounts of solar irradiance from about 1932 to 1970 at meteorological stations around the world.

weather-resistant metal case (Figure 3.9). Prior to the 1970s, the measurement uncertainties introduced by a variety of undesirable response characteristics resulted in daily total irradiation within ± 10% of the best performing instruments (Coulson, 1975). More recent design improvements have addressed these shortcomings:

- Calibration of the rectangular receiver varied greatly with solar azimuth and elevation.
- Pen zero-point depended on the varying thermal exchanges with the day/night longwave atmospheric conditions; receiver deflections varied with ambient temperature.
- Time response was relatively slow at about 15 minutes due to the large thermal mass of the receiver.
- There were optical distortions caused by the glass dome.

3.4.7 Eppley 180° Pyrheliometer (1930)

To address the needs of agriculture and to advance the meteorological observing capabilities in the United States in the early 1920s, the Weather Bureau and the National Bureau of Standards developed a new instrument for measuring total hemispheric

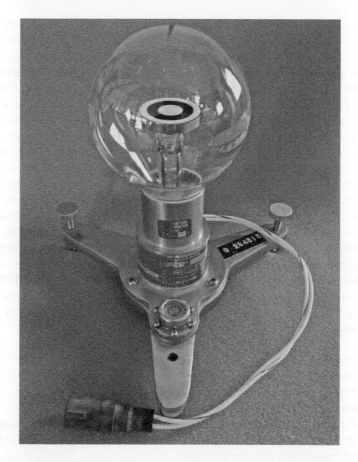

FIGURE 3.10 Eppley Laboratory, Inc., model 50 "light bulb" pyranometer used from 1951 to 1975 by the National Weather Service in their SOLRAD network of 61 stations. (Courtesy of Eppley Laboratory, Inc.)

solar irradiance on a horizontal surface (Kimball and Hobbs, 1923). Marvin suggested the black and white annular ring concept for the thermopile detector used in the "180° pyrheliometer" design as shown in Figure 3.10. Eppley Laboratory, Inc. introduced the first pyranometer based on the *Weather Bureau thermopile* at a meeting of the American Meteorological Society in May 1930 and continued to produce the radiometer with improvements for more than 30 years (Coulson, 1975). The National Weather Service (formerly the U.S. Weather Bureau) used variants of this pyranometer from 1951 to 1975 in their SOLRAD network (NREL, 1992).

The thermopile detector consisted of gold-palladium and platinum-rhodium alloys with 10 (early model) or 50 (later model) thermojunctions in good thermal contact but electrically insulated from the concentric rings of the annular design. Measuring about 29 mm in diameter, the detector was made from two thin silver rings and a small inside disk. The outer ring and the small disk were coated with magnesium oxide, a material with high reflectance in the solar spectrum. Depending on the year of manufacture, the inner ring was coated with either lamp black or Parson's optical black lacquer to provide for low reflectance at solar wavelengths (SOLMET, 1979). The glass bulb was about 76 mm in diameter, made of soda-lime glass about 0.6 mm thick, and filled with dry air at atmospheric pressure at the time of assembly (Coulson, 1975). This glass envelope was the basis for the more common reference to this pyranometer as the Eppley light bulb. Instrument performance was affected by the degradation of the soda-lime glass transmittance, the degradation of black detector paint's absorptivity with exposure, the lack of detector temperature compensation (except for later models), and the optical distortions associated with the blown-glass envelope. The Parson's optical black lacquer's absorptivity deteriorated significantly under exposure to sunlight, turning from black to gray to green. The last production version of this pyranometer was named model 50 to reflect the number of thermojunctions in the detector. In response to the energy crisis of the 1970s, considerable effort was applied to rehabilitating the historical measurement records based on this pyranometer at the National Climatic Data Center to reduce the effects of these instrument performance characteristics on data uncertainty (SOLMET, 1978). The resulting data archive provides the original measured value with additional corresponding "Engineering Correction" data that account for known instrumentation deficiencies:

1. Calibration changes
2. Solar radiation measurement scale differences
3. Midscale chart recorder setting
4. Pyranometer detector paint degradation
5. Calibration "cross-match problem" due to the uses of Parson's black and lamp black paints among the network instruments and calibration references
6. Temperature response characteristics

In total, the amount of adjustment to these hourly solar irradiance data could exceed 20%, depending on the pyranometers and chart recorders deployed over time at a SOLMET station.

3.4.8 Eppley Model PSP (1957)

The first model of the precision spectral pyranometer (PSP) developed by the Eppley Laboratory, Inc. was introduced in 1957 (Marchgraber and Drummond, 1960). The PSP has been widely accepted for deployment on land, sea, and aircraft platforms (Figure 3.11). The advanced design improvements of the PSP at the time it was introduced included

- Electrical compensation of the detector response to ambient temperature changes between –20°C and +40°C
- Precision optics protecting a smaller thermopile detector than previously used in black-and-white designs resulting in better angular response (±1% over incidence angles of 0° to 80°)

FIGURE 3.11   Eppley Laboratory, Inc., model precision spectral pyranometer. (Courtesy of Eppley Laboratory, Inc.)

- A single Parson's optical black painted receiver to eliminate spectral reflectance and durability issues associated with the white magnesium oxide coating previously used in black-and-white designs
- Improved heat transfer characteristics of the detector, domes, and instrument body
- Faster time response for measuring rapid changes of irradiance (1 sec for 1/e)
- Rugged design for improved long-term operation
- Replaceable desiccant for maintaining dry air within the body
- Replaceable outer hemispheres for optional color filter measurements of spectral bands

The wire-wound copper–constantan thermopile produces about 10 μvolts per Wm^{-2} and is mounted under two precision-ground Schott glass hemispheres as shown in Figure 3.12. The double-dome design addresses the need to minimize the convective heat losses between the thermopile and the surrounding environment. The white sunshade protects the instrument's cast bronze body from thermal gradients caused by solar heating. An optional ventilator is available for providing forced air over the outer hemisphere to reduce the effects of dew, snow, dust, and other potential contaminants from accumulating on the outer dome, hence improving the irradiance measurements (Figure 3.13).

3.4.9   Yanishevsky Pyranometer (1957)

The Yanishevsky (now Yanishevsky-Savinov) series of pyranometers served as the principal instrument for solar irradiance measurements in the former Soviet Union and continues that tradition in Russia and Central Asia today. The thermopile-type detector used in early models was constructed in a square checkerboard

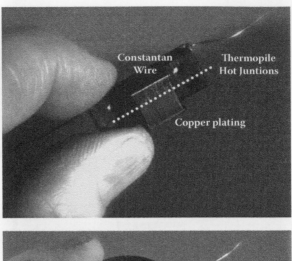

FIGURE 3.12   (a) Wire-wound F3 thermopile developed by the Eppley Laboratory, Inc., for use in their model precision spectral pyranometer. (b) Final assembly of Eppley F3 thermopile ready for installation in a model PSP. (Courtesy of Eppley Laboratory, Inc.)

pattern of alternate black and white squares and rectangles (Figure 3.14). Later models have a radial pattern of alternate black and white segments. The latest version, Yanishevsky-Savinov M115-M, pyranometer detector has alternating black and white squares forming a planar detector surface. The thermocouples are composed of alternating strips of Manganin and constantan. The hot junctions are blackened with carbon soot, and cold junctions are whitened with magnesium oxide. A single glass hemisphere protects the black-and-white receiver. Early models also had an auxiliary opaque metal hemisphere for providing a measurement of zero offsets. The measurement performance of the M115-M has been compared with other pyranometers deployed as part of the International Polar Year (Ivanov et al., 2008). Results of the 24-day comparison of hourly irradiances measured by the Norwegian Polar Institute's reference CMP 11 pyranometer show agreement within accepted measurement uncertainty limits at 6.3 ±5.6 Wm^{-2} over a measurement range of 0 to 500 Wm^{-2} (Ivanov, 2008).

FIGURE 3.13 Two model PSPs installed in Eppley ventilators to reduce effects of outer hemisphere contamination by dew, frost, snow, dust, and other sources. PSP with clear WG295 outer dome in foreground and PSP with RG780 outer dome in background. (Courtesy of Eppley Laboratory, Inc.)

FIGURE 3.14 The Yanishevsky pyranometer served the needs of the former Soviet Union for solar irradiance measurements. The black-and-white thermopile detector measures 3 cm × 3 cm.

3.4.10 Eppley Model NIP (1957)

The normal incidence pyrheliometer (NIP) designed and manufactured by the Eppley Laboratory, Inc. measures the direct normal ("beam") solar irradiance. As shown in Figure 3.15, the blackened wire-wound copper–constantan thermopile receiver is at the base of a view-limiting tube providing the 1 to 10 aperture-opening-to-tube-length

(a)

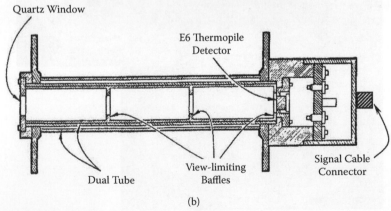

Quartz Window

E6 Thermopile Detector

Dual Tube

View-limiting Baffles

Signal Cable Connector

(b)

FIGURE 3.15 (a) Eppley Laboratory, Inc., model normal incidence pyrheliometer mounted on an automatic solar tracker for measuring direct normal irradiance. (Courtesy of Eppley Laboratory, Inc.) (b) NIP design elements shown in cross section.

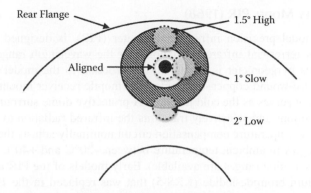

FIGURE 3.16 Diopter target used on the rear flange of a NIP to determine alignment with solar disc. The detector is fully illuminated when the sun spot image is fully within the white area of the target.

geometry responsible for the NIP's 5.7° field of view. In 1956, consistent with the existing solar tracker alignment capabilities and the need to provide instrument design standards, the International Radiation Commission established pyrheliometer design recommendations that included this viewing geometry (Bolle, 2008). Over the years, the double-walled tube has been constructed with an outer wall made of chrome-plated brass, white-painted brass, or stainless steel with the inner wall made of brass and fitted with light-baffle diaphragms to limit stray light from reaching the thermopile receiver. The viewing end of the tube is fitted with a quartz window of 1 mm thickness (currently Infrasil II material). The sealed double-walled construction is filled with dry air at atmospheric pressure at the time of manufacture. This design improves measurement stability, especially during windy conditions. The pyrheliometer has two flanges. The rear flange has three holes previously used for mounting on older solar trackers. The front flange has a sighting diopter arrangement for alignment with the target on the rear flange (Figure 3.16).

Recommended as one of the instruments for use during the 1957–1958 International Geophysical Year (IGY), the NIP continues to be produced today with these performance characteristics:

- Sensitivity: approx. 8 μvolts/Wm^{-2}
- Impedance: approx. 200 ohms
- Temperature dependence: ±1% over ambient temperature range –20 to +40°C (Compensation can be supplied over other temperature ranges)
- Linearity: ±0.5% from 0 to 1400 Wm^{-2}
- Response time: 1 second (1/e signal)
- Mechanical vibration: tested up to 20 g's without damage
- Calibration: referenced to Eppley primary standard group of pyrheliometers
- Size: 11 inches long
- Weight: 5 pounds

3.4.11   EPPLEY MODEL PIR (1968)

The Eppley model precision infrared radiometer (PIR), is designed to measure atmospheric or terrestrial infrared radiation over the wavelength range of 3.4 μm to 50 μm. This pyrgeometer is similar in construction to the model PSP with a single black wire-wound copper–constantan thermopile receiver mounted in a cast bronze body that serves as the cold junction. A protective dome surrounds the thermopile hot junctions and selectively transmits the infrared radiation to the receiver (Figure 3.17). A temperature compensation circuit nominally adjusts the thermopile output for changes in ambient temperature between –20°C and +40°C (other temperature compensation ranges are available). Early models of the PIR used a dome made of thallium bromide-iodide (KRS-5) that was replaced in the 1980s with a silicon hemisphere coated with tellurium and zinc selenide to improve thermal conductance and wavelength response of the instrument. As described in more detail in Chapter 10, the PIR receiver responds to these sources of infrared radiation:

- Incoming atmospheric or terrestrial (depending on mounting orientation)
- Effective instrument case temperature
- Effective hemispheric dome temperature

Therefore, the PIR design includes thermistors in contact with the hemispherical "solar blind" filter and the pyrgeometer case to provide temperature data coincident with the thermopile voltage generated during irradiance measurements. The irradiance can be computed from the calibration coefficients determined for the thermopile, case emittance, and dome correction factor (see Chapter 10). Historically, these calibration coefficients have been primarily determined from measurements made with laboratory blackbody calibrators. Since 2003, pyrgeometer calibrations are traceable to the interim World Infrared Standard Group (WISG) developed by

FIGURE 3.17   Eppley Laboratory, Inc., model precision infrared radiometer used for total hemispheric longwave radiation measurements (foreground) with model PSP (background). (Courtesy of Eppley Laboratory, Inc.)

the World Radiation Center (WRC; Gröbner and Los, 2007). The WISG is a group of pyrgeometers that form a reference standard that includes two Eppley PIRs.

3.4.12 PRIMARY ABSOLUTE CAVITY RADIOMETER (PACRAD) (1969)

This radiometer is historically significant for several reasons. Developed at the Jet Propulsion Laboratory to support NASA's Deep Space Network Energy Program, the PACRAD was needed to improve the measurement of simulated solar output in space chambers. Nearing retirement, J. M. Kendall, Sr. was tasked with designing a new radiometer after the thermal control system of at least one space probe did not function as predicted by space simulator testing. The simulator design problem was linked to the inability to accurately measure the simulated solar flux in the space chamber. The resulting *JPL standard total-radiation absolute radiometer* designed by Kendall was used to measure the absolute total radiation intensity with an absolute accuracy of 0.5% and to experimentally determine the Stefan-Boltzmann constant to within 0.5% of the theoretical value (Kendall, 1968). About this time, the scientific community was in need of an absolute solar irradiance measurement reference. A series of pyrheliometer measurement comparisons had produced solar irradiance measurement differences approaching 5% among the Ångström Scale of 1905, the Smithsonian Scale of 1913, and the International Pyrheliometric Scale of 1956 (Latimer, 1973). Clearly, an improved international reference for solar radiation measurement was needed.

Working from first principles of measurement, Kendall applied several new design concepts to develop the PACRAD instrument (Figure 3.18). First, he introduced the cavity-shaped receptor to enhance the detector absorptivity. Earlier pyrheliometer designs used flat plate detectors (i.e., disks or rectangular plates) that were prone to reflecting photons regardless of surface coating and collimator design. Second, the PACRAD had a built-in electrical calibrating heater to provide cavity heating accurately and equivalent to radiation heating. As a result, PACRAD irradiance measurements were based on the ability to make accurate and traceable electrical power measurements (i.e., voltage and resistance) and accurate dimensional measurements (i.e., aperture area) to produce solar irradiance data in Wm^{-2}. A third critical element in Kendall's design was the attention to heat transfer within the instrument. The materials selection, component design and placement, including the compensation cavity, were based on the need to absorb the incoming solar radiation that is converted to heating of the cavity receiver. The heat flows through a metallic thermal resistor to the massive heat sink to produce a temperature difference of a fraction of a Kelvin. The temperature difference is measured by a thermopile that is electrically calibrated as part of the measurement procedure. Key to the PACRAD design and operation was the thorough characterization of the effects of the various types of heat transfer within the radiometer and the resulting correction factors for all heat transfers that are not equivalent to the heating of the cavity by electrical substitution.

Three models of the PACRAD were built and compared with Eppley Ångström pyrheliometers (Kendall, 1969). Solar measurements taken at Table Mountain, California, on April 23, 1969, with the three PACRAD models agreed with one another within 0.11% and the Ångström pyrheliometers agreed with each other within 0.18%. However, solar irradiance measurements from the pyrheliometers

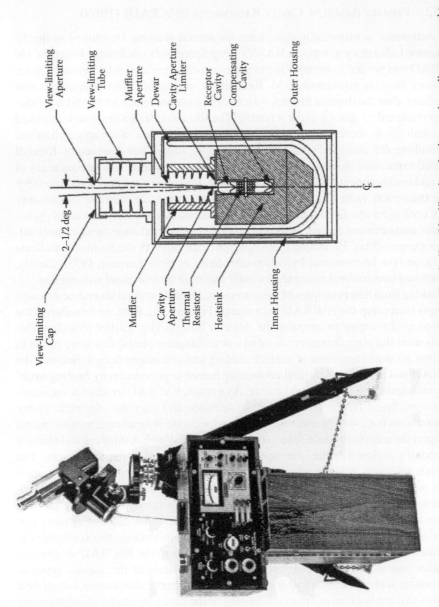

FIGURE 3.18 The primary absolute cavity radiometer became the basis for the electrically self-calibrating absolute cavity radiometers used to define the World Radiometric Reference. (After Kendall, J.M., *Primary Absolute Cavity Radiometer*, Technical Report 32-1396, Pasadena, CA, 1969.)

View-limiting Aperture

View-limiting Tube

Muffler Aperture

Dewar

Cavity Aperture Limiter

Receptor Cavity

Compensating Cavity

Outer Housing

View-limiting Cap

2–1/2 deg

Muffler

Cavity Aperture

Thermal Resistor

Heatsink

Inner Housing

were 2.3% lower than the PACRADs. This discrepancy was later resolved with the establishment of the World Radiometric Reference (WRR) based on a series of International Pyrheliometer Comparisons (IPCs) conducted by the World Radiation Center in Davos, Switzerland.

During IPC-III in 1970, Kendall's third model, PACRAD III, was observed to be very stable and was the only absolute cavity radiometer in the comparison. By 1975, five other absolute cavity designs had been developed and participated in IPC-IV (Fröhlich, 1991; Zerlaut, 1989):

- Active cavity radiometer (ACR) by R. C. Willson (Willson, 1973)
- Eppley-Kendall PACRAD-type (EPAC) by the Eppley Laboratory, Inc. and J. Kendall (Drummond and Hickey, 1968)
- Model PMO developed by the Physikalisch-Meteorologisches Observatorium
- Model PVS by Yu. A. Skliarov at the Saratov University in the former Soviet Union
- Model TMI, PACRAD-type, by Technical Measurements, Inc.

Because of its performance in IPC-III, the PACRAD III was chosen as the reference instrument for IPC-IV. The weighted average of all instruments for all observations during IPC-IV was 1.0017 with a standard error of only 0.18% among the absolute cavity instruments (Zerlaut, 1989). The subsequent establishment of the absolute WRR measurement scale is a result of the excellent agreement among all of the absolute cavity instruments at IPC-IV and later comparisons. In 1979, based on the results of IPC-IV, WMO adopted the WRR as the replacement for the International Pyrheliometric Scale of 1956 for the measurement of direct normal irradiance. Thus, the PACRAD instrument served an important role in establishing the WRR and providing for more accurate and consistent solar irradiance measurements around the world.[*]

3.4.13 Eppley Model 8-48 (1969)

Developed as a lower-cost replacement for the 180° pyrheliometer or Eppley light bulb pyranometer described earlier, this instrument offered improved measurement performance characteristics with a more robust design than its predecessor. In place of a rather large and delicate glass envelope surrounding the detector, a precision-ground optical Schott glass WG295 hemisphere protects the black-and-white segments of the differential thermopile detector from the elements (Figure 3.19). The black (hot) junctions were originally painted with Parson's optical black and the white (cold) junctions coated with barium sulfate. The relatively large receiver produced about 10 μvolts per Wm^{-2} and a time response of 5 seconds for a change in signal of 1/e. The challenge of constructing a planar receiver from black and white segments resulted in angular response specifications of ±2% for incidence angles 0–70°

---

[*] During this same period, other researchers were developing advanced pyrheliometer designs for solar irradiance measurements from space. A summary of the results from Willson's active cavity radiometer, NASA's eclectic satellite pyrheliometer, and others is available from R. C. Willson (1993).

FIGURE 3.19 Eppley Laboratory, Inc., model 8-48 pyranometer exhibits minimal thermal offset due to equal exposure of thermopile junctions to the effective sky temperature. (Courtesy of Eppley Laboratory, Inc.)

and ±5% for incidence angles 70–80°. The balanced black-and-white receiver design, however, results in minimal thermal offset on the order of one Wm^{-2} (Dutton et al., 2001) and negates the need for a protective sunshade over the body and an inner dome to limit the convective heat losses to the surroundings. In the course of production, these pyranometers have had bodies made from machined aluminum (early years) and painted cast aluminum (current production). A temperature compensation circuit nominally adjusts the thermopile output for changes in ambient temperature between –20°C and +40°C (other temperature compensation ranges are available).

3.4.14 LI-COR MODEL LI-200SA (1971)

The origin of this now ubiquitous pyranometer was based on the need for a low-cost radiometer to support the work of the Department of Soil and Water Sciences at the University of Wisconsin (Kerr, Thurtell, and Tanner, 1967). The silicon photodiode detector used in the original design was mounted below a plastic diffuser to provide adequate angular response and coupled with a purpose-built solid-state integrator for recording daily sums of global irradiance. The prototype radiometer was deployed in a 17-station mesoscale network in Wisconsin to study the temporal and spatial variations of solar radiation for applications in bioclimatology (Kerr, Thurtell, and Tanner, 1968). Lambda Instruments Corporation produced the first commercial version of the design in 1971. The name was changed to LI-COR, Inc. in 1978.

As shown in Figure 3.20, the LI-200SA pyranometer is compact (2.38 cm diameter × 2.54 cm tall) and lightweight (28 g). Because the detector is based on the

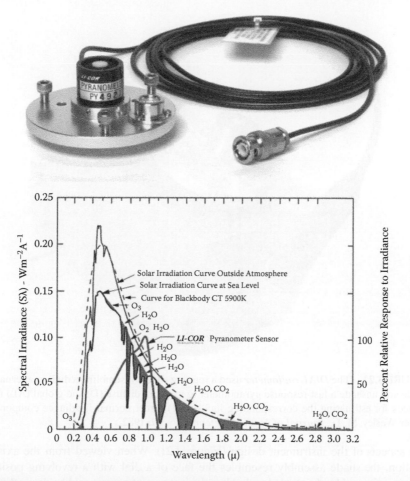

FIGURE 3.20 (Upper) LI-COR, Inc. model LI-200SA photodiode-based pyranometer and (Lower) the instrument's solar spectral response between 350 nm and 1100 nm compared with the spectra for extraterrestrial, sea level, and black-body radiation corresponding to a temperature of 5900 K. (From LI-COR Environmental, Inc. Used by permission.)

photoelectric effect rather than the thermoelectric effect used for thermopile radiometers, the LI-200SA has a very fast time response (10 μsec). The silicon photodiode, however, does not respond uniformly to all wavelengths and does not respond at all to wavelengths greater than about 1100 nm (Figure 3.20). This limited spectral response increases the measurement uncertainty when compared with thermopile-based detectors (King and Myers, 1997).

3.4.15 ROTATING SHADOWBAND RADIOMETER (1975)

Using a single LI-COR LI-200 pyranometer to measure global and diffuse solar irradiance for computing the coincident direct irradiance was the design basis for the *DIAL radiometer* (Wesely, 1982). The original DIAL radiometer prototype got its name from

FIGURE 3.21   The *DIAL radiometer* used a rotating shade (resembling a dial) to alternately shade and unshade a fast-response pyranometer to measure diffuse (D) and global (AL) irradiances for estimating the corresponding incident (I) direct normal irradiance component. (After Wesley, M. L., *Journal of Applied Meteorology, 21,* 1982.)

two aspects of the instrument design (Figure 3.21). When viewed from the axis of rotation, the shade assembly resembles the face of a dial with a revolving position indicator formed by the shade band. Also, the instrument name could be derived from its abilities to measure diffuse (D), estimate direct (I), and measure global (AL) solar irradiance components. The design innovations addressed the needs for a low-cost radiometer system that would be simple to install, operate, and maintain for purposes of estimating the broadband solar irradiances at the earth's surface and the corresponding atmospheric turbidity. The innovations included the following:

- No mechanical adjustment to the shading device
- Compact and lightweight mounting arrangement
- Commercially available pyranometer using a silicon photodiode mounted beneath a cosine-correction diffuser
- Integrated assembly, suitable for field applications in remote areas

The DIAL radiometer concept was later commercialized for applications in solar resource assessment and atmospheric science. Commercial versions based on the DIAL radiometer concept are now known as a rotating shadowband radiometer (RSR), a rotating shadowband pyranometer (RSP), a thermopile shadowband radiometer (TSR), or a multifilter rotating shadowband radiometer (MFRSR). The DIAL radiometer design used a small pyranometer positioned under a motorized shading band that completed one rotation about every 4 minutes. The band was ~0.9 cm wide and ~6.0 cm away from the 0.8 cm diameter diffuser on the pyranometer. With a shadowband rotation speed of about 0.25 rpm, the pyranometer detector was shaded for about 0.7 s. This was a sufficient shading interval for strip-chart recorders of the day to respond and display the minimum signal from the fast-response photodiode detector used in the pyranometer. The diffuse and global irradiance data were based on the minimum (shaded) and maximum (unshaded) pyranometer signals, respectively.

Initial comparisons of vertically integrated estimates of atmospheric turbidity made with a DIAL radiometer and two broadband thermopile pyranometers (Eppley Laboratory, Inc. model 8-48—unshaded and shaded) suggested single reading accuracy of ±30% for turbidity measurements for solar zenith angles less than 65°. When data from the DIAL radiometer were corrected for angular response and excessive shading of diffuse sky radiance due to shade bandwidth, averaged results compared within ±2%. This comparison missed the spectral response problem seen on clear days with modern rotating shadow band radiometers.

Since the introduction of the rotating shadowband concept, considerable effort has been made to characterize and improve the system design and data correction algorithms (Augustyn et al., 2004; Harrison, Michalsky, and Berndt, 1994; King and Myers, 1997; Rosenthal and Roberg, 1994; Stoffel, Riordan, and Bigger, 1991; Vignola, 2006). Specifically, these response characteristics of an RSR continue to be studied:

- **Spectral response**: From the outset of DIAL radiometer design, the need to provide fast response to changing irradiance levels has been met by the use of a pyranometer with a silicon photodiode detector. The time response for such a detector is on the order of 10 μsec compared with 1 to 5 seconds (or longer) for thermal detectors. Because the photodiode detector responds nonuniformly to the solar radiation between about 300 nm and 1100 nm, various correction schemes have been developed to simulate the corresponding broadband measurements. Because a silicon photodiode is less responsive near the 400 nm spectral region (i.e., blue), so-called spectral mismatch errors of up to 30% can occur for measurements of clear-sky diffuse irradiance. The advent of small thermopile detectors has provided the opportunity for fast-acting and broadband spectral response when used in an RSR (http://www.nrel.gov/midc/SRRL_TSR/).

- **Temperature response**: The electrical current produced by the photoelectric effect in a photodiode detector increases with increasing temperature. Corrections to the measured solar irradiance have been made with coincident measurements of air temperature near the RSR and from detector temperature measurements from a thermistor placed near the photodiode detector. RSR irradiance measurement uncertainties due to temperature response can be on the order of 3% for a 10°C temperature change (King and Myers, 1997).

- **Angular response**: Computing accurate direct normal irradiance from RSR measurements of global and diffuse components requires adequate design and characterization of the pyranometer for angular or "cosine" response. Typically, commercially available pyranometers have adequate angular response characteristics between incidence angles of 0° and 60° as measured normal to the detector surface. Pyranometer angular response is commonly undefined for incidence angles greater than 85°. Accounting for the specific angular response characteristics of a pyranometer can reduce the measurement uncertainty of direct normal irradiance as estimated from RSR measurements by a few percent for solar zenith angles less than about 70°.

Coming full circle, the original intent of the DIAL radiometer design has been fulfilled by the MFRSR with its capabilities to measure solar spectral irradiance at selected wavelengths allowing for the retrieval of atmospheric optical depths of aerosols, water vapor, and ozone (Harrison et al., 1994). The MFRSR uses six independent interference filter-silicon photodiode combinations, mounted in a temperature-controlled enclosure, to detect spectral irradiance using six narrowband filters (10 nm wide). One unfiltered channel is used to obtain broadband solar irradiance estimates (Figure 3.22). As a result, the MFRSR design provides for the spectral measurement of diffuse and global irradiance and estimates of direct normal irradiance.

FIGURE 3.22 Multifilter rotating shadowband radiometer detector arrangement as positioned beneath the diffuser.

FIGURE 3.23 (See color insert.) The WRR is determined by a group of absolute cavity radiometers named the World Standard Group. At the moment, the WSG is composed of six instruments: PMO-2, PMO-5, CROM-2L, PACRAD-III, TMI-67814, and HF-18748.

3.4.16 World Standard Group (1979)

The Physikalisch-Meteorologisches Observatorium Davos (PMOD) established the World Radiation Center (WRC) to guarantee international homogeneity of solar radiation measurements by maintaining the World Standard Group (WSG) of electrically self-calibrating absolute cavity radiometers. The WSG defines the World Radiometric Reference (WRR) as a detector-based measurement reference (Fröhlich, 1978). Presently, the WSG comprises six well-characterized instruments: PMO-2, PMO-5, CROM-2L, PACRAD-III, TMI-67814, AND HF-18748 (Figure 3.23). The WSG serves as the reference measurements for the IPC held every five years at the WRC.

3.5 RECENT ADVANCES IN SOLAR MEASUREMENTS

3.5.1 Automatic Hickey–Frieden Cavity Radiometer

The Eppley Laboratory, Inc. model automatic Hickey–Frieden (AHF) absolute cavity pyrheliometer is the automated version of the model HF radiometer that first participated in an IPC in 1980 (Zerlaut, 1989). Similar to the PACRAD and in collaboration with JPL, the HF cavity radiometer was developed to serve as a reference standard for the measurement of DNI. In fact, HF-18748 is one of the WSG absolute cavity radiometers defining the WRR at the WRC in Davos, Switzerland.

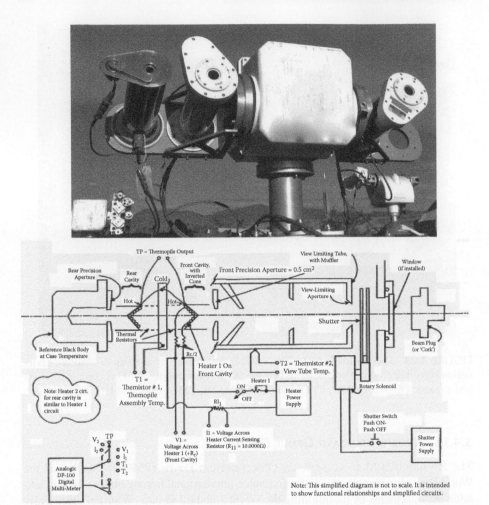

FIGURE 3.24 Eppley Laboratory, Inc., model automatic Hickey–Frieden electrically self-calibrating absolute cavity radiometer. (Top: Courtesy of Eppley Laboratory, Inc.; bottom: after Reda, I., NREL Report No. TP-463-20619, 1996.)

An electrically self-calibrating instrument, the sensor consists of a balanced cavity receiver pair attached to a circular wire-wound and plated thermopile (Figure 3.24). The blackened cavity receivers are fitted with heater windings that allow for absolute operation using the electrical substitution method, which relates radiant power to electrical power in SI units. The forward cavity views the DNI through a precision aperture immediately in front of the cavity receiver and behind a view-limiting aperture. The precision aperture is made of Invar, an alloy of 63.8% iron, 36% nickel, and 0.2% carbon with a very low coefficient of thermal expansion. The area is nominally 50 mm^2 and is measured for each unit. The rear receiver views an ambient temperature black body. The HF radiometer element with a baffled view-limiting tube and black body are fitted into an outer tube that acts as the enclosure for the instrument and reduces thermal gradients in the instrument due to wind and rapid changes to ambient temperature. The model AHF has an automatic shutter attached to the outer tube for simplifying the self-calibration and measurement controls.

The operation of the cavity radiometer and the measurements of the required calibration parameters are performed using digital electronics (AHF and later versions of the HF). The control functions include setting of the calibration heater power level, activation of the calibration heater, selection of the signals to be measured, and control of the digital multimeter measurement functions and ranges. The measured parameters include the thermopile signal, the heater voltage, and the heater current that is measured as the voltage drop across a 10-ohm precision resistor. The instrument temperature may also be measured using an internally mounted thermistor. The digital multimeter resolution of 100 nanovolts allows for the measurement of a thermopile signal change equivalent in radiation to approximately 0.1 Wm^{-2}.

Although these are absolute devices, the radiometers are compared with the Eppley reference cavity radiometers that have participated in multiple IPCs and other intercomparisons to provide direct calibration traceability to the WRR.

3.5.2 TOTAL IRRADIANCE MONITOR (TIM)

The Laboratory for Atmospheric and Space Physics at the University of Colorado at Boulder designed, built, and operates the solar total irradiance monitor (TIM) as one of the instruments developed for NASA as part of the SOlar Radiation and Climate Experiment (SORCE) launched in January 2003 (Kopp, Lawrence, and Rottman, 2003). The instrument continues the 25-year record of space-based measurements of total solar irradiance (TSI) that began in 1979 with the earth radiation budget (ERB) instrument and continues with the active cavity radiometer irradiance monitor (ACRIM) and variability of solar irradiance and gravity oscillations (VIRGO) radiometers. With an estimated absolute accuracy of 350 ppm (0.035%), the TIM can detect relative changes in solar irradiance to less than 10 ppm/yr (0.001%/yr), allowing determination of possible long-term variations in solar output. The TIM has four electrical substitution cavity radiometers (ESRs) to provide for measurement redundancy and for monitoring sensor degradation via duty cycling. The cavity interiors are etched nickel phosphorus (NiP) producing solar reflectances in the range of 10^{-4}. Each ESR can be independently shuttered from sunlight. The electrical power needed to maintain fixed cavity temperature during shutter cycles determines the radiative power absorbed by the cavity due to the entering solar power. The ESRs are balanced as pairs, one acting as a thermal reference while the other is actively driven to the temperature of the corresponding reference cavity. The use of phase-sensitive detection provides for the high measurement precision of the instrument (Gundlach, Adelberger, Heckel, and Swanson, 1996).

Comparing TSI measurements during 2008 from the TIM (~1361 Wm^{-2}) coincident with those from ACRIM III (~1363 Wm^{-2}) and VIRGO (~1366 Wm^{-2}) suggests a negative bias of about 5 Wm^{-2} for the TIM (http://www.pmodwrc.ch/pmod.php?topic=tsi/composite/SolarConstant). The TSI measurements resulted in a workshop hosted by NIST and NASA in 2005 to discuss instrument design,

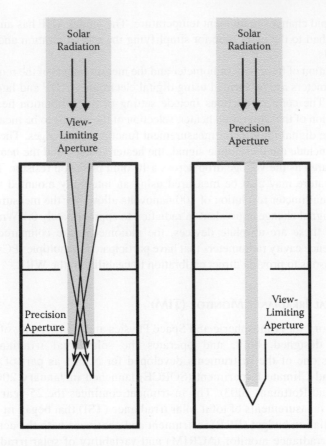

FIGURE 3.25 The relative position of the view-limiting and precision apertures is different between the designs of absolute cavity radiometers (left) and the total irradiance monitor. The optical layout of the TIM (right) places the small precision aperture at the front of the instrument so that only the solar radiation to be measured enters the cavity detector (left). With the precision aperture near the cavity detector, excess radiation is allowed into the instrument interior that can lead to erroneously high irradiance measurements from scatter (indicated by the arrows). (After Kopp, G., and J. L. Lean, *Geophys. Res. Letters Frontier article,* 38, 2011.)

measurement uncertainties, and stabilities in detail (Butler, Johnson, Rice, Shirley, and Barnes, 2008). Notable among the findings was the relative placement of the view-limiting and precision apertures relative to the cavity receiver in the instrument designs (see Figure 3.25). The TIM positions the smaller precision aperture at the front of the instrument so that only the light that is intended for measurement enters the cavity. All other cavity radiometers place the view-limiting aperture at the front, allowing excess light into the instrument interior that can lead to erroneously high irradiance measurements from scattered light.

The TIM value of total solar irradiance is 1360.8 ± 0.5 Wm⁻² during the 2008 solar minimum period (Kopp and Lean, 2011). The discrepancy between the canonical minimum value of TSI of 1365.4 ±1.3 Wm⁻² and the TIM value remains under investigation.

3.5.3 PMOD Model 8 (PMO8)

During 2009–2010, the PMOD developed a new prototype model in its series of model PMO cavity radiometers. The PMO8 design goals were to improve the performance of the existing model PMO6 by reducing or eliminating these measurement issues:

- Static nonequivalence
- Temperature dependency
- Dynamic nonequivalence
- Aperture heating

Like the PMO6 production unit, the PMO8 prototype is an electrically self-calibrating absolute cavity pyrheliometer designed to accurately substitute radiant power with electrical power for measuring DNI. During operation of the instrument, an electronic circuit maintains a constant heat flux from the cavity heater by adjusting the input electrical power. The heat flux is measured differentially between the receiving cavity and the reference cavity to compensate for the rate of change of the heat sink's temperature.

In comparison, the standard mode of PMO6 operation consists of alternate measurements of cavity heating with the shutter closed and open. The duration of each shutter cycle (closed and open) is selectable with a default value of 60 seconds. The operator determines the number of cycles and the integration time. The start time of the measurements can be set to a predefined time. After every open–closed cycle the irradiance is shown on the display and sent out to the RS232 port for recording.

The new PMO8 prototype features the following:

- Synchronous shutter control
- Two heater power levels
- Continuous sampling (50 ms)
- Serial communications via RS232 or USB interface

Results of the PMO8 participation in IPC-XI in 2010 will be used to verify the design and determine the readiness for production.

SUMMARY

The measurement of radiant energy remains a difficult and challenging science in spite of the fact that fundamental radiometry has a long history of steady improvements. Beginning with the need to satisfy basic curiosities of physicists and astronomers, the present interests in solar energy conversion and global climate change continue to motivate the development of more accurate, reliable, and economical radiometers for ground- and satellite-based observations. The current internationally accepted measurement reference for solar irradiance, the WRR, is a detector-based standard with an estimated measurement uncertainty of ±0.3% and a precision of ±0.1%. Improvement of the WRR appears imminent as scientific and engineering advances are applied to the measurement of solar energy.

QUESTIONS

1. Who is credited with developing the first pyrheliometer for measuring direct normal irradiance?
2. What thermopile design facilitated the development of the modern pyranometer?
3. When was the World Radiometric Reference established as the international measurement standard for solar irradiance?
4. What is the currently accepted value of the total solar irradiance?
5. Name the response characteristic contributing most to the estimated measurement uncertainty of a photodiode-based pyranometer.
6. Name the type of radiometer designed to measure atmospheric or terrestrial infrared radiation over the wavelength range of 3.4 to 50 μm.
7. In what year did the U.S. Weather Bureau begin the routine measurement of solar radiation using a pyranometer?

REFERENCES

Abbot, C. G. 1911. The silver disk pyrheliometer. *Smithsonian Miscellaneous Collection* 56:7.

Abbot, C. G. 1929. *The sun and the welfare of man*. In the Smithsonian Series Volume 2, ed. Charles Greenley Abbot. New York: The Series Publishers, Inc.

Abbot, C. G. and L. B. Aldrich. 1916. The pyranometer—an instrument for measuring sky radiation. *Smithsonian Miscellaneous Collection* 66(7).

Abbot, C. G. and L. B. Aldrich. 1932. An improved water-flow pyrheliometer and the standard scale of solar radiation. *Smithsonian Miscellaneous Collection* 3182(87):8.

Ångström, K. 1893. Eine elegtrische Kompensationsmethode zur quantitativen Bestmmung strahlender Warme. *Nova Acta Regiae Societatis Scientiarum Upsala* 3(16) (Translated in *Physical Review* 1, 1894).

Ångström, K. 1899. The absolute determination of the radiation of heat with the electric compensation pyrheliometer, with examples of the application of this instrument. *Astrophysical Journal* 9:332–346.

Augustyn, J., T. Geer, T. Stoffel, R. Kessler, E. Kern, R. Little, F. Vignola, and B. Boyson. 2004. Update of algorithm to correct direct normal irradiance measurements made with a rotating shadow band pyranometer. In *Proceedings of Solar-2004*, American Solar Energy Society, Portland, Oregon, July 10–14.

Beaubien, D. J., A. Bisberg, and A. F. Beaubien. 1998. Investigations in pyranometer design. *Journal of Atmospheric and Oceanic Technology* 15:677–686.

Bolle, Hans-Jürgen, 2008. International Radiation Commissions 1896 to 2008: Research into atmospheric radiation from IMO to IAMAS. *International Association of Meteorology and Atmospheric Sciences, IAMAS Publication Series* No. 1, ISBN 978-3-00-024666-1. Available at: http://www.irc-iamas.org/resources/

Butler, J., B. C. Johnson, J. P. Rice, E. L. Shirley, and R. A. Barnes. 2008. Sources of differences in on-orbital total solar irradiance measurements and description of a proposed laboratory intercomparison. *Journal of Research of the National Institute of Standards and Technology* 113:187–203.

Callendar, H. L. and A. Fowler. 1906. The horizontal bolometer. *Proceedings of the Royal Society Series A*. 77:15–16.

Coulson, K. L. 1975. *Solar and terrestrial radiation: Methods and measurements*. New York: Academic Press.

Covert, R. N. 1925. Meteorological instruments and apparatus employed in the U.S. Weather Bureau. *J. Opt. Soc. Amer. Rev. Sci. Inst.* 10:299–425.

Crova, M. A. 1876. New pyrheliometer. *American Journal of Science and Arts* Volume XI, Third Series (January–June), eds. J. D. Dana, B. Silliman, and E. S. Dana, 220–221. New Haven: Editors.

Dorno, C. 1923. Progress in radiation measurements. *Monthly Weather Review* 50(10). U.S. Weather Bureau.

Drummond, A. J. 1970. A survey of the important developments in thermal radiometry. In A. J. Drummond, ed., *Advances in geophysics* (Vol. 14). New York: Academic Press.

Drummond, A. J. and J. R. Hickey. 1968. The Eppley-JPL Solar Constant Measurement Program. *Solar Energy* 12:217–232.

Drummond, A. J., W. J. Scholes, J. J. H. Brown, and R. E. Nelson. 1968. *Radiation including Satellite Techniques*. Proceedings of the WMO/IUGG Symposium held in Bergen, August 1968. WMO No. 248, 549 pp.

Dutton, E. G., J. J. Michalsky, T. Stoffel, B. W. Forgan, J. Hickey, D. W. Nelson, T. L. Alberta, and I. Reda. 2001. Measurement of broadband diffuse solar irradiance using current commercial instrumentation with a correction for thermal offset errors. *Journal of Atmospheric and Oceanic Technology* 18:297–314.

Eddy, J. A. 1976. The Maunder Minimum. *Science* 192:1189-1202. doi: 10.1126/science.192.4245.1189

Foote, P. D. 1918. Some characteristics of the Marvin pyrheliometer. *Scientific Papers of the Bureau of Standards*, No. 323. Department of Commerce. Issued June 28, 1918 (Government Printing Office).

Fröhlich, C. 1991. History of solar radiometry and the world radiometric reference. *Metrologia* 28:111–115.

Glickman, T. S. (ed.). 2000. *Glossary of meteorology,* 2d ed. American Meteorological Society. Available at: http://amsglossary.allenpress.com/glossary

Gröbner J. and A. Los. 2007. Laboratory calibration of pyrgeometers with known spectral responsivities. *Applied Optics* 46:7419–7425.

Gundlach, J. H., E. G. Adelberger, B. R. Heckel, and H. E. Swanson. 1996. New technique for measuring Newton's constant G. *Phys. Rev. D* 54:R1256–R1259.

Hand, I. F. 1941. A summary of total solar and sky radiation measurements in the United States. *Monthly Weather Review* 69:95–125.

Harrison, L. and J. Michalsky. 1994. Objective algorithms for the retrieval of optical depths from ground-based measurements. *Applied Optics* 33:5126-5132.

Harrison, L., J. Michalsky, and J. Berndt. 1994. Automated multifilter rotating shadow-band radiometer: An instrument for optical depth and radiation measurements. *Applied Optics* 33:5118–5125.

Iqbal, M. 1983. *An introduction to solar radiation*. New York: Academic Press.

Ivanov, B., P. Svyaschennikov, J.-B. Orbaek, C.-P. Nillsen, N. Ivanov, V. Timachev, A. Semenov, and S. Kaschin. 2008. Intercomparison and analysis of radiation data obtained by Russian and Norwegian standard radiation sensors on example of Barentsburg and Ny-Ålesund research stations. *Journal of the CNR's Network of Polar Research Polarnet Technical Report, Scientific and Technical Report Series*, ISSN 1592-5064. Roberto Azzolini (ed.).

Jensenius Jr., J. S. and G. M. Carter. 1979. *Specialized agricultural weather guidance for South Carolina*. TDL Office Note 79-15, Silver Spring, MD: National Weather Service, National Oceanic and Atmospheric Administration, U.S. Department of Commerce.

Jensenius Jr., J. S. and G. F. Cotton. 1981. The development and testing of automated solar energy forecasts based on the model output statistics (MOS) technique. In *Proceedings of the first workshop on terrestrial solar resource forecasting and on the use of satellites for terrestrial solar resource assessment*, Newark, DE, Boulder, CO: American Solar Energy Society, pp. 22–29.

Kendall, J. M. 1968. *The JPL standard total-radiation absolute radiometer*. Technical Report 32-1263, Jet Propulsion Laboratory, Pasadena, CA, May 15, 1968.

Kendall, J. M. 1969. *Primary absolute cavity radiometer*. Technical Report 32-1396, Jet Propulsion Laboratory, Pasadena, CA, July 15, 1969.

Kerr, J. P., G. W. Thurtell, and C. B. Tanner. 1967. An integrating pyranometer for climatological observer stations and mesoscale networks. *Journal of Applied Meteorology* 6:688–694.

Kerr, J. P., G. W. Thurtell, and C. B. Tanner. 1968. Mesoscale sampling of global radiation analysis of data from Wisconsin. *Monthly Weather Review* 96:237–241.

Kimball, H. H. and H. E. Hobbs. 1923. A new form of thermoelectric recording pyrheliometer. *Monthly Weather Review* 51:239–242.

King, D. L. and D. R. Myers. 1997. *Silicon-photodiode pyranometers: Operational characteristics, historical experiences, and new calibration procedures.* Paper presented at the 26th IEEE Photovoltaic Specialists Conference, September 29–October 3, Anaheim, CA.

Kopp, G., G. Lawrence, and G. Rottman. 2003. The total irradiance monitor design and on-orbit functionality. In *SPIE Proc.* 5171-4, 2003, pp. 14–25.

Kopp, G. and J. L. Lean. 2011. A new, lower value of total solar irradiance: Evidence and climate significance. *Geophys. Res. Letters Frontier article* 38: L01706. doi: 10.1029/2010GL045777

Latimer, J. R. 1973. On the Ångström and Smithsonian absolute pyrheliometric scales and the International Pyrheliometric Scale 1956. *Tellus* 25:586–592.

Löf, G. O. G., J. A. Duffie, and C. O. Smith. 1966. World distribution of solar radiation. *Solar Energy* 10:27–37.

Marchgraber, R. M. and A. J. Drummond. 1960. *A precision radiometer for the measurement of total radiation in selected spectral bands*. Monograph No. 4, 10–12, Paris: Int. Union Geodesy Geophys.

Marty, C., R. Philipona, J. Delamere, E. G. Dutton, J. Michalsky, K. Stamnes, R. Storvold, T. Stoffel, S. A. Clough, and E. J. Mlawer. 2003. Downward longwave irradiance uncertainty under arctic atmospheres: Measurements and modeling. *J. Geophys. Res.* 108. doi: 10.1029/2002JD002937

Michalsky, J., E. G. Dutton, D. Nelson, J. Wendell, S. Wilcox, A. Andreas, P. Gotseff, D. Myers, I. Reda, T. Stoffel, K. Behrens, T. Carlund, W. Finsterle, and D. Halliwell. 2011. An extensive comparison of commercial pyrheliometers under a wide range of routine observing conditions. *Journal of Atmospheric and Oceanic Technology* 28:752–766.

Michalsky, J. J., R. Perez, R. Stewart, B. A. LeBaron, and L. Harrison. 1988. Design and development of a rotating shadowband radiometer solar radiation/daylight network. *Solar Energy* 41:577–581.

Moll, W. J. H. 1923. A theromopile for measuring radiation. *Proc. Phys. Soc. London Sect. B* 35:257–260.

National Renewable Energy Laboratory (NREL). 1992. User's manual—1961–1990 National Solar Radiation Data Base, Version 1.0. NSRDB—Volume 1. NREL/TP-463-4859, National Renewable Energy Laboratory, 243 pp. Also available as TD-3282, National Climatic Data Center, Asheville, NC.

Reda, I. 1996. Calibration of a solar absolute cavity radiometer with traceability to the world radiometric reference. NREL Report No. TP-463-20619, 79 pp.

Rosenthal, A. L. and J. M. Roberg. 1994. Twelve month performance evaluation for the rotating shadowband radiometer. Contractor Report, Sandia National Laboratories, SAND94-1248.

SOLMET. 1978. *User's manual—hourly solar radiation—surface meteorological observations*, Vol. 1, TD-9724. Asheville, NC: National Climatic Data Center.

SOLMET. 1979. *Final report—hourly solar radiation—surface meteorological observations*, Vol. 2, TD-9724. Asheville, NC: National Climatic Data Center.

Stoffel, T., C. Riordan, and J. Bigger. 1991. Joint EPRI/SERI Project to evaluate solar radiation measurement systems for electric utility solar radiation resource assessment, *The conference record of the twenty-second IEEE photovoltaic specialists conference*, 2992, 533.

Vignola, F. 2006. Removing systematic errors from rotating shadowband pyranometer data. In *Proceedings of the 35th ASES annual conference*, Denver, CO.

Waldmeier, M. 1961. *The sunspot-activity in the years 1612-1960*. Zurich: Schulthess.

Wesely, M. L. 1982. Simplified techniques to study components of solar radiation under haze and clouds. *Journal of Applied Meteorology* 21:373–383.

Willson, R. C. 1973. Active cavity radiometer. *Applied Optics* 12:810–817.

Willson, R. C. 1993. Solar irradiance. In R. J. Gurney, J. L. Foster, and C. L. Parkinson, eds., *Atlas of satellite observations related to global change*. New York: Cambridge University Press.

World Meteorological Organization (WMO). 1965. Measurement of radiation and sunshine. Guide to Meteorological Instrument and Observing Practices, 2d ed., WMO No. 8, TP3.

World Meteorological Organization (WMO). 2008. Measurement of radiation and sunshine. Guide to Meteorological Instruments and Observing Practices, 7th ed., WMO-No. 8. Available at: http://www.wmo.int/pages/prog/www/IMOP/IMOP-home.html

Zerlaut, G. 1989. Solar radiation instrumentation. In R.L. Hulstrom, ed., *Solar resources*. Cambridge, MA: MIT Press.

4 Direct Normal Irradiance

We receive from the sun perfectly enormous quantities of radiation.

Charles Greeley Abbot
(1872–1973)

4.1 OVERVIEW OF DIRECT NORMAL IRRADIANCE

Solar radiation that arrives at the earth's surface having come directly from the sun is defined as direct normal irradiance (DNI). Even if the sky is clear, DNI is smaller than would be measured at the top of the earth's atmosphere because DNI has undergone scattering (by molecules and aerosols) and absorption (by gases and aerosols) within the earth's atmosphere. If clouds are between the sun and the observer, and they are optically thick, then no direct normal irradiance reaches the earth's surface. The global horizontal irradiance (GHI) observed at the surface is a mixture of DNI that reaches the earth's surface without being scattered or absorbed and diffuse horizontal irradiance (DHI), the irradiance resulting from molecules, aerosols, and clouds scattering of the DNI. This partitioning is ever changing because the atmosphere is not static. The governing equation is

$$GHI = DNI \cdot \cos(sza) + DHI \tag{4.1}$$

where *sza* is the solar zenith angle, the angle between the zenith and solar directions. This partitioning is illustrated in Figure 4.1. It is easy to see that the solid dark line GHI is the sum of the direct normal component on the horizontal in gray (the first term in Equation 4.1) plus the dotted line DHI (the second term of the equation). This is consistent throughout the day until the sun is completely blocked and the first term on the right-hand side in Equation 4.1 goes to zero, leaving GHI equal to DHI between 16:00 and 17:00 and after 17:30 where the solid dark line and the dotted line coincide.

The sun's output at the top of the earth's atmosphere, called the total solar irradiance (TSI), has been measured by a variety of space-based radiometers over several decades. The latest is the total irradiance monitor (TIM) on the SOlar Radiation and Climate Experiment (SORCE) satellite (http://lasp.colorado.edu/sorce). TIM measurements of the TSI began in March 2003 and indicate stable solar output with the largest variation of around 0.3%, lasting just a few days, but with typical variations of TSI under 0.1% (Kopp and Lean, 2011). The sun's output is so stable that it goes by the misnomer "solar constant." Even this almost constant value at the top of the earth's atmosphere changes for the simple reason that the distance between the sun and the earth changes as explained in Chapter 2. The earth's orbit around the sun causes a change in the amount of solar radiation received at the top of the earth's atmosphere by a little over 6.7% between early January when it

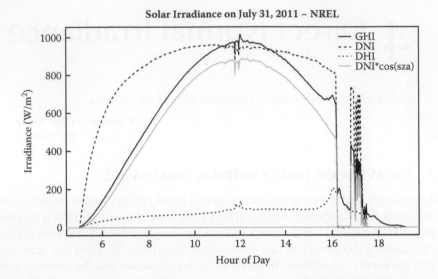

FIGURE 4.1 A demonstration of the partitioning of the GHI into the horizontal component of DNI and DHI using actual data from NREL's Solar Radiation Research Laboratory. DNI can be higher than GHI when it is clear (dashed). When it is cloudy, GHI is equal to DHI (e.g., between 16:00 and 17:00).

has its highest value (minimum earth–sun distance) and early July when it has its lowest value (maximum earth–sun distance). This change is entirely predictable from simple orbital mechanics and must be taken into account when computing the extraterrestrial radiation (ETR).

The amount of direct solar radiation that reaches the earth's surface is much more uncertain and difficult to forecast. Direct normal ("beam") irradiance is the only form of sunlight that can be used for concentrating solar energy conversion technologies such as concentrating solar power (CSP) thermal systems, which produce steam to generate electricity, and concentrating photovoltaic (CPV), which generate electricity directly. Flat plate collectors such as those that heat water and flat photovoltaic modules that produce electricity, often on the roofs of houses or businesses, can use any incident solar radiation: direct sunlight, sunlight that has been scattered by clouds, molecules, aerosols in the atmosphere, or sunlight reflected by the surface. DNI plays an important role for solar flat plate collectors because it offers the highest energy density for conversion. As shown in Figure 4.2, the area under the DNI time-series curve can represent the largest amount of solar energy available for a day. Flat plate solar collectors can be mounted with a fixed, adjustable, or tracking mode for optimum energy conversion. Under clear and partly cloudy sky conditions, DNI often is the predominant contributor to the plane of array (POA) solar irradiance.

Measuring DNI costs the most in the field of broadband solar and infrared radiation measurement. Although a good pyrheliometer for measuring DNI is typically less expensive than a good pyranometer used to measure GHI, the pyrheliometer must be accurately pointed at the sun from sunrise to sunset. Good automatic solar

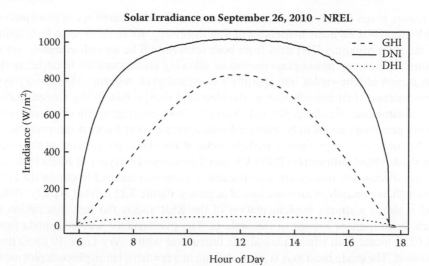

FIGURE 4.2 Demonstration of the integrated DNI far exceeding the integrated GHI on a very clear day.

trackers are much more costly than the typical thermopile pyrheliometer, usually 5 to 10 times higher. Less expensive clock-drive trackers have been used but require regular adjustments to keep the pyrheliometer pointed at the sun. Because of expense or the requirement to regularly align a manual tracker, there are relatively few continuous, long-term measurements of DNI with a pyrheliometer mounted on an automatic solar tracker. Globally, there are perhaps 100 times more pyranometer installations measuring GHI than pyrheliometers measuring DNI (http://www.nrel. gov/rredc/pdfs/solar_inventory.pdf; Stoffel et al., 2010). However, long-term direct beam data are crucial for selecting a CSP or CPV site, for estimating the expected annual output from a functioning plant, and for developing an operating strategy for the solar plant. Given the millions of dollars in capital costs to establish a large solar power plant, it is surprising that a reliable pyrheliometer and tracker are not used more often in prospecting for appropriate sites. Once operations begin, direct beam data are useful to estimate the plant's efficiency and to spot problems with plant operations. Although surface measurements are the most accurate, it is reasonable to use DNI estimated from satellite images to make initial surveys of the direct solar resource (Vignola, Harlan, Perez, and Kmiecik, 2005). Satellites can prospect over a wide area, and analysis of these satellite-based surveys can help narrow the search for the optimum CSP or CPV location in an area. When anchored by accurate ground-based data from the same general area, systematic errors in the satellite estimates can be accurately assessed and a more precise determination of the solar resource can be made. Another advantage of satellite estimates is that geostationary satellites have been flying for decades, so even a year's worth of surface-based data can be used to validate satellite-estimated irradiance. Satellite-derived estimates can then be used to evaluate the interannual variability and long-term trends of solar radiation at the site. Again, ground-based measurements can be used to reduce biases in the satellite-derived estimates. Further, satellites in combination with ground-based measurements will find increasing use in short-term forecasting of the solar resource. DNI data from both sources will be an indispensable aid in future solar energy power plant operation, allowing plant operators to optimize the integration of renewables into a utility's electrical grid. As renewable energy systems increase their contribution to the electrical energy mix in the United States, the transmission of solar-generated electricity from geographically dispersed solar power generators needs to be managed with accurate solar forecast information.

An alternative to pointing a pyrheliometer at the sun is the concept of the rotating shadowband radiometer (RSR). Chapter 7 describes two types of RSRs in detail, but we discuss this instrument here because it is another method to obtain the DNI, although, necessarily, with some loss of accuracy. Figure 3.21 is from Wesely (1982), and it shows a simple implementation of the RSR and is the first description of such an instrument known to the authors. The pyranometer was a Lambda (now LI-COR model 200) silicon photodiode instrument with a very fast (~10 μsec) time constant. The shade band axis is pointed south in a northern hemisphere deployment, and it is rotated with constant speed. When the band is beneath the motor housing, a measurement is made of the GHI, and with the LI-COR diffuser completely shaded by the band in front of the sun, a measurement is made of the DHI. If the skies are clear and stable, the difference of these two measurements is the *horizontal* component of the DNI that would fall on this pyranometer. The DNI can be obtained by dividing by cos (sza). If the cosine response of the LI-200 pyranometer is measured, a correction for the imperfect cosine response can be made to further improve the results. Other factors that add to the estimated DNI uncertainty include the temperature dependence of the photodiode pyranometer; the spectral response of silicon, which is neither flat nor responsive over the whole solar spectrum; and the need to correct for excess diffuse skylight blocked by the relatively wide band. With these corrections properly made, a reasonable estimate of DNI can be obtained, although it will be inferior to a good tracking pyrheliometer. This instrument eliminates the necessity of an expensive tracker, but, as will be shown in Chapter 7, it requires considerable effort to correct the results.

4.2 PYRHELIOMETER GEOMETRY

Figure 4.3 is a photograph of an Eppley model normal incidence pyrheliometer (NIP) mounted on an Eppley model SMT-3 automatic tracker. Figure 4.4 is a schematic of the optical geometry within a pyrheliometer with a circular field of view like the NIP. Sunlight enters from the right and is incident on the receiver behind the aperture stop on the left. The Gershun tube, not shown, that surrounds and holds this assembly, has its interior blackened. A window in front of the field stop (not shown) is typically selected from a material that transmits direct sunlight uniformly over all solar wavelengths to the interior of the instrument. The window also serves to seal the instrument from the elements. Sets of blackened baffles (three in this example) are positioned to minimize scattered light within the instrument. The detector sits directly behind and fills the aperture stop (or precision aperture) at the bottom of the tube that, along with the window and field stop at the top of the tube, defines

FIGURE 4.3 The Eppley NIP pyrheliometer mounted on an Eppley SMT-3 tracker; the insert is a blowup of the target on the back flange of the NIP used for alignment (see text for explanation of the numbering).

the acceptance angles for the instrument. Figure 4.4 illustrates the angles that are defined by the field and aperture stops. The angle designated z_0 is called the opening angle and is referred to when designating the field of view for a pyrheliometer (actually one-half of the total field of view). The limit angle z_l is denoted by this term because beyond this angle no sunlight is detectable. However, since the sun has a finite diameter, solar radiation is actually detectable up to 0.26° beyond this limit. The slope angle z_s is the most important angle with regards to tracking accuracy. If the center of the sun is at this angular distance from the center of the aperture,

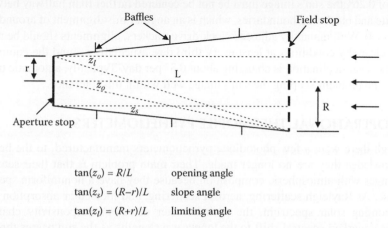

$$\tan(z_o) = R/L \qquad \text{opening angle}$$
$$\tan(z_s) = (R-r)/L \qquad \text{slope angle}$$
$$\tan(z_l) = (R+r)/L \qquad \text{limiting angle}$$

FIGURE 4.4 Schematic and definitions of the angles used to define the field of view of pyrheliometers.

TABLE 4.1

Optical Geometry (Defined in Figure 4.4) for a Few Pyrheliometers

Pyrheliometer	Slope Angle z_s (°)	Opening Angle z_o (°)	Limit Angle z_l (°)
Eppley NIP	1.78	2.91	4.03
Eppley AHF	0.804	2.50	4.19
Kipp & Zonen CHP 1	1.0	2.5	4.0

Source: Gueymard, C., *Journal of Applied Meteorology*, 37, 414–435, 1998.

vignetting occurs (radiation from the full disk of the sun does not strike the detector surface). Since the sun has a maximum radius of about 0.26°, the sun center must be closer to the aperture center than an angle equal to the slope angle minus 0.26° to avoid vignetting. The data in Table 4.1 are from Gueymard (1998) and list these angles for a few common pyrheliometers.

As an aside, LI-COR, Inc. (1982) published an instruction manual for the LI-2020 solar tracker that is no longer sold. However, in it they give a useful explanation of the alignment for the Eppley NIP, which we summarize here. The Eppley NIP has a top flange with a hole and a bottom flange with a target (see Figure 4.3 inset; the gray dots will be explained momentarily). The instrument is aligned when the sun's image is on the central black dot of the target as seen in the inset. A white ring that is surrounded by a black ring surrounds the central black dot. If the sun's image center is on the edge of the central black dot, its misalignment is 0.33° (position 1); if the sun's image center is on the edge of the white ring, its misalignment is 1.00° (position 2); and if the sun's image center is on the edge of the outer black ring, its misalignment is 2.00° (position 3). Therefore, for the Eppley NIP with a slope angle of 1.78° given the sun's radius of 0.26°, the sun's image must be not be centered farther than halfway between the white and black ring boundaries, which is an angular misalignment of around 1.5° (position 4). With manually adjusted clock-drive trackers, alignments should be made during clear-sky conditions at least every third day, especially around the equinoxes when the solar declination is changing about 0.3° per day. Typically, automatic trackers have no problem keeping the sun's image well within this tolerance.

4.3 OPERATIONAL THERMOPILE PYRHELIOMETERS

Although there were a few photodiode pyrheliometers manufactured, to the best of our knowledge they are no longer made. Their main problem is that their sensitivity changes with atmospheric composition because they have a nonuniform spectral response. As Rayleigh scattering, aerosol scattering, and molecular absorption alter the incoming solar spectrum, the pyrheliometer detector's responsivity changes. There is a marked spectral shift to the longer wavelengths as the sun passes through larger air masses. More and more of the blue light is scattered from the beam irradiance (DNI) as the sunlight travels longer distances through the atmosphere. This change in the spectral distribution of the DNI affects the responsivity of the photodiode sensor, making them unsuitable for accurate and consistent DNI measurements.

The most accurate pyrheliometers are electrically self-calibrating absolute cavity radiometers (ACRs). ACRs are not usually used in the field because they are a factor of 10 times more expensive than thermopile pyranometers; the frequent self-calibrations for maintaining accurate DNI values interrupt the data stream; and in their normal operating mode they are used without a window, which means that they must be attended to ensure that they are not damaged by exposure to weather and insects. Occasionally, ACRs are used with special windows to allow accurate DNI values to be obtained under a variety of conditions, but in almost all cases thermopile pyrheliometers are used for operational direct normal irradiance measurements.

The geometry that is used for the typical thermopile pyrheliometer is described in Figure 4.4. The thermopile detector is positioned at the bottom of the Gershun tube behind and thermally in contact with a black absorbing surface that fills the aperture stop. The hot junction is exposed to the direct sun, and the cold junction is attached to a heat sink in the pyrheliometer. The temperature difference creates a voltage that is proportional to the received direct solar irradiance. Voltages produced by the DNI on a clear day with little haze peak near 8 millivolts for typical commercial pyrheliometers and irradiances near 1000 Wm^{-2}.

Table 4.2 contains some operational characteristics of the type of pyrheliometers most often employed for long-term DNI measurements. This information is taken from

TABLE 4.2
Operational Characteristics of High-, Good-, and Secondary-Quality Pyrheliometers

Characteristic WMO/ISO	High-quality/ Secondary Standard	Good-quality/ First Class	Second Class (ISO Only)
Response time (95%)	<15 sec	<30 sec/ <20 sec	<30 sec
Zero offset (signal for 5 K h^{-1} change in ambient temperature)	±2 W m^{-2}	±4 W m^{-2}	±8 W m^{-2}
Resolution (smallest detectable change)	±0.5 W m^{-2}	±1 W m^{-2}	±5 W m^{-2}
Stability (% of full scale change/year)	±0.1% / ±0.5%	±0.5% / ±1%	±2%
Temperature response (maximum % error for 50 K change in ambient temperature)	±1%	±2%	±10%
Nonlinearity (% deviation from response at 500 W m^{-2} within the 100–1100 W m^{-2} range)	±0.2%	±0.5%	±2%
Spectral sensitivity (% deviation of transmission times absorption from this product mean between 300 and 3000 nm)	±0.5%	±1.0%	±5%
Tilt response (% deviation from response in horizontal position to a 1000 W m^{-2} beam between 0 and 90°)	±0.2%	±0.5%	±2%

WMO (2008) and ISO (1990) with ISO requirements if the second entry is different from WMO. Note that there are no comparable WMO entries for second-class pyrheliometers.

To evaluate most operational pyrheliometers, an extensive comparison was conducted over 10 months in 2008–2009, including all winter months and the hottest summer months at a midlatitude, midcontinent, high-elevation site (Michalsky et al., 2011). All manufacturers of thermopile pyrheliometers at the time were invited to send instruments, and all with pyrheliometers on the market at the time were represented. Since fall 2008, newer models may be available from these same manufacturers or others new to the market, but no evaluation of these pyrheliometers on the scale of this study has been scheduled. The comparison included more than the traditional calibration conditions that require clear, stable skies with direct irradiances greater than 700 Wm^{-2}. The conditions sampled included calm, windy, cold, hot, and thin to partly cloudy conditions. All instruments were calibrated over the 10-month study on 10 different occasions using unwindowed absolute cavity radiometers. Calibrations are discussed in Section 4.5 after the description of the cavity radiometer. Outside the 10 calibration periods, three cavity radiometers with windows that allowed for near-continuous operations were the standard to which the thermopile pyrheliometers were compared. All instruments during the study were stable in the sense that their calibrations were repeatable to within the uncertainty of the calibration process.

The variable conditions pyrheliometer comparison (VCPC), which was the title and acronym attached to this study, revealed three levels of performance for the instruments in the comparison. This ignores two outliers that performed so poorly that further discussion of them in the paper was not considered. One of these was a photodiode pyrheliometer that was not expected to perform well, and the other was a thermopile instrument that was available, but it is uncertain that it was used correctly as there was no instruction manual for operating this instrument. The best performers, as expected, were the absolute cavity radiometers with windows that transmitted all solar wavelengths. The estimated 95% confidence level uncertainty in the measurements made by the unwindowed cavity radiometers is 0.45%. A 95% confidence level means that 95% of the time, the measurements will be made within the uncertainty level expressed. The 95% confidence level uncertainty will be referred to as 95% uncertainty in the rest of this chapter. Transferring the calibration to windowed cavities produced a total 95% uncertainty of only 0.5% or just slightly more than the unwindowed cavities' calculated uncertainty. Several thermopile pyrheliometers performed with a 95% uncertainty between 0.7% and 0.8%; these were considered to be in the second tier of performance. One manufacturer supplied four versions of the same pyrheliometer model that had 95% uncertainties between 1.0% and 1.7%. Surprisingly, all of these 95% uncertainties are better than that claimed by their respective manufacturers.

These results assume that calibrations using unwindowed absolute cavity radiometers are performed on a regular basis, temperature corrections are applied, and, most importantly, instrument windows are cleaned on a routine basis, at least twice weekly or more often in particularly dusty or dirty sites. The results for most instruments in the study are given in Table 4.3. Temperature corrections in the fourth column refer to mathematical corrections suggested by the manufacturer that are in addition to those achieved by circuit design.

TABLE 4.3
Instrument Identification and Uncertainties for Clear-Sky Conditions from the Various Conditions Pyrheliometer Comparison

Manufacturer	Model	Serial No.	Temperature Correction (Yes/No)	95% Uncertainty
Eppley	All-weather cavity	31114	N	±0.5%
Eppley	All-weather cavity	29219	N	±0.5%
Eppley	All-weather cavity	32452	N	±0.5%
Kipp & Zonen	CH 1	970147	N	0.70%
				−0.60%
Kipp & Zonen	CHP 1	80011	Y	0.70%
				−0.60%
Kipp & Zonen	CHP 1	80010	Y	0.70%
				−0.60%
Kipp & Zonen	CHP 1	80009	Y	0.70%
				−0.60%
Kipp & Zonen	CH 1	30346	Y	±0.7%
Kipp & Zonen	CH 1	930039	N	±0.7%
Kipp & Zonen	CH 1	30340	N	±0.7%
Kipp & Zonen	CH 1 calcium fluoride window	30347	Y	0.70%
				−0.80%
Middleton	DN5	5094	N	0.70%
				−0.80%
Middleton	DN5	5027	N	0.70%
				−0.80%
Middleton	DN5	5029	N	0.70%
				−0.80%
Hukseflux	DR01	8029	Y	0.70%
				−0.80%
Hukseflux	DR01	8028	Y	0.70%
				−0.80%
Hukseflux	DR01	8027	Y	0.70%
				−0.80%
Kipp & Zonen	CHP 1	OPROTO1	N	±0.9%
Eppley	NIP ventilated	28322	Y	1.00%
				−0.70%
Eppley	NIP new brass	34504	Y	1.10%
				−0.90%
Eppley	NIP new brass	34507	Y	1.10%
				−0.90%

(continued)

TABLE 4.3 (CONTINUED)
Instrument Identification and Uncertainties for Clear-Sky
Conditions from the Various Conditions Pyrheliometer Comparison

Manufacturer	Model	Serial No.	Temperature Correction (Yes/No)	95% Uncertainty
Eppley	NIP new brass	34129	Y	1.10% −0.90%
Hukesflux	DR01P	8026	N	1.10% −1.30%
Eppley	NIP old brass	16229	Y	±1.2%
Eppley	NIP Old Brass	16319	Y	±1.2%
Eppley	NIP old brass	16521	Y	±1.2%
Eppley	NIP stainless steel	31139	Y	1.40% −1.20%
Eppley	NIP stainless steel	25791	Y	1.40% −1.20%
Eppley	NIP stainless steel	31144	Y	1.40% −1.20%
Eppley	NIP stainless steel	28260	Y	1.70% −1.30%
Matrix	MK-III (photodiode)	2457	N	−
Cimel	183A7	501657	N	−

Source: Adapted from Michalsky, J., E. G. Dutton, D. Nelson, J. Wendell, S. Wilcox, A. Andreas, P. Gotseff, D. Myers, I. Reda, T. Stoffel, K. Behrens, T. Carland, W. Finsterle, and D. Halliwell, *Journal of Atmospheric and Oceanic Technology, 28,* 752–766, 2011.

4.4   ABSOLUTE CAVITY RADIOMETERS

Before a discussion of pyrheliometer calibration, it is necessary to review design and performance characteristics of the absolute cavity radiometer. It is important to note that the measurement of DNI cannot be made from the first principles of physics. Instead, the concept of an absolute cavity radiometer is based on the ability to measure electrical power and equate the data to solar power. That is, an absolute cavity radiometer is considered an electrically self-calibrating instrument. The electrical substitution can take place in an *active* or *passive* mode of operation. In the active mode, the electrical substitution is performed as part of the irradiance measurement. In the passive mode, irradiance measurements are taken between electrical calibration events. A very simple schematic of one type of absolute cavity radiometer operating in the passive mode is shown in Figure 4.5. Only the passive mode of operating an absolute cavity radiometer will be described here.

Direct sunlight enters from the left in Figure 4.5 and encounters an aperture with a very precisely measured area. Sunlight hits the blackened cone where it is almost

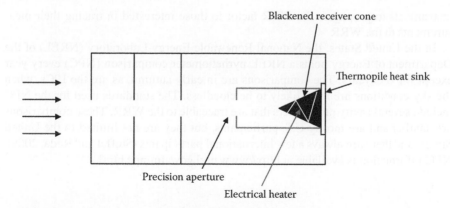

FIGURE 4.5 Schematic of an absolute cavity radiometer.

completely absorbed, thus heating the cone to which is attached the hot junction of a thermopile. The cold junction of the thermopile is attached to the heat sink. The temperature differential produces a voltage in the thermopile that is precisely measured. The field stop is then covered to prevent any sunlight from reaching the cone. The electric heater is activated to raise the temperature of the cone to the same temperature and produce the same voltage in the thermopile as it had with the sun striking it. The power to heat the cone electrically is divided by the area of the field stop to yield power per unit area in milliwatts cm^{-2} or, equivalently, Wm^{-2}. In the passive mode of operation, this calibration is made before and after DNI measurements over a selected interval (usually, sub-hourly).

While the process seems simple, in fact, there are several small correction factors that have to be considered to perform this measurement accurately. The major considerations (Reda, 1996) are the losses caused by reflection from the cone, which is nearly but not perfectly absorbing, and the losses resulting from the heated cone cooling by emitting infrared radiation; the nonequivalence of heating the cone with electrical power versus solar power; the scattered light from the precision aperture; light diffraction into the cavity; the electrical losses in the wire leads that heat the cavity; and, most importantly, the accuracy of the measurement of the area of the aperture.

The Physikalisch-Meteorologisches Observatorium Davos and World Radiation Center (PMOD/WRC) developed and maintains the World Radiometric Reference (WRR). The WWR is defined by the World Standard Group (WSG), which consists of a number of different manufacturers' cavity radiometers donated by research laboratories and manufacturers. In the tenth International Pyrheliometer Comparison (IPC-X), for example, six cavity radiometers constituted the WSG (Finsterle, 2006). Their average is defined to be the WRR. The 95% uncertainty level for the WRR is 0.3%. Every five years an IPC is held in Davos, Switzerland, the last in 2010. The comparisons are conducted in early autumn at the PMOD. In 2005, at IPC-X, 73 participants calibrated 89 pyrheliometers, mostly cavity radiometers. This group included 16 regional and 23 national radiation centers as well as other institutions and manufacturers. These participants then used their newly calibrated or recalibrated instruments to propagate the WRR factor to those interested in tracing their measurements to the WRR.

In the United States, the National Renewable Energy Laboratory (NREL) of the Department of Energy holds a NREL pyrheliometer comparison (NPC) every year except for IPC years. The comparisons are in early autumn, as are the IPCs, when the sky conditions are most likely to be cloudless. The standards used for the NPC include several cavity radiometers that are traceable to the WRR. These comparisons are smaller and are mostly U.S. participants, but they are not limited to the United States, and there are always a few international participants (Stoffel and Reda, 2009). NPC information is available at http://www.nrel.gov/aim/npc.html.

4.5 UNCERTAINTY ANALYSIS FOR PYRHELIOMETER CALIBRATION

The Joint Committee for Guides in Metrology (JCGM, 2008) published the *Guide to the Expression of Uncertainty in Measurements (GUM)* with minor changes from the 1995 version. This guide unifies the approach to expressing uncertainty for the measurement sciences internationally. Two practical guides (Cook, 2002; Taylor and Kuyatt, 1994) are used in the discussion that follows in an attempt to make the *GUM* procedures of uncertainty analysis in radiometry somewhat more accessible. The application of *GUM* to the calibration of a pyranometer explained in Reda, Myers, and Stoffel (2008) is a clear, published example of the practical application of the *GUM* approach in radiometry.

A clear statement of the reliability of a measurement in terms of its uncertainty should accompany every measurement. An important change in *GUM* from earlier descriptions of uncertainty analyses was the separation of uncertainty associated with repeated measurements of the same quantity that allows for a statistical treatment of the random behavior of the measurement (type A) and the uncertainty from all other causes that cannot be treated statistically (type B). For example, assume that a measurement of the DNI is made on a clear day near solar noon using a pyrheliometer. The pyrheliometer outputs a voltage proportional to the solar radiation incident upon its detector. Assume that the data logger recording the voltage output can sample every second and average 60 samples to get a 1-minute mean value and a standard deviation. This standard deviation is an example of a type A standard uncertainty. The data logger has a stated uncertainty of how well the voltage can be measured on a particular voltage range. That statement of the uncertainty for the voltage scale used is provided in the data logger manufacturer's specifications and is, therefore, a type B uncertainty since it was not obtained from a series of measurements.

A summary of the steps that should be followed in performing an uncertainty analysis using *GUM* include the following:

- Write down the measurement equation.
- Calculate the standard uncertainty for each variable in the measurement equation either as a type A or a type B uncertainty.
- Calculate sensitivity factors associated with each variable.

- Use the root sum of the squares of the *standard* uncertainties modified by sensitivity factors to determine the *combined* standard uncertainty.
- Calculate the *expanded* uncertainty by multiplying the combined uncertainty by a coverage factor found from the student's *t*-distribution to obtain the chosen standard level of confidence, often 95% representing a coverage factor (*k*) approximately equal to two.

The measurement equation that applies to the calibration of a pyrheliometer is simply

$$R = \frac{V}{DNI} \tag{4.2}$$

where R is the calibrated response to be determined in $\mu volts/Wm^{-2}$, V is the voltage output of the thermopile in $\mu volts$, and *DNI* is the reference direct normal irradiance. The most accurate DNI results come from an electrically self-calibrating absolute cavity radiometer that is traceable to the WRR in Wm^{-2}. For example, assume a measured 1-minute average voltage in full direct sunlight of 7993 $\mu volts$. The digital logging system used to read the voltage has a resolution of 1 $\mu volt$ and has an auto-zero feature that ensures that there is no offset. The standard deviation of the 60 1-second samples that result in the 1-minute average of 7993 $\mu volt$ is 111 $\mu volt$. The cavity radiometer, which was calibrated at an IPC, has a 95% uncertainty of 0.41%, including the 0.3% uncertainty of the WRR. The average cavity reading for the measurements is 932.3 Wm^{-2} with a calculated standard deviation of 13.7 Wm^{-2} over the same minute that the pyrheliometer samples were collected. Table 4.4 lists the standard uncertainties associated with calibrating the pyrheliometer; both type A and type B uncertainties are listed that arise from both the voltage measurement and the irradiance measurement.

The resolution of the data logger on the scale used to read the voltage is 1 $\mu volt$; this suggests that the voltage can lie anywhere between -0.5 and $+0.5$ $\mu volts$ of the reading, implying a standard uncertainty of $0.5/\sqrt{3}$ for a rectangular distribution. The standard sample deviation reported by the data logger is divided by $\sqrt{60}$ to get the standard uncertainty of the mean, where 60 is the number of samples in the 1-minute average. In a similar manner the DNI standard uncertainties were

TABLE 4.4
Derivation of Standard Uncertainty for Sample Pyrheliometer Calibration

Variable	Value	Semirange (B) or Std Dev (A)	Type	Distribution	DF	Std Uncertainty
V resolution	7993 $\mu volts$	0.5 $\mu volts$	B	Rectangular	1000	0.29 $\mu volts$
V measurements	7993 $\mu volts$	111 $\mu volts$	A	Normal	59	14.33 $\mu volts$
DNI cavity	932.3 Wm^{-2}	0.205% = 1.91 Wm^{-2}	B	Rectangular	1000	1.10 Wm^{-2}
DNI measurements	932.3 Wm^{-2}	13.7 Wm^{-2}	A	Normal	59	1.77 Wm^{-2}

calculated. The root sum of the squares of the standard uncertainties of the voltage is 14.33 μvolts to two decimal places implying that, for this case, the resolution of the meter does not add significantly to the uncertainty. The square root of the sum of the squares of the standard deviations of the DNI is 2.08 Wm^{-2}. The sensitivity factors for V and DNI are calculated by taking partial derivatives of Equation 4.2 with respect to V and DNI. Adding in quadrature, the combined standard uncertainty of the responsivity (R) is

$$u_{RC}^2 = \left(\frac{\partial R}{\partial V}\right)^2 (\Delta V)^2 + \left(\frac{\partial R}{\partial DNI}\right)^2 (\Delta DNI)^2 = \frac{(\Delta V)^2}{DNI^2} + \frac{(V \cdot \Delta DNI)^2}{DNI^4} \qquad (4.3)$$

Taking the square root yields a standard uncertainty u_{RC} of 0.025, which implies that 68% of the determinations of R would fall within ±0.025. Multiplying by a coverage factor of two, results in a responsivity and expanded uncertainty (U_E) of 8.574 ± 0.050 μvolts/(Wm^{-2}) with 95% confidence. The 95% confidence level implies that if this exercise were repeated, 95 of 100 times the calibration would fall within this range of ±0.050 μvolts/(Wm^{-2}).

4.6 UNCERTAINTY ANALYSIS FOR OPERATIONAL THERMOPILE PYRHELIOMETERS

The previous section covered an uncertainty analysis for the calibration of the pyrheliometer that will be deployed in the field. When the pyrheliometer is operated unattended in the field, other factors affect the measurement and add uncertainty to those measurements. Some of the contributors to this added uncertainty are temperature dependence, nonlinearity, pointing accuracy, solar zenith angle dependence, and the data logger voltage measurement, which may be of lower quality than the calibration data logging system. A huge uncertainty that can easily overwhelm everything just described is that caused by dirty optics (i.e., the window used to seal the pyrheliometer). Dirty optics result when instruments are not routinely maintained, with the frequency of cleaning dependent on the site and weather.

In Table 4.5 the listed uncertainties for estimating overall operational uncertainty are handled as percent uncertainties to demonstrate a different approach from that used in Table 4.4.

The combined standard uncertainty is calculated by taking the square root of the sum of the squares of the standard uncertainties from the right-hand column of Table 4.5. Small uncertainties that would not contribute significantly to the overall uncertainty have been intentionally neglected, and some contributions to uncertainty may have been inadvertently forgotten, but all contributions of consequence are thought to be included. The total standard uncertainty is given in Table 4.5, implying that with the coverage factor of two the uncertainty of the field measurement with 95% confidence in the result is 1.268% of the direct normal irradiance measurement. This is with the assumption of a clean pyrheliometer window; remember that even a modest dust veil on the window can reduce the measured DNI and easily double or triple this uncertainty systematically.

TABLE 4.5

Estimating Standard Uncertainty for an Operational Pyrheliometer

Variable	Value	Semirange (B) or Std Dev (A)	Type	Distribution	DF	Std Uncertainty
Calibration Table 4.4	R	0.583%	B	Rectangular	1000	0.337%
Nonlinearity		0.1%	B	Rectangular	1000	0.058%
Temperature dependence		0.5%	B	Rectangular	1000	0.289%
V resolution for 100 Wm^{-2}		0.33%	B	Rectangular	1000	0.191%
Window transmittance		0.60%	B	Rectangular	1000	0.364%
V measurements	7993 μvolts	111 μvolts	A	Normal	59	0.179%
Combined standard uncertainty						0.634%

4.6.1 WINDOW TRANSMITTANCE, RECEIVER ABSORPTIVITY, AND TEMPERATURE SENSITIVITY

The largest uncertainties in Table 4.5, besides that associated with the calibration of the pyrheliometer, are those due to the window transmittance and the temperature sensitivity of the instrument. If the window transmission were constant throughout the solar spectrum, this term could be reduced or even neglected. Some windows have nearly uniform transmittance, such as calcium fluoride, which has a tendency to scratch and is hygroscopic (attracts and absorbs water). Sapphire is another excellent choice, but it is probably too expensive to use on an operational pyrheliometer. Fused silica is the most popular choice for operational pyrheliometers because it transmits the ultraviolet well; however, the transmittance is not constant to 3000 nm but shows some dips in transmission, to low values near 2800 nm. This gives rise to the size of the transmission uncertainty used in Table 4.5. Another requirement for spectral insensitivity is an absorbing paint for the hot junction of the thermopile that is wavelength independent. Fortunately, there are several paints that are satisfactory in that they are nearly wavelength independent and durable.

Temperature sensitivity in thermopile pyrheliometers is usually corrected to the first order using a temperature compensation circuit. Manufacturers supply the temperature-dependent correction for any first-order mathematical temperature correction if there is no temperature compensation circuitry, and if there is electronic temperature compensation then a second-order mathematical temperature compensation may be suggested.

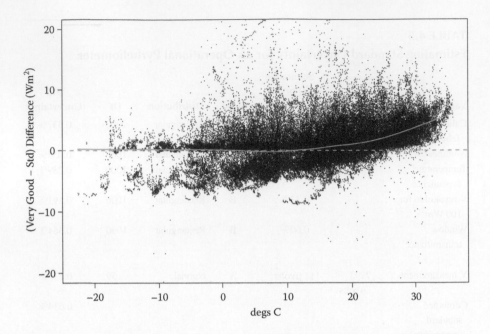

FIGURE 4.6   The temperature dependence of one very good pyrheliometer after all manufacturer's corrections applied.

Figures 9 and 10 in Michalsky et al. (2011) illustrate the effects of correcting and not correcting the temperature dependence of pyrheliometers. Figure 4.6 is a plot of the difference in irradiance measurements between a very good thermopile pyrheliometer and a cavity radiometer after all known temperature corrections have been made. The cavity radiometer has no temperature dependence; therefore, the figure indicates that, although most of the temperature dependence is removed, there is room for improvement in the thermopile pyrheliometer at the higher ambient temperatures. Algorithms can be developed from data using ambient air temperature as a proxy for the correction. The bottom line is that every appropriate attempt to correct for this effect will reduce the uncertainty of the direct normal irradiance measurement.

4.6.2   SOLAR ZENITH ANGLE DEPENDENCE

This uncertainty is not included in Table 4.5 but is of significant concern in some models of pyrheliometers. This effect is not understood but is clearly illustrated, for example, in Figure 4.7, which gives data from a broadband outdoor radiometer calibration (NREL, 2010) at the Southern Great Plains Atmospheric Radiation Measurement Facility of the U.S. Department of Energy. The figure is a plot of the calibrated responsivity over a day with a responsivity change of over 2% from 80° to 20° solar zenith angle during the morning and almost no change from 20° to 80° in the afternoon. Sometimes this change is smaller, and sometimes it is more symmetric about solar noon. This behavior is not uncommon for some pyrheliometers, and it exceeds the uncertainty of any contribution listed in Table 4.5.

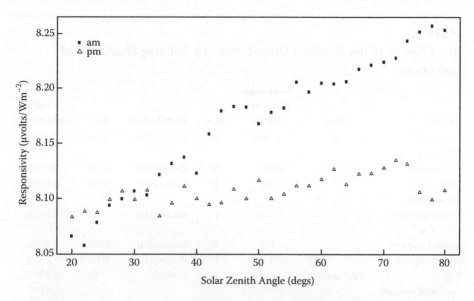

FIGURE 4.7 Solar zenith angle dependence of the Eppley NIP 29008E6 pyrheliometer during a 2010 broadband outdoor radiometer calibration (BORCAL). Note that the morning and afternoon responsivities differ as a function of solar zenith angle.

4.7 UNCERTAINTY ANALYSIS FOR ROTATING SHADOWBAND RADIOMETER MEASUREMENTS OF DIRECT NORMAL IRRADIANCE

Rotating shadowband radiometers derive direct normal irradiance by subtracting measured diffuse horizontal from measured global horizontal irradiance and dividing by the cosine of the calculated solar zenith angle at the time of measurement. Since the uncertainty associated with the GHI is larger than the uncertainty associated with the DNI measured by a pyrheliometer, the uncertainty of the calculated DNI, which includes the uncertainties of the GHI and DHI, will necessarily be even larger. Further, there is a small uncertainty associated with the solar position calculation, which is negligible with the best algorithms, and there is an additional uncertainty associated with the horizontal leveling of the pyranometer. Since most rotating shadowband radiometers use a photodiode pyranometer, there is a significant additional uncertainty associated with the nonuniform spectral response between 400 and 1100 nm and the nonresponse between 300 and 400 nm and beyond 1100 nm. Other significant uncertainty issues are the temperature dependence of the silicon cell's response function and the angular (cosine) response of the photodiode pyranometer (see chapter 5 for more details on photodiode pyranometers). Table 4.6 lists some of the major contributors to uncertainty for a rotating shadowband radiometer that uses a photodiode pyranometer for the calculation of the direct irradiance. The expanded uncertainty for a 95% confidence in the direct irradiance calculation is 3.226%. Certain factors have very little effect on the overall uncertainty of an instrument. For example, if we eliminate

TABLE 4.6

One Estimate of the Standard Uncertainty of a Rotating Shadowband Radiometer

Variable	Value	Semirange (B) or Std Dev (A)	Type	Distribution	DF	Std Uncertainty
Calibration Table 4.4	R	0.583%	B	Rectangular	1000	0.337%
Nonlinearity		0.1%	B	Rectangular	1000	0.058%
Temperature dependence		1.5%	B	Rectangular	1000	0.866%
V resolution for 100 W m^{-2}		0.33%	B	Rectangular	1000	0.191%
Spectral response		1.0%	B	Rectangular	1000	0.577%
Cosine response		2%	B	Rectangular	1000	1.155%
V	7993 μvolts	111 μvolts	A	Normal	59	0.179%
Combined standard uncertainty						1.613%

from Table 4.6 the uncertainties associated with the nonlinearity, the voltage resolution, and the voltage measurement standard deviation, then the combined standard uncertainty decreases by only 0.022%. Therefore, it is important to account for the large contributors to uncertainty, and many of the minor contributors to uncertainty can safely be ignored. One should first be certain that the contribution is small, however, before neglecting it.

The assumptions made for Table 4.6 were that the calibration was derived for the calculated direct normal irradiance. Had one calibrated the photodiode pyranometer for global horizontal irradiance, the uncertainty of the direct would have been greater. Most RSRs use the GHI calibration factor. For example, Michalsky, Augustine, and Kiedron (2009) demonstrated that the calibration for the global horizontal and for the direct normal irradiances are different by about 2.5% for the photodiode pyranometer as used in that paper. There would be similar differences for other photodiode pyranometers. In Table 4.6, the temperature dependence is not corrected; there is about a 3% change over typical annual operating temperatures. If this temperature is measured and the correction is made, then this contribution to the uncertainty drops significantly to around 0.3%. Similarly, if the cosine response is measured and a correction applied to the direct beam calculation, this large contribution to the uncertainty can be reduced to a much smaller number, perhaps as low as 0.4%. In Table 4.6 it is assumed that there has been a correction applied to reduce the uncertainty of the spectral correction to 1.0%. However, this contribution to the uncertainty will dominate rotating shadowband radiometer measurement uncertainty if not corrected properly. For more discussion of RSRs and photodiode pyranometers see Chapters 5 and 7.

4.8 DIRECT NORMAL IRRADIANCE MODELS

The term *model* has different meanings. Two types of models are briefly discussed here. First, we discuss models that are radiative transfer codes that require inputs to these codes to describe the atmospheric extinction sources in the path of the direct solar beam. The second type of model uses satellite images or atmospheric soundings as observed from satellites to estimate the direct solar radiation that strikes the earth's surface.

4.8.1 GROUND-BASED MODELING

Clear-sky direct irradiance radiative transfer models perform very well if model inputs are accurately specified. Michalsky et al. (2006) evaluated six direct irradiance models of varying complexity. For the 30 cases studied, it was demonstrated that the estimated clear-sky DNI values differed from measurements by less than 1%. The mean difference for the six models was 0.5%. This implies that radiative transfer models are well developed, and, given the correct input parameters, one can estimate clear-sky DNI values to an accuracy of better than field pyrheliometer measurements. So, why are models not preferred over measurements? The reason is that good models need several accurately specified input parameters to yield correct broadband irradiances. Aerosol optical depth (AOD) and some indication of its wavelength dependence are needed. The wavelength dependence can be a simple Ångström expression such as $\tau = \beta\lambda^{-\alpha}$, which will be discussed in Chapter 12. The next most important contributor to direct beam extinction is water vapor, so one requires some estimate of this quantity, and finally, an ozone column measurement or estimate is needed. Daily column ozone can be obtained from the NASA website (http://toms.gsfc.nasa.gov/teacher/ozone_overhead_v8.html). Water vapor can be crudely estimated from surface relative humidity or, better yet, using nearby radiosondes, if available. Aerosol optical depths are measured with sun radiometers, which are not widely available, and will be discussed in Chapter 12. If one's goal is to measure or model direct irradiance, AOD or direct normal irradiance measurements present different but similar measurement challenges. The more complex models of the spectral distribution of direct normal irradiance, which can be integrated to calculate broadband solar irradiance, are discussed in Chapter 12.

4.8.2 SATELLITE MODEL ESTIMATES

Satellite estimates of surface solar irradiance use models that range in complexity from those based on regressions of ground-based measurements to satellite pixel brightness (i.e., completely empirical) to those based on radiative transfer models using satellite-retrieved or climatological information on cloud properties, aerosols, humidity, and temperature profiles.

Perez et al. (2002) published their paper on a well-known operational model for deriving surface irradiance from satellite measurements that has a hybrid approach. The technique uses the pixel brightness as measured by the Geostationary Operational Environmental Satellites (GOES) but takes into account the site's elevation, a climatological estimate of the monthly average aerosol loading, snow cover, and a specular reflectance correction factor. Clean Power Research uses this model and its modifications of it for its *SolarAnywhere* service, which provides estimates of surface irradiance (http://www.cleanpower.com/SolarAnywhere), including direct normal irradiance. 3TIER (http://www.3tier.com/en/products/) processes satellite data to generate surface irradiance using its own modified version of the Perez et al. model.

A popular site for solar direct normal irradiance information based on satellite retrievals of the input parameters is the NASA Surface meteorology and Solar Energy (SSE) website (http://power.larc.nasa.gov). This is also known as the Prediction Of Worldwide Energy Resource (POWER) project website. NASA's SSE approach is a more physics-based approach in that it uses a radiative transfer code with cloud optical property inputs from satellite retrievals and a chemical transport model to estimate aerosol extinction to calculate surface solar direct normal irradiance.

The uncertainties associated with satellite retrievals of direct normal irradiances are about twice as large as those for the retrievals of global horizontal irradiance from satellites. Satellite estimates of hourly DNI using the Perez et al. (2002) model have biases of only a few percent and root mean square errors (RMSEs) in the 35–40% range for hourly data (e.g., Vignola, Harlan, Perez, and Kmiecik, 2007). In Vignola et al. the RMSE and the mean bias error (MBE) of DNI are calculated from hourly averages of frequently sampled ground-based (gnd_i) measurements and one satellite image (sat_i) per hour. The MBE in percent is calculated using

$$MBE = 100 \cdot \left[\frac{\Sigma_{i=1}^{i=n}(gnd_i - sat_i)}{n} \right] \cdot \frac{1}{\left(\Sigma_{i=1}^{i=n} gnd_i \right)/n} \tag{4.4}$$

where the sum is ideally formed over all the daylight hours in at least one year to determine a typical MBE for a site. For the RMSE Vignola et al. (2007) use

$$RMSE = 100 \cdot \left(\frac{\Sigma_{i=1}^{i=n}(gnd_i - sat_i)^2}{(n-1)} \right)^{\frac{1}{2}} \cdot \frac{1}{\left(\Sigma_{i=1}^{i=n} gnd_i \right)/n} \tag{4.5}$$

where again the sum is ideally formed over all of the daylight hours in at least one year to determine a typical RMSE for a site.

As an aside, it may seem that a more straightforward way to calculate MBE in percent would be to use a formula such as

$$MBE = 100 \cdot \left[\Sigma_{i=1}^{i=n} \left(\frac{gnd_i - sat_i}{gnd_i} \right) \right] \Big/ n \tag{4.6}$$

where the individual percent differences are averaged; however, if a single gnd_i value were equal to zero, MBE becomes undefined. If zeros were screened from the sum, small nonzero denominators would still produce large fractions that would have a large and detrimental influence on the result. The same arguments apply to calculating the RMSE using

$$RMSE = 100 \cdot \left[\Sigma_{i=1}^{i=n} \left(\frac{gnd_i - sat_i}{gnd_i} \right)^2 \right]^{\frac{1}{2}} \Big/ n \qquad (4.7)$$

The hourly RMSEs of 35–40% may seem high because MBEs and RMSEs are often quoted for mean *monthly* daily totals, which are much smaller numbers. For example, Myers, Wilcox, Marion, George, and Anderberg (2005) quote uncertainties around 14% for the sites in their study using Perez et al.'s (2002) model. The average monthly daily total (AMDT) is the total solar energy per square meter per day summed over all days in the month divided by the number of days in the month. Every month is likely to have direct sunlight; therefore, this calculation of MBE

$$MBE = 100 \cdot \left[\Sigma_{i=1}^{i=m} \frac{(gnd_i - sat_i)}{gnd_i} \right] \Big/ m \qquad (4.8)$$

is likely to give a valid and defined MBE. In this equation gnd_i and sat_i are AMDTs for one of the m months in the sample size. Note this has the same form as Equation 4.6, but instead of summing over individual hours the sum is over monthly averages of daily totals. The same argument applies to the RMSE for AMDTs. Alternately, it would be as informative to calculate and compare MBE and RMSE using equations of the form in (4.4) and (4.5). It is useful to state just how MBE and RMSE are calculated when quoting biases and uncertainties, and it should be required for publications.

4.9 HISTORICAL AND CURRENT SURFACE-MEASURED DIRECT NORMAL IRRADIANCE DATA

An excellent review of the historical solar irradiance data and accompanying metadata regarding these can be found in Stoffel et al. (2010) and need not be repeated here. Years of calibrated and well-maintained instrumentation that captures as complete a record of direct normal irradiance data as possible are critical in understanding the annual and interannual variability of radiation. Climate research and central concentrating solar power generation require long records to understand the long-term variability of DNI.

In this section, only current (as of June 2011) measurements of direct normal irradiance with expectations of continuance will be discussed. Much of the focus, but not all, is on the United States and on measurements made with first-class instruments. Rotating shadowband radiometers will be discussed because of their proliferation.

The Baseline Surface Radiation Network (BSRN) was started by the World Climate Research Program in 1992 to measure solar and infrared surface radiation with the highest-quality instruments and techniques available (Ohmura et al., 1998). The data produced have been used for validating and improving satellite retrievals of surface radiative fluxes, for comparisons to climate model calculations, and

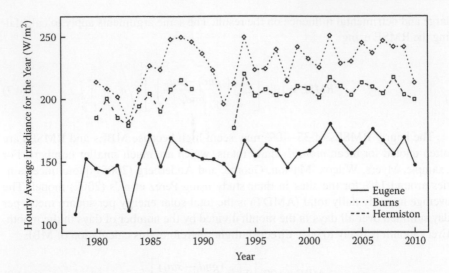

FIGURE 4.8 Hourly average DNI for Burns, Eugene, and Hermiston, Oregon, for the years from 1978 through 2010; monthly average hourly values are used to determine the annual average. The monthly average values exclude hours or days with missing data or misaligned measurements.

for monitoring subtle long-term changes in the surface radiative environment. The database is maintained at the Alfred Wegener Institute in Bremerhaven, Germany. Data are available from about 50 stations worldwide at http://www.bsrn.awi.de.

The World Radiation Data Center is in St. Petersburg, Russia. It was established by the World Meteorological Organization to archive and publish solar radiation data. Most of the data from the more than 1000 sites included in the archive are GHI measurements representing daily total irradiation values. Some sites submit hourly data, and some sites submit DNI[*].

In the United States, three small networks continue to provide moderate- to high-quality direct normal irradiance data: the University of Oregon Solar Radiation Monitoring Laboratory (UO SRML), the NOAA Surface Radiation (SURFRAD) network, and the NOAA Integrated Surface Irradiance Study (ISIS) network.

The UO SRML has run the Pacific Northwest (United States) Solar Radiation Data Network since 1977. The annual average DNI from the three longest running stations is shown in Figure 4.8. Currently, 12 stations operate in Idaho, Montana, Oregon, Utah, Washington, and Wyoming to measure direct normal irradiance; five

---

[*] The contact information is:
Voeikov Main Geophysical Observatory
World Radiation Data Centre
7, Karbyshev Str.
194021, St. Petersburg, Russian Federation
Voice: (812) 297-43-90 Fax: (812) 297-86-61
Director: Dr. Anatoly Tsvetkov (812) 295-04-45 t
svetkov@main.mgo.rssi.ru or wrdc@main.mgo.rssi.ru

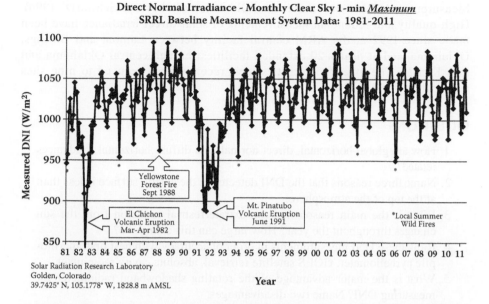

Direct Normal Irradiance - Monthly Clear Sky 1-min _Maximum_
SRRL Baseline Measurement System Data: 1981-2011

Solar Radiation Research Laboratory
Golden, Colorado
39.7425° N, 105.1778° W, 1828.8 m AMSL

FIGURE 4.9 Maximum observed DNI for Golden, Colorado, based on 1- and 5-minute measurements from 1981 through 2011. The data have not been adjusted for sun-earth distance, showing the intra-annual and inter-annual variability. The effects of volcanic eruption and local forest fires are illustrated by the local minima.

stations use pyrheliometers, and seven stations use RSRs. Eugene and Hermiston, Oregon, have colocated RSRs and pyrheliometers to evaluate the accuracy of the RSRs. Data are taken at 5-minute intervals at most sites, although 1-minute and 15-minute data are taken at a few stations. More information can be found at http://solardata.uoregon.edu/.

The NOAA SURFace RADiation (SURFRAD) network has operated since 1994. All SURFRAD sites are part of the BSRN. As such the sites acquire high-quality direct solar radiation data as a subset of more extensive radiation measurements. The locations of the seven sites, their data record, and much more can be found at http://www.srrb.noaa.gov/surfrad/index.html.

The NOAA Integrated Surface Irradiance Study (ISIS) is a remnant of the NOAA National Weather Service solar network known as SOLRAD in the 1950s through 1980s. There are currently seven NWS stations that continue to operate solar equipment that includes direct irradiance sensors. The stations and the data can be found at http://www.srrb.noaa.gov/isis/index.html.

There are other medium- to high-quality direct solar irradiance measurements made in the United States, but typically their operation is short-term; the National Renewable Energy Laboratory's (NREL) web page lists several of these (http://www.nrel.gov/midc/). NREL's high-quality data from the Solar Radiation Research Laboratory (SRRL) in Golden, Colorado, dates from 1981 (see Figure 4.9). The U.S. Department of Energy operates the Atmospheric Radiation

Measurement (ARM) site in northern Oklahoma (Stokes and Schwartz, 1994). High-quality and redundant measurements of solar direct irradiance have been made since 1992 at the ARM central facility between Lamont and Billings, Oklahoma, and at nearly 20 extended facilities through central Oklahoma and into Kansas. The extended facilities have recently been reduced to nine sites nearer the central facility.

QUESTIONS

1. How are global horizontal, direct normal, and diffuse horizontal irradiances related?
2. Name three reasons that the DNI detected at the earth's surface is less than at the top of the atmosphere.
3. What is the main reason that the extraterrestrial radiation from the sun changes throughout the year? How large can this change be?
4. Rank these DNI measurements in order of accuracy with best first: thermopile pyrheliometer, GOES satellite retrieval, absolute cavity radiometer.
5. What is the major advantage of the rotating shadowband radiometer for measuring DNI? Name two disadvantages.
6. How would one obtain a pyrheliometer calibration that is linked to the WRR?
7. What is a good reference for choosing a pyrheliometer with proven accuracy?
8. Did the direct normal irradiance reach 700 Wm^{-2} in Eugene, Oregon, on January 1, 2009? Hints: This can be checked in five clicks. How about January 2, 2009? One more click.
9. How many Dobson units of ozone were over Denver (40°, −105°) on March 31, 2010?
10. What should accompany every measurement of DNI?

REFERENCES

Cook, R. R. 2002. Assessment of uncertainties of measurement for calibration and testing laboratories. National Association of Testing Authorities, Australia. Available from: http://www.nata.asn.au/publications/uncertainty

Finsterle, W. 2006. *WMO International Pyrheliometer Comparison IPC-X, final report.* IOM report No. 91. WMO/TD 1320, Geneva.

Gueymard, C. 1998. Turbidity determination from broadband irradiance measurements: A detailed multicoefficient approach. *Journal of Applied Meteorology,* 37:414–435.

ISO. 1990. *Specification and classification of instruments for measuring hemispherical solar and direct solar radiation.* ISO 9060, Geneva. Available from: http://www.iso.org/

Joint Committee for Guides in Metrology (JCGM). 2008. *Evaluation of measurement data— Guide to the expression of uncertainty in measurement. GUM 1995 with minor revisions,* Bureau International des Poids et Mesures. Available from: http://www.bipm.org/en/publications/guides/gum.html

Kopp, G. and J. L. Lean, 2011. A new, lower value of total solar irradiance: Evidence and climate significance. *Geophysical Research Letters,* 38:L01706. doi:10.1029/2010GL045777

LI-COR, Inc. 1982. *LI-2020 automatic solar tracker instruction manual.* Publication No. 8203-27. Lincoln, NE.

Michalsky, J. J., G. P. Anderson, J. Barnard, J. Delamere, C. Gueymard, S. Kato, P. Kiedron, A. McComiskey, and P. Ricchiazzi. 2006. Shortwave radiative closure studies for clear skies during the atmospheric radiation measurement 2003 aerosol intensive observation period. *Journal of Geophysical Research,* 111:D14S90. doi:10.1029/2005JD006341

Michalsky, J. J., J. A. Augustine, and P. W. Kiedron. 2009. Improved broadband solar irradiance from the multi-filter rotating shadowband radiometer. *Solar Energy* 83:2144–2156.

Michalsky, J., E. G. Dutton, D. Nelson, J. Wendell, S. Wilcox, A. Andreas, P. Gotseff, D. Myers, I. Reda, T. Stoffel, K. Behrens, T. Carlund, W. Finsterle, and D. Halliwell. 2011. An extensive comparison of commercial pyrheliometers under a wide range of routine observing conditions. *Journal of Atmospheric and Oceanic Technology,* 28:752–766.

Myers, D. R., S. Wilcox, W. Marion, R. George, and M. Anderberg. 2005. *Broadband model performance for an updated national solar radiation database in the United States of America.* Proceedings of the ISES Solar World Conference, Orlando, FL.

NREL. 2010. *Broadband outdoor radiometer calibration* (BORCAL 2010-02). August 5, National Renewable Energy Laboratory.

Ohmura, A., H. Gilgen, H. Hegner, G. Müller, M. Wild, E. G. Dutton, B. Forgan, C. Fröhlich, R. Philipona, A. Heimo, G. König-Langlo, B. McArthur, R. Pinker, C. H. Whitlock, and K. Dehne. 1998. Baseline Surface Radiation Network (BSRN/WCRP): New precision radiometry for climate research. *Bulletin of the American Meteorological Society* 79:2115–2136.

Perez, R., P. Ineichen, K. Moore, M. Kmiecik, C. Chain, R. George, and F. Vignola. 2002. A new operational satellite-to-irradiance model. *Solar Energy* 73:307–317.

Reda, I. 1996. *Calibration of a solar absolute cavity radiometer with traceability to the World Radiometric Reference.* National Renewable Energy Laboratory NREL/TP-463-20619, Golden, CO.

Reda, I., D. Myers, and T. Stoffel. 2008. Uncertainty estimate for the outdoors calibration of solar pyranometers: A metrologist perspective. *Journal of Measurement Science,* 3:58–66.

Stoffel, T. and I. Reda. 2009. NREL Pyrheliometer Comparisons, September 22–October 3 (NPC-2008). NREL/TP-550-45016, February, National Renewable Energy Laboratory, Golden, CO. http://www.nrel.gov/publications

Stoffel, T., D. Renné, D. Myers, S. Wilcox, M. Sengupta, R. George, and C. Turchi. 2010. Concentrating solar power; best practices handbook for the collection and use of solar resource data. Technical Report NREL/TP-550-47465. National Renewable Energy Laboratory, Golden, CO. http://www.nrel.gov/docs/fy10osti/47465.pdf

Stokes, G. M. and S. E. Schwartz. 1994. The Atmospheric Radiation Measurement (ARM) program: Programmatic background and design of the cloud and radiation test bed. *Bulletin of The American Meteorological Society,* 75:1201–1221. doi:10.1175/1520-0477

Taylor, B. N. and C. E. Kuyatt. 1994. *Guidelines for evaluation and expressing the uncertainty of NIST measurement results.* NIST Technical Note 1297, National Institute of Standards and Technology, Gaithersburg, MD. Available from: http://physics.nist.gov/cuu/Uncertainty/basic.html

Vignola, F., P. Harlan, R. Perez, and M. Kmiecik. 2005. Analysis of satellite derived beam and global solar radiation data. Proceedings of the ISES Solar World Conference, Orlando, FL.

Vignola, F., P. Harlan, R. Perez, and M. Kmiecik. 2007. Analysis of satellite derived beam and global solar radiation data. *Solar Energy* 81:768–772.

Wesely, M. L. 1982. Simplified techniques to study components of solar radiation under haze and clouds. *Journal of Applied Meteorology* 21:373–383.

WMO. 2008. *WMO guide to meteorological instruments and methods of observation.* WMO-No. 8 (7th ed.), Chapter 7. Available from: http://www.wmo.int/pages/prog/www/IMOP/publications/CIMO-Guide/CIMO_Guide-7th_Edition-2008.html

5 Measuring Global Irradiance

This instrument, as its name (from the Greek, fire, up, a measure) indicates, is intended to measure the heat equivalent of radiation received from or going out toward the complete hemisphere above the plane of the measuring surface.

C. G. Abbot and L. B. Aldrich
The Pyranometer: An Instrument for Measuring Sky Radiation, 1916

5.1 INTRODUCTION TO GLOBAL HORIZONTAL IRRADIANCE MEASUREMENTS

Global horizontal irradiance (GHI) is the total solar flux available from the hemispheric sky dome that is incident on a horizontal surface. By convention global irradiance implies a hemispheric, 2π steradian, field of view on any surface. GHI is the sum of the solar radiation coming directly from the solar disk, also called the direct normal irradiance (DNI) or "beam irradiance," which is normal to the rays from the sun, projected (Equation 5.1) onto a horizontal plane and the solar radiation coming from all other directions of the sky dome, or "sky irradiance" (Figure 5.1). The solar radiation coming from all parts of the sky dome other than directly from the sun is called diffuse horizontal irradiance (DHI). Planar surfaces of any orientation receive solar radiation directly from the sun plus diffuse radiation from the sky and reflected radiation from the ground. Historically, this has been termed total solar radiation on a tilted surface. Those involved with photovoltaic systems refer to this total irradiance as plane of array (POA) irradiance. Studies of the thermal performance of buildings rely on the amount of solar irradiance on various parts of a building envelope and refer to this total irradiance as global irradiance on tilted surfaces. Traditionally, the study of nonconcentrating thermal collectors has also relied on global irradiance available to tilted flat plate collectors. In this book global horizontal irradiance (GHI) will refer to the total solar irradiance on a horizontal surface. When discussing irradiance on a surface oriented at any angle other than horizontal, the irradiance will be referred to as global tilted irradiance GTI.

Global irradiance is measured by a pyranometer. The root of the word *pyrano* is derived from the Greek "pyr" meaning fire or heat and "ano" meaning sky. Therefore, a pyranometer is a meter for measuring heat from the sky. Earlier versions of the pyranometer were referred to as a 180° pyrheliometer, but the terminology was changed to pyranometer to help avoid the confusion with the term pyrheliometer that

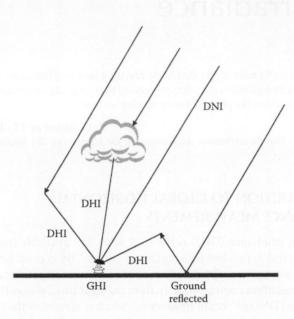

FIGURE 5.1 Drawing showing the sources of DHI as the DNI passes through the atmosphere. DHI can come from scattering by the constituents of the atmosphere, clouds, and ground-reflected irradiance that is rescattered by the clouds or atmospheric constituents.

is used to describe the instrument for measuring the DNI or beam irradiance coming directly from the sun (see Chapter 4).

As described in Chapter 2, GHI is equal to the DNI projected onto the horizontal surface plus the DHI that is from all other areas of the sky.

$$GHI = DNI * \cos(sza) + DHI \qquad (5.1)$$

where *sza* is the solar zenith angle of the sun as measured from the vertical (zenith) at the time of interest. Near solar noon, *sza* is smallest (determined by the location and date), and at sunrise or sunset, *sza* is 90°. Note that when using angles in the cosine function, most software programs interpret the value in radians instead of degrees. For example, one of the most common errors in computing GHI is to use degrees instead of radians when calculating the cosine of the solar zenith angle.

On a clear day, the GHI is mainly composed of the DNI projected onto the horizontal surface, and the DHI contribution is often a small fraction of GHI in Equation 5.1. The relationship over a clear day between GHI, DNI, and DHI is illustrated in Figure 5.2. The direct *horizontal* irradiance is the projection of DNI on a horizontal surface (i.e., DNI times the cosine of the solar zenith angle) as shown by the dashed line in Figure 5.2, and this produces the bell-shaped curve

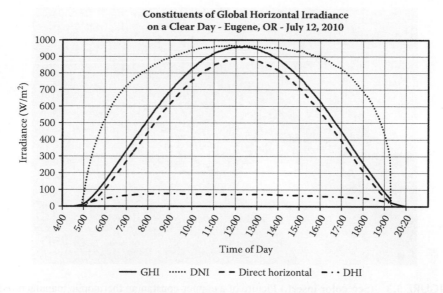

FIGURE 5.2 Plot of 1-minute data for GHI, DNI, and DHI irradiance on a clear day in Eugene, Oregon. GHI is the solid line, the dotted line is DNI, and the dot-dash line is DHI. The dashed line is the DNI projected onto a horizontal surface. Note that DHI is less than 100 Wm⁻² and decreases in the afternoon as the water vapor in the atmosphere decreases. At 19:12, the direct sunlight is blocked by an obstruction near the horizon.

seen in time-series plots of GHI. Note that DNI, shown by the dotted gray line, can be greater than GHI, especially in the early morning and late afternoon hours. Depending on the location and season, the DNI can exceed the GHI near solar noon. The cosine of the solar zenith angle is smallest during the morning and evening hours, and the direct *horizontal* component should always be smaller than GHI, as seen in Figure 5.2.

As described in Chapter 3, GHI has been measured in many ways during the past 100 years. Abbot and Aldrich (1916) discussed a design for pyranometers. A variety of pyranometers have been developed and used in the field from the Robitzsch actinograph to the more sophisticated Eppley bulb-type pyranometer up to today's modern designs. A description of these pyranometers from a historical perspective was presented in Chapter 3. This information is useful because a considerable amount of irradiance data has been and continues to be gathered using these historical pyranometers. However, pyranometers in wide use today are much more accurate, and they output their data to automated data-logging equipment instead of strip-chart recorders that need to be digitized for analysis.

Based on their detector design, three types of pyranometers will be discussed in more detail in this chapter:

- Black-disk thermopile pyranometers
- Black-and-white thermopile pyranometers
- Photodiode-based pyranometers

FIGURE 5.3   (See color insert.) Picture of a copper-constantan thermopile manufactured by Eppley Laboratory, Inc. Constantan wire is wrapped around a solid core and then dipped into a copper bath. (Photograph by Warren Gretz, NREL staff photographer.)

The basis for all thermopile radiometers is the *thermoelectric effect*, the generation of voltage from temperature differences in two dissimilar metals. The thermoelectric effect was discovered in 1821 by Thomas Seebeck, a German physicist. The *Seebeck Effect* predicts 41 μvolts from junctions of copper and constantan (a copper-nickel alloy) per Kelvin (as determined by $T_2 - T_1$ in Figure 2.16) at room temperature. A number of thermocouples in series form a thermopile. A copper constantan thermopile is shown in Figure 5.3. Constantan wire is wrapped around a solid core and plated with copper. The resulting black-disk thermopile pyranometer has the hot junction of the thermopile attached to a black-disk that absorbs incoming solar radiation, and the voltage produced is the result of the heat flow between the black-disk and the cold junction that is attached to the body of the pyranometer. Some pyranometers have a pair of thermopiles in parallel, called "bucking" thermopiles, with one thermopile being heated by the sun and the other thermopile being shaded from the sun and typically looking at the body of the pyranometer. This combination is purported to compensate for the thermal offset experienced by a single thermopile, but the exact details on these instruments are proprietary. The black-and-white or star-type pyranometer has thermopiles attached to alternating black (hot junction) and white (cold junction) segments forming the receiver surface, and the difference in voltages produced by the different temperatures of the colored surfaces (i.e., thermopiles connected in series) is proportional to the incident radiation (see Figure 5.15 and Figure 5.16).

In 1905, Albert Einstein explained the photoelectric effect as the emission of electrons by a surface, typically a metal, in response to the absorption of photons (i.e., individual quanta of light). The development of photovoltaic (PV) devices, also called solar cells, is based on the photoelectric effect. The photodiode pyranometer uses a solar cell to generate a current across a precision resister to produce a voltage that is proportional to the incident radiation. Each type of pyranometer will be described in more detail in subsections of this chapter.

The International Organization for Standardization (ISO) and the World Meteorological Organization (WMO) have undertaken the classification of pyranometers based on their measurement performance characteristics. These standards organizations classify pyranometers into three classes: (1) a secondary standard, (2) first class, and (3) second class (Table 5.1). While most of the ISO and WMO standards are identical, the characterizations specified are not the same in all cases. When ISO and WMO specifications differ, the ISO specifications will be listed first, followed by a "/", and then the WMO characteristics.

The main difference between the ISO and the WMO standards is in the specification of spectral sensitivity. The WMO standard requires about twice the spectral range that the ISO standard requires. Silicon photodiode–based pyranometers do not respond to solar irradiance with wavelengths beyond 1100 nm, so while

TABLE 5.1
Specifications for Pyranometer Classification

ISO Specifications/WMO Characteristics	Secondary Standard/ High-Quality	First Class/ Good Quality	Second Class/ Moderate Quality
Response time (to 95% of final value)	<15 s	<30 s	<60 s
Zero offset response: to 200 Wm⁻² net radiant loss to sky (ventilated)	7 Wm⁻²	15 Wm⁻²	30 Wm⁻²
Zero offset response: to 5°C/hr change in ambient temperature	±2 Wm⁻²	±4 Wm⁻²	±8 Wm⁻²
Resolution (smallest detectable change)	±1 Wm⁻²	±5 Wm⁻²	±10 Wm⁻²
Stability (change in sensitivity per year)	±0.8%	±1.6/1.5%	±2.0/3.0%
Nonlinearity (deviation from sensitivity at 500 Wm⁻² over 100 to 1000 Wm⁻² range)	±0.2%/0.5%	±0.5%/1.0%	±2.0%/3.0%
Directional response for beam radiation (error due to assuming that the normal incidence response at 1000 Wm⁻² is valid for all directions)	±10 Wm⁻²	±20 Wm⁻²	±30 Wm⁻²
Spectral selectivity (deviation of the product of spectral absorptance and transmittance from the mean)	±2%	±5%	±10%
Temperature response (error due to 50°C ambient temperature change)	±2%	±4%	±8%
Tilt response (deviation from horizontal responsivity due to tilt from horizontal to vertical at 1000 Wm⁻²)	±0.5%	±2%	±5%
Suitable applications	Working standard	Network operations	Low-cost network

some photodiode pyranometers perform better than second-class pyranometers, they technically do not qualify under these WMO or ISO standards. The optical domes for the thermopile pyranometers are used to isolate the thermopile sensor from the environment. The material used in these domes transmits a limited portion of the solar spectrum to the sensing thermopile. Ideally, the spectral transmittance would be the same for all solar irradiance wavelengths (generally 300 to 3000 nm). The domes of several first-class and second-class thermopile pyranometers are made of materials that meet the ISO spectral standards but do not transmit the longer wavelengths and hence do not meet WMO specifications.

The rational for characterizing the performance of pyranometers in specific categories will become more apparent from the more detailed descriptions found later in this chapter. Four main performance characteristics will help determine the quality of pyranometers:

- Deviation from true cosine response of the detector to receive radiation within a 2π steradian field of view
- Change in responsivity as a function of ambient temperature
- Dependence of pyranometer responsivity on the wavelength of incident solar radiation
- Thermal offsets of the pyranometer

The characteristic response of the pyranometers to changes in these factors will be discussed for each of the pyranometer types along with other important factors, such as changes in response as the solar azimuth angle changes. These factors all affect the accuracy of pyranometer measurements, and the magnitude of these effects will be used to determine the uncertainty in the irradiance values obtained using the instrument. The specifications are one way to differentiate between the capabilities of each pyranometer and to provide guidance in matching the application's particular solar radiation measurement requirements with suitable a pyranometer.

5.2   BLACK-DISK THERMOPILE PYRANOMETERS

Major design features of black-disk thermopile pyranometers are as follows:

- A single all-black thermopile measures the heat flow between the black-disk of the receiver and the body of pyranometer.
- Two quartz domes reduce convective losses.
- Design minimizes azimuth angle dependence.
- GHI and DHI measurements suffer from measureable thermal offsets.

Secondary-standard and first-class instruments use the single black thermopile design. Single black-disk thermopile pyranometers measure the heat flow between a black-disk receiver exposed to the incident radiation and the body of the pyranometer. Radiant energy is absorbed by the black receiver that is thermally attached to the "hot junction" of the thermopile, while the "cold junction" is connected to the body of the pyranometer. The temperature difference between the hot junction of the thermopile and the cold junction produces a voltage typically on the order of 10 mV for a flux of 1000 Wm^{-2} incident on the detector. This thermoelectric effect was observed by Seebeck, Peltier, and Thomson (Velmre, 2007; Bunch and Hellemans, 2004). Danish physicist Hans Christian Ørsted helped explain this thermoelectric effect (Velmre, 2007).

The voltage created in the circuit is on the order of several μV per Kelvin difference in temperature. In some pyranometers a copper–constantan thermopile is used because copper–constantan has a relatively large Seebeck coefficient of 41 μV K^{-1} at room temperature. Many thermocouples are connected in series to form the thermopile that produces a larger, more easily measured output voltage.

For accurate solar irradiance measurements, the heat flow should be only between the disk and the body of the pyranometer that has considerable thermal mass compared to the disk and acts as a heat sink. Heat losses by other means such as convection, advection, conduction, and radiation reduce the accuracy of the measurements. Convection is the movement of heat in a medium such as water or air. An example of convention is using a ceiling fan to circulate warm air near the ceiling of a room to the lower part of the room. Advection is the process that transfers heat from a surface to a gas or liquid. An example of advection is when a fan is used to blow air across hot elements in a heater, resulting in the transfer of heat to the air. Conduction is the movement of heat in a material. An example of conduction is when a spoon is put in a hot cup of coffee and the handle of the spoon becomes hot as heat travels up the spoon. Heat is also transferred through electromagnetic radiation. An example of radiative heating is microwave heating of water or sunlight heating a solar collector to heat water (see Section 2.12 for more information on thermodynamic fundamentals).

To reduce heat losses by convection and advection from wind blowing across the pyranometer receiver, high-quality pyranometers use two glass domes. One glass dome prevents advective heat losses from wind blowing across the pyranometer receiver being heated by the sun. A second glass dome acts much like a double-pane window to reduce the heat loss with a layer of air between the domes acting as a buffer to further reduce convective heat losses.

Conductive heat losses are reduced by isolating the thermopile hot junction from other areas of the pyranometer. The better the thermal isolation, the slower the heat transfers from the pyranometer's receiver to unintended heat sinks. Ideally only the heat flow from the receiver disk to the body of the pyranometer should be allowed.

Stopping radiative heat loss from a pyranometer is difficult because an accurate pyranometer should pass all solar radiation through the glass domes, including the near-infrared (NIR) radiation to be measured. Therefore, the best pyranometers have precision-ground glass domes that uniformly transmit irradiance from around 300 nm to 3000 nm. However, the receiver disk that absorbs solar radiation also emits thermal infrared (IR) radiation (wavelengths longer than 3000 nm). The glass domes that allow the IR portion of the solar spectrum to impinge on the receiver disk also allow thermal IR radiation to radiate to the dome and then to the sky; this causes the receiver disk to become cooler as it radiates to the cooler sky. This is the basis for the so-called thermal offset found in pyranometers designed with a single black disk (Dutton et al., 2001).

As described previously, thermopile-type pyranometers produce a voltage proportional to the energy flow from the receiver disk to the cold junction or body of the instrument. At night, the energy flow reverses as the receiver disk and protective domes cool due to exposure to the cold sky, and a negative output voltage from the thermopile results. During the day, there is a similar heat flow (thermal IR radiation) from the sensor disk to the cooler sky. This path to the sky reduces the energy flow through the thermopile and, hence, reduces the voltage that would be produced otherwise. This radiation to the sky is called the *IR radiative loss* or *thermal offset*, resulting in a negative bias in the pyranometer readings (Dutton et al., 2001; Gulbransen, 1978).

Because the effective sky temperature varies with the amount and vertical distribution of clouds, atmospheric water vapor, and aerosol loading, the pyranometer thermal offsets vary with location, time of day, and season, making it difficult to account for this thermal offset. It can range anywhere from zero during foggy conditions to as much as -30 Wm^{-2} for high-altitude clear-sky conditions, depending on the pyranometer construction. The use of ventilators will also affect the thermal offset contributions to uncertainties in pyranometer calibration and field measurement performance.

Examples of three black-disk thermopile pyranometers in common use are examined here. The Kipp & Zonen model CM 22 pyranometer (Figure 5.4) is considered a secondary standard. The factory specifications for the CM 22 are given in Table 5.2. The CM 22 is similar to the Kipp & Zonen CM 21 except the CM 22 uses an Infrasil II quartz dome and the CM 21 uses a Schott K5 optical glass dome that transmits a smaller range of the solar spectrum in a less uniform manner. The CM 21 is classified

FIGURE 5.4   (See color insert.) Picture of Kipp & Zonen CM 22 pyranometers in ventilators at SRRL in Colorado.

TABLE 5.2
Specifications of the Kipp & Zonen CM 22 and CM 21 Pyranometers

Characteristic	CM 22 (Secondary Standard)	CM 21 (First Class)
Spectral range	200–4000 nm (50% transmission points)	285–2800 nm (50% transmission points)
	290–3500 nm (95% transmission points)	335–2200 nm (95% transmission points)
Typical sensitivity	10 μV/Wm^{-2}	10 μV/ Wm^{-2}
Typical impedance	10–100 ohms	10–100 ohms
Linearity	±0.25% < 1000 Wm^{-2}	±0.25% < 1000 Wm^{-2}
Temperature dependence	±0.5% (–20°C to +50°C)	±0.5% (–20°C to +50°C)
Stability	±0.5% sensitivity change per year	±0.5% sensitivity change per year
Cosine response	±1% deviation at 60° solar zenith angle	±1% deviation at 60° solar zenith angle
	±3% deviation at 80° solar zenith angle	±3% deviation at 80° solar zenith angle
Response time	<5 seconds for 95%	<5 seconds for 95%
Quartz dome	Infrasil II	Schott K5 optical glass 2 mm thick, 30 mm inner and 50 mm outer dome diameter

as a first-class pyranometer, and its specifications are also given in Table 5.2. Older versions of the Kipp & Zonen CM 11 and CM 21 pyranometers have domes with nonuniform transmission values at the short-wavelength end of the solar spectrum. The 50% transmission point in the older pyranometers started around 310 nm instead of 285 nm for the current versions of these pyranometers.

The Eppley Laboratory, Inc. precision spectral pyranometer (PSP) (Figure 5.5) is considered a first-class pyranometer, and its factory specifications are given in

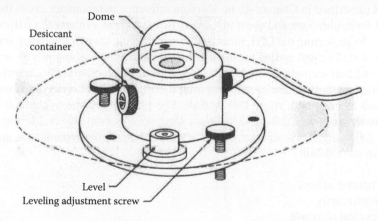

FIGURE 5.5 Schematic drawing of an Eppley PSP pyranometer. Sun shield represented by the dotted oval.

TABLE 5.3

Specification of an Eppley PSP Pyranometer (First Class)

Characteristic	Specification
Spectral range	280–2800 nm
Typical sensitivity	9 μV/Wm^{-2}
Typical impedance	650 ohms
Linearity	±0.5% from 0 to 2800 Wm^{-2}
Temperature dependence	±1.0% (−20°C to +50°C)
Stability	±0.5% sensitivity change per year
Cosine response	±1% from 0° to 70° solar zenith angle
	±3% deviation from 70° to 80° solar zenith angle
Response time	1 second (1/e signal)
Glass domes	clear WG295 glass

Table 5.3. Note that only some of the factory specifications are described in terms that translate directly into ISO or WMO specifications. In fact, specifications among manufacturers are not uniform. There are many reasons for this diversity in specification, and much of this variance results from the way the manufacturers conduct their standardized tests. Many performance specifications are best determined indoors under controlled conditions. Manufacturers also perform outdoor tests to determine the responsivities under conditions experienced in the field. There are many ways to perform tests on pyranometers, and descriptions of factory specifications often relate directly to the testing method that is used. Therefore, if manufacturers use different testing methods, it is not always easy to compare specifications because the details in the specifications are different. In addition, the performance of various pyranometer models is affected differently by environmental conditions such as air temperature and wind speed. The best measurement comparisons are done side by side under conditions likely to be experienced by the pyranometer in the field. Unlike pyrheliometers (described in Chapter 4), no absolute reference pyranometer exists that can be used for calibrations and accurate comparisons. At best a reference GHI value is obtained by projecting the DNI, measured by an absolute cavity radiometer, onto the horizontal surface and adding the diffuse horizontal irradiance obtained by using an offset-corrected secondary standard pyranometer or a second-class black-and-white type pyranometer (because of their minimal thermal offset characteristics) with the direct sun blocked by a shade ball or disk. The black-and-white pyranometer will be discussed in Section 5.3.

The following characteristics of a black-disk thermopile pyranometer are discussed in more detail:

- Thermal offsets
- Nonlinearity
- Spectral response
- Angle of incidence response
- Response degradation over time

- Temperature dependence
- Ice and snow on dome—ventilator section
- An optical anomaly
- Care and maintenance

5.2.1 THERMAL OFFSETS

Pyranometer thermal offset is a subtle thermal loss mechanism that can affect the performance of single black detector pyranometers. By design, the black disk is a good absorber of solar radiation as well as a good emitter in the IR portion of the radiation spectrum. The body of the pyranometer is generally warmer than the sky. Therefore, there is IR radiative transfer from the pyranometer to the sky. The transfer actually consists of several steps with the outer dome of the pyranometer radiating to the sky, the inner dome radiating to the outer dome, and the receiver radiating to the inner dome, lowering its temperature with respect to the body and producing a negative signal (Dutton et al., 2001; Gulbrandsen, 1978; Long, Younkin, and Powell, 2001). Thermal offsets usually vary from minus a few Wm^{-2} to as much as -20 to -40 Wm^{-2}. During the day, thermal offsets are larger (more negative) than the nighttime values as the ground and the body of the pyranometer are heated by solar radiation (Figure 5.6). This causes the thermal offsets to increase during the day. The magnitude of the thermal offset is proportional to the net IR at the time of measurement. The net IR is a function of the effective sky and ground temperatures. Since the amount of atmospheric precipitable water vapor affects the net IR, and in

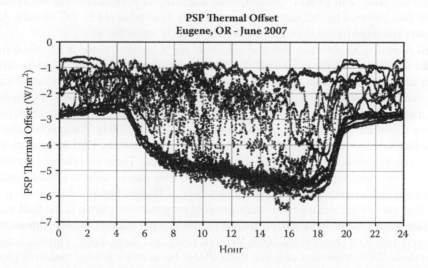

PSP Thermal Offset
Eugene, OR - June 2007

FIGURE 5.6 Calculated thermal offset for an Eppley PSP pyranometer in Eugene, Oregon, during June 2010. One-minute data for all hours in June 2010 are plotted. Using nighttime readings and pyrgeometer data, the methodology of Younkin and Long (2003) is employed to calculate the thermal offset of the PSP during the day. During cloudy periods there is minimal thermal offset as the sky temperature is similar to the ambient temperature. During clearer weather, the thermal offset reaches its maximum values.

turn the thermal offset, the offset is related to the station elevation. That is, under cloudless sky conditions, lower thermal offsets can be expected from pyranometers used in Eugene, Oregon, at 150 m (-2 Wm^{-2} to -10 Wm^{-2}) than the offsets found in measurements made in Golden, Colorado, at 1829 m (-15 Wm^{-2} to -40 Wm^{-2}). The magnitude of the thermal offsets depends on the difference between the ambient air and sky temperature, and thermal offsets are significantly less during cloudy periods with warmer skies as opposed to clear periods with cooler sky temperatures. The body of the black-disk thermopile pyranometer is the cold reference junction and is usually at or near ambient temperature, especially when installed in a ventilator. During clear weather, sky temperatures are much colder that ambient surface air temperatures. Because different sites exhibit different ranges of sky temperatures due to elevation and amounts of atmospheric water vapor and aerosols, the thermal offsets will be more severe at some sites than at others. The fact that thermal offsets vary from site to site makes it difficult to clearly determine the proper calibration constant of a pyranometer, especially ones that have large thermal offsets. Starting around 2005, some pyranometers have been calibrated to determine both the broadband responsivity and the IR responsivity (R_{net}) (Reda et al., 2005).

Thermal offsets in pyranometer measurements can most easily be explored with nighttime data. Predominately the negative nighttime values result from thermal offsets. The thermal offsets differ depending on whether the pyranometer is ventilated and whether a direct current (DC) or alternating current (AC) motor powers the ventilator. Most ventilators heat the body of the pyranometer exacerbating the thermal offset. For an Eppley PSP with an Eppley ventilator, the thermal offset at night was observed to be -2 to -3 Wm^{-2} greater, more negative, for pyranometers in ventilators with fans powered by AC motors compared with those powered by DC motors. AC motors use more power than DC motors and slightly warm the air.

It is easiest to correlate thermal offsets at night because there is no incoming solar present to heat the receiver and the pyranometer output is expected to be zero. Key to modeling thermal offsets is the use of pyrgeometers from which net IR can be determined from the pyrgeometer thermopile signal. Correlations between net IR and thermal offsets, relative humidity, and sky temperature have been developed using nighttime pyranometer values (Younkin and Long, 2003). These correlations can then be used with data gathered during the day to adjust the GHI measurements based on pyrgeometer thermopile signals during the day. These correlations are specific to pyranometer and pyrgeometer pairs, and they can be used to produce more accurate measurements of GHI and, in particular, DHI (see Figure 5.6).

Unfortunately, there are a limited number of pyrgeometers paired in the field with pyranometers. Some work has been done to estimate the thermal offsets without the pyrgeometer data (Reda et al., 2005; Vignola, Long, and Reda, 2007; Vignola, Long, and Reda, 2008; Vignola, Long, and Reda, 2009), but to date a general meteorological model that can be associated with a pyranometer has not been developed.

With thermal offsets changing from site to site, it is very difficult to determine the appropriate pyranometer responsivity without correcting for the thermal offset. Systematic errors result if calibration values are obtained from sites where large thermal offsets occur and are then used at locations where the thermal offsets are much less.

Three options are available to address this calibration issue:

1. Use calibration values determined without consideration of thermal offsets.
2. Use calibration values determined with mean nighttime offsets subtracted during the day.
3. Use calibration values determined with thermal offsets corrected.

If one does not have a pyrgeometer and meteorological data necessary to calculate the pyranometer thermal offsets, one should calibrate the instrument and use it at the same location to minimize thermal offset issues.

An alternative that accounts for some of the thermal offsets is to subtract the average nighttime values from the night before and the night after from the daytime values (Dutton et al., 2001). This has been shown to account for roughly half of the thermal offsets during the day (Vignola et al., 2007). If this method is used, the calibration constant should be determined from a calibration that subtracts the nighttime values from daytime measurements. In other words, the calibration value used for a pyranometer should be determined in a manner similar to how the pyranometer is used.

The best option is if pyrgeometer and other meteorological data are available and one can calculate thermal offsets. Then the calibration constant determined with the thermal offsets subtracted should be used.

5.2.2 Nonlinearity

Nonlinearity is the deviation of the radiometer's responsivity ($\mu V/Wm^{-2}$) as the solar irradiance level increases or decreases. The ISO specification for nonlinearity is less than $\pm0.5\%$ for a first-class instrument. Nonlinearity is defined as the deviation from the responsivity measured at 500 Wm^{-2} over a range of 100 Wm^{-2} to 1000 Wm^{-2}. For example, if the measured responsivity at 500 Wm^{-2} is 8.00 $\mu V/Wm^{-2}$, then the responsivity from 100 Wm^{-2} to 1000 Wm^{-2} has to be between 7.96 $\mu V/Wm^{-2}$ and 8.04 $\mu V/Wm^{-2}$. Factory specifications for an Eppley PSP indicate a linear response better than $\pm0.5\%$ from 0 to 2800 Wm^{-2}, whereas Kipp & Zonen models CM 22 and CM 21 claim a linear response better than $\pm0.25\%$ for irradiances below 1000 Wm^{-2}.

At high levels of solar irradiance, black-disk thermopile pyranometers have a good linear response. It is at lower levels of irradiance that the thermopile pyranometers may exhibit nonlinearity problems. For example, if a pyranometer exhibits a 20 Wm^{-2} thermal offset at a solar irradiance of 1000 Wm^{-2}, then its readings are offset by 20/1000, or 2%. If the thermal offset is 20 Wm^{-2} and the incident irradiance is 100 Wm^{-2}, then the readings are off by 20%.

Consequently, nonlinearity tests are usually performed under indoor laboratory conditions with the instrument and surroundings at a stable (room) temperature. Under such conditions, there are no thermal offsets caused by the detector radiating to a colder target. A stable lamp with a chopper wheel is used to test for response change with intensity. The intensity of light is proportional to the area blocked by the chopper wheel. The National Institute of Standards and Technology has developed a more exacting way to test for linearity (see Walker, Saunders, Jackson, and McSparron, 1987). Under laboratory conditions, ISO classification specifies a pyranometer as first class if it deviates from a linear response by less than 0.5%.

5.2.3 SPECTRAL RESPONSE

Pyranometer spectral response is largely determined by the spectral transmission of the material used for the protective and insulating domes and the spectral absorption of the detector coating. Ideally, the pyranometer output should be proportional to the sum of the solar flux across all solar wavelengths. The Kipp & Zonen model CM 22 uses two Infrasil II quartz domes to protect the detector. The transmission of light through the Infrasil II domes changes with wavelength by about 2% from 300 nm to 3000 nm (see Figure 5.7 for transmission through an Infrasil II window that is used on the Eppley normal incidence pyrheliometer [NIP]). The spectral distribution of solar energy changes as sunlight takes a longer path through the atmosphere. The amount of atmosphere through which sunlight passes is called air mass (see Section 12.5.1 for a more detailed discussion of air mass). As air mass increases from 1.5 to 6.0, the percentage of incident solar radiation that is transmitted through each Infrasil II dome increases by ~0.1% to ~0.3%. The CM 21 and the PSP use other optical glass whose transmission is less uniform with wavelength. The transmittance

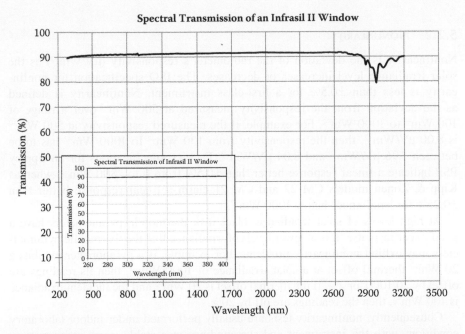

FIGURE 5.7 Spectral transmission of light through an Infrasil window from 280 nm to 3200 nm. Inset shows transmission from 280 nm to 400 nm. The percent transmission varies from around 89% at 280 nm to around 92% at 1100 nm to 2600 nm. The transmission drops to 80% around 2950 nm and returns to around 90% around 3200 nm. Measurements were made by Fuding Lin at the University of Oregon Support Network for Research and Innovation in Solar Energy.

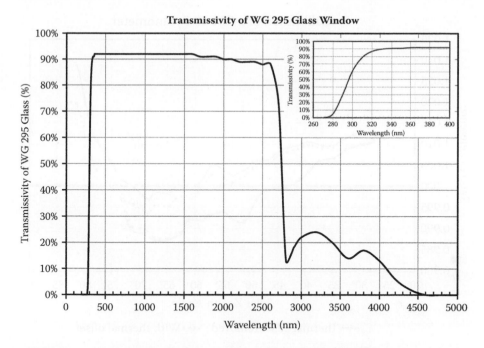

FIGURE 5.8 Spectral transmission of light through a Schott WG295 glass window from factor specifications. Inset shows spectral transmission form 260 nm to 400 nm.

of WG295 Schott glass used in the PSP is shown in Figure 5.8. The WG295 glass cuts off some ultraviolet and some IR radiation. Similar transmission properties are seen in the Schott K5 domes.

5.2.4 Angle of Incidence Response

Ideally, the responsivity of a pyranometer to the solar radiation from any direction within its hemispherical field of view, 2π steradians, is independent of incidence angle. Lambert's cosine law states that the irradiance received at a surface should be proportional to the cosine of the incident angle. Therefore, a pyranometer with an ideal angular response, or Lambertian response, would respond to incident radiation in accordance with Lambert's cosine law. When the pyranometer measurements are not Lambertian the instrument's responsivity is said to vary with incident angle. It is difficult, if not impossible, to produce a pyranometer with a perfect angular response, especially for a pyranometer with a relatively large receiver that is covered by glass domes. In most literature and in this book, angular response is referred to as cosine response and a perfect or Lambertian angular response is called a true angular or cosine response. When describing an instrument's angular response, the comparison is normalized against the true angular response to emphasize the difference from a perfect response. By making the comparison against an ideal Lambertian response, deviations of a few percent become easier to detect. Optical leveling of the receiving disk is very important, especially at large solar zenith angles where

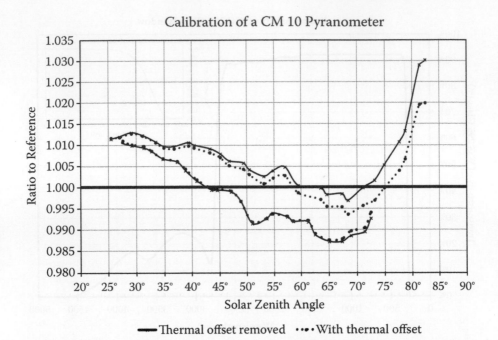

FIGURE 5.9 Cosine response of Kipp & Zonen CM 10 pyranometer. The reference GHI is calculated using the summation technique where the direct-horizontal component, obtained from the DNI by multiplying by the solar zenith angle, is added to the DHI. The DNI is measured using an AHF cavity radiometer and the DHI is measured with a Schenk Star pyranometer mounted on a tracker using a shade ball. The responsivities are shown with and without taking the thermal offset into account (see Section 5.2.1). The results are for morning and afternoon measurements.

small differences in the angle of incidence can translate into large errors in angular response. The spirit levels used to orient a pyranometer are not always consistent with the optical plane defined by the detector surface. Imperfections in the glass domes and even refraction of light as it passes through the domes can cause some deviations from true cosine (Lambertian) response. For solar irradiance measurements, the sun is considered an idealized point source with the collimated light coming from a given direction.

Results from a calibration of a Kipp & Zonen CM 10 pyranometer are shown in Figure 5.9. The reference GHI is calculated from the DNI measured with an absolute cavity radiometer (an Eppley automatic Hickey-Friedan [AHF] in this example) projected onto a horizontal surface plus the diffuse irradiance obtained by a pyranometer with minimal thermal offset (a Schenk star or Eppley 8-48 pyranometer) shaded by a shade ball. In this book, the reference GHI is obtained by this method using either an absolute cavity radiometer or a thermopile pyrheliometer for the DNI. The source of the DNI will be referenced for each example. The GHI readings from the CM 10 were divided by the reference GHI measurements and normalized to 1.00 at a solar zenith angle of 45°. During this calibration, there were a limited number of data points near 45° in the afternoon, hence the average at 45° is heavily weighted by the morning data. Since the absolute accuracy of these reference measurements are less than 1%, it will be assumed for these discussions that they are precise; any deviation from the ratio of 1.00 represents deviation from a calibrated instrument with a perfect angular response.

In Figure 5.9, the ratio of the pyranometer output to reference values is plotted against solar zenith angle. While some of the deviation from 1.00 may result from changes in temperature and possibly tilt of the instrument, much of the deviation is associated with a nonperfect Lambertian or cosine response. In Figure 5.9, the correction for thermal offset is up to 0.5%, a small fraction of the deviation from an ideal angular response. Of course, the thermal offset in Eugene is only about -6 Wm^{-2} due to the amount of atmospheric water vapor.

Understanding the error introduced by imperfect leveling of the pyranometer illustrates the difficulty in obtaining a true Lambertian response. When the sun is at 30° solar zenith angle, a 0.5° error in leveling will change the results by about 0.1%, at 70° angle of incidence it leads to about a 0.6% error in the measurement, and at 80° it leads to a 1.2% error. Therefore, it is important that the pyranometer is leveled at the site and that the platform supporting the pyranometer is securely anchored. It is also prudent to occasionally check the level of the pyranometer as determined by the integral spirit level.

Most of the deviation from true cosine response comes from the detector failing to absorb incident irradiance uniformly at all angles of incidence. The thickness of the glass domes, the variation in the thickness, the concentric alignment of the domes with respect to the receiver disk, the uniformity of the receiver disk coating, the "flatness" of the receiver surface, and optical leveling of the receiver disk within the instrument body all contribute to a non-Lambertian response. Dust or moisture on the dome can significantly contribute to the apparent angular response problem.

Ideally, the receiver should be flat or shaped to give the sensor a better cosine response. One problem that has been observed in a few pyranometers is that the circular receiver disk glued to the thermopile will become deformed and will take on the rectangular shape of the thermopile below. This deformation can be caused during manufacture, by vibrations, or by rough treatment during shipping. If the receiver disk takes on the shape of the underlying thermopile, deviation from a true cosine response becomes pronounced, and the pyranometer should be repaired (see Figure 5.10).

5.2.5 Response Degradation

The black receiver of the pyranometer is protected from direct contact with the elements but is exposed to global irradiance, and, like paint, the exposure to sunlight, especially ultraviolet (UV) radiation, will change the "color" of the paint and the absorptivity of the receiver over time. In the case of a thermopile pyranometer, the change in detector color is not easily perceived even after many years, but the change in responsivity typically decreases between 0.5 and 1.0% per year. The greater the exposure to sunlight, especially ultraviolet light, the faster the change in responsivity. Wilcox, Myers, Al-Abbadi, and Bin Mahfoodh (2001) showed that the formula used to measure paint aging also works for modeling the change in pyranometer responsivity. Using analyses of clear-sky records and long-term calibration records, Riihimaki

FIGURE 5.10 Figure of an distorted disk on an Eppley PSP pyranometer. The disk should be flat for a good cosine response. This disk has been bent to take on the shape of the thermopile to which the disk is attached.

and Vignola (2008) confirmed Wilcox et al.'s model and showed this effect continues at a linear rate over the decades that a pyranometer may be in the field.

Reference pyranometers usually have limited exposure to sunlight between calibrations. This greatly reduces the degradation rate of the pyranometer and again confirms the paint-aging hypothesis.

5.2.6 TEMPERATURE DEPENDENCE

All pyranometers exhibit some change in responsivity as ambient temperature changes. Many pyranometer manufacturers provide a temperature response curve for the specific model or model type. Thermopile pyranometers have a temperature dependence even if first-order temperature-compensating circuits are used to minimize the primary temperature effects. However, correcting the temperature response at one temperature will affect the temperature response in a different temperature range. Most first-class pyranometers have a responsivity that changes by less than 1% over the −20°C to 40°C range after temperature compensation circuitry is added. Temperature tests are performed in the laboratory because the responsivity of pyranometers outdoors is dependent on many meteorological variables that are not under the control of those making the test. Examples of the pyranometer temperature dependence are shown in Figure 5.11 for a Kipp & Zonen CM 22 and an Eppley PSP pyranometer.

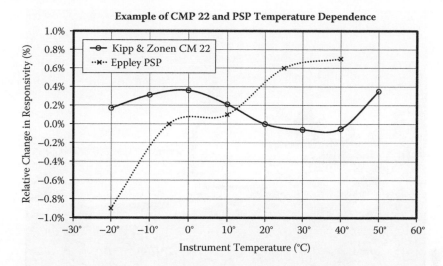

FIGURE 5.11 Example of the temperature dependence of a Kipp & Zonen CM 22 and an Eppley PSP. These measurements were done in the laboratory under controlled conditions. Each instrument has its own unique temperature dependence. Note that the temperature dependence of both instruments meets the factory specifications.

5.2.7 ICE AND SNOW ON DOME—VENTILATORS

Ventilators are extremely useful for keeping the domes of pyranometers free of moisture, frost, snow, and dust. If obtaining a complete and accurate measurement record is important, a ventilator is a necessary addition for any pyranometer installation.

A ventilator is basically a short cylinder with a fan mounted near the bottom to blow air under the pyranometer shield and across the dome. Three ventilators are shown in Figure 5.12. The ventilator on the left is a standard Eppley model ventilator (VEN) with a clear acrylic plenum surrounding the pyranometer. The ventilator in the middle is a Schenk ventilator, and the ventilator on the right is a Physicalisch-Meteorologisches Observatorium Davos (PMOD) design installed on an Eppley VEN base. The ventilator fans can be AC or DC powered, and some ventilators include a resistive element to heat the air slightly just before it blows across the dome, such as the PMOD and the Schenk ventilators shown in Figure 5.12. This arrangement may be helpful in reducing the temperature difference between the dome and body that is responsible for the infrared losses (thermal offsets) discussed in Section 5.2.1. AC fans generate and dissipate more heat from their motors than DC fans. In conclusion, ventilators can enhance or reduce the thermal offsets experienced by the instrument.

Ventilators usually have a screen beneath the fan to prevent larger debris from being added to the airflow. This screen should be cleaned at least once a year because it can become blocked and hinder the airflow.

FIGURE 5.12   (See color insert.) Three ventilators mounted on a Sci-Tech tracker. Going from right to left, the ventilator on the left is a Swiss PMOD ventilator holding a Kipp & Zonen CM 22 pyranometer. The ventilator is mounted on an Eppley ventilator base so the height of the dome matches the height of the domes of the other instruments. PMOD also makes a base for leveling. The middle ventilator is a Schenk ventilator for Schenk Star pyranometers. The ventilator on the left is an Eppley ventilator holding an Eppley precision infrared radiometer or pyrgeometer.

5.2.8   An Optical Anomaly

One anomaly observed with some pyranometers occurs when the sun is at a specific location in the sky. Over a period of about an hour, the irradiance readings first drop and then increase to an anomalously high reading before falling back to normal behavior. This is the result of reflections from the domes that will create a bright spot or "caustic" near the receiver disk. During a specific time of year, this intense spot of light will move across the ring surrounding the disk and illuminate a small portion of the absorbing disk. As the sun moves across the sky, the spot moves off the detector and crosses the metal ring surrounding the detector. When the spot is on the ring, the reading from the pyranometer is low and when the spot is on the detector, the reading can increase by 50% or more. The caustic on an Eppley 8-48 pyranometer is illustrated in Figure 5.13, and Figure 5.14 for a PSP.

Some pyranometers don't exhibit this effect, while others by the same manufacturer will exhibit this effect at different but specific times of the year at certain times of the day. Dust or film on the dome can cause or enhance this effect. Therefore, it is important to keep the domes as clean as possible. Pyranometers mounted on trackers usually do not exhibit this effect because the angle between the sun and the pyranometer

FIGURE 5.13   Photo of a caustic on an Eppley 8-48 pyranometer. Picture is taken in the lab and an image of the light source can be seen in the reflection on the top. The caustic is the array of light on the bottom of the white wedge.

FIGURE 5.14   Photo of the caustic on an Eppley PSP pyranometer. Moisture on the dome focused the light on the receptor disk.

domes never reaches the critical combination of zenith and azimuth angles at which the internal reflections occur. When analyzing pyranometer data to identify problems, one should watch for the occasional clear days that exhibit this anomalous behavior.

5.3 BLACK-AND-WHITE PYRANOMETERS

Major features of black-and-white pyranometers are as follows:

- Use of thermopiles with alternating black and white surfaces
- Performance specifications consistent with second-class pyranometers
- Responsivity has a measureable temperature dependence
- Responsivity varies with azimuth
- Responsivity affected by pyranometer tilt
- Limited thermal offset

Black-and-white type pyranometers consist of alternating black and white surfaces that are attached to the hot and cold junctions, respectively, of a thermopile. When exposed to solar radiation, the resulting temperature difference between the hot and cold junctions of the thermopile produces a voltage.

5.3.1 CHARACTERISTICS OF BLACK-AND-WHITE PYRANOMETERS

The specifications of the Eppley Laboratory, Inc. model 8-48 (more commonly referred to as a black-and-white pyranometer) (Figure 5.15) and the Schenk Star pyranometer (Figure 5.16) are given in Table 5.4 and Table 5.5, respectively.

Black-and-white pyranometers are classified as second-class pyranometers because the responsivity of the rather large detector surface changes significantly

FIGURE 5.15   Image of Eppley black-and-white type pyranometer.

FIGURE 5.16 Image of Schenk Star type pyranometer.

as a function of azimuth angle and tilt. One of the main reasons for this is that their alternating black and white surfaces make them sensitive to the sun's azimuth position. The responsivity of the pyranometer changes if the sun is directly aligned with a white surface as opposed to being directly aligned with a black surface. This problem can be addressed to some extent by making the surface areas smaller or using other geometric patterns. In addition, if the pyranometer is mounted on a tracker,

TABLE 5.4

Eppley Model 8-48 Pyranometer Specifications

Characteristic	Specification
Spectral sensitivity	0.285 to 2.8 μm
Typical sensitivity	10 μV/Wm^{-2}
Typical impedance	350 ohms
Temperature dependence	± 1.5% over ambient temperature range –20 to +40 °C
Linearity	± 1.0% from 0 to 1400 Wm^{-2}
Response time	5 seconds (1/e signal)
Typical cosine accuracy	± 2% from normalization 0–70° incident angle
	± 5% from normalization 70–80° incident angle
Mechanical vibration	tested up to 20 g's without damage
Calibration method	integrating hemisphere
Size	5.75 inch diameter, 2.75 inches high
Weight	2 pounds

TABLE 5.5
Schenk Star Pyranometer Specifications

Characteristic	Specification
Measuring range	0 to 1500 Wm^{-2}
Spectral sensitivity	0.3 to 3 μm
Typical sensitivity	15 μV/Wm^{-2}
	or 4 … 20 mA = 0 … 1500 Wm^{-2}
Typical impedance	35 ohms
Temperature dependence	< 1 % of the value between –20°C and +40°C
Linearity	< 0.5 % in the range 0.5 to 1330 Wm^{-2}
Response time	< 25 sec (95%), < 45 sec (99%)
Typical cosine accuracy	< 3 % of the value, incident angle 0°–80°
Azimuth response	< 3 % of the value
Stability	< 1% per year
Resolution	< 1 Wm^{-2}
Operational ambient temperature range	–40°C to +60°C
Weight	1.0 kg
Cable	2-polar shielded, 3 m length

the pyranometer would always have the same orientation to the sun and thus be less susceptible to azimuth affects.

Like other pyranometers, black-and-white pyranometers have imperfect angular responses. The receiver tends to be much larger than those found in single black-disk designs, making it more difficult to have a planar surface. The angular response of a Schenk Star type pyranometer is illustrated in Figure 5.17. This particular instrument

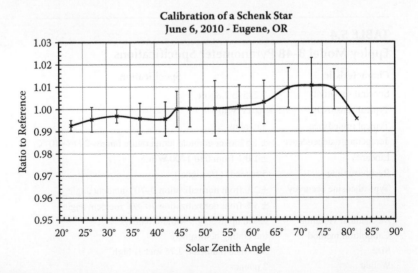

FIGURE 5.17 Cosine response of a Schenk Star pyranometer.

has an excellent cosine response, increasing by only 2% from a solar zenith angle of 22.5° to 82° with peak responsivity around 72°.

As with other black-and-white receiver designs, the Schenk pyranometer has a single glass dome. Having a single dome also leaves the pyranometer subject to more convective and advective thermal losses.

A major concern with black-and-white pyranometers is that their responsivity changes when they are tilted. Gravity starts to affect the convective heat flow as the pyranometer is tilted. As the pyranometer tilt increases, its responsivity decreases. Therefore, black-and-white pyranometers were not recommended for measuring irradiance on tilted surfaces. Design changes have been made to Eppley 8-48 pyranometers, and these redesigned pyranometers are reported to have a smaller tilt effect (Tom Kirk and John Hickey, personal communication).

The Schenk star is often difficult to level because Schenk's bubble level is not easy to read. It is often easier to put a level on the metal ring that holds the dome in place. The desiccant is in a container under the pyranometer. This requires lifting the pyranometer from its platform or ventilator to check and replace the desiccant. When the desiccant is examined, the pyranometer has to be releveled. The Eppley black-and-white pyranometer (model 8-48) has its desiccant on the front side of the pyranometer and is therefore easier to view and change.

As with other types of pyranometers, responsivity of black-and-white type pyranometers also changes with temperature (Figure 5.18). Temperature compensation is typically available between –20°C and +40°C.

Black-and-white pyranometers should be mounted in a ventilator or with an air gap between the instrument base and the mounting surface to maintain thermal isolation from the mounting platform.

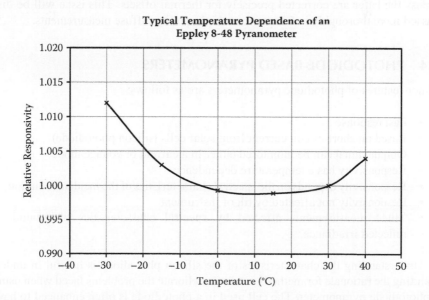

Typical Temperature Dependence of an Eppley 8-48 Pyranometer

FIGURE 5.18   Typical temperature response of an Eppley 8-48 (black-and-white) pyranometer.

5.3.2 Lack of Thermal Offset

For accurate measurements, the heat flow should be between the thermopiles behind the black and white surfaces. Other heat losses, such as from wind blowing across the pyranometer surfaces, reduce the accuracy of the measurements. In addition, temperature-compensating components are added to the detector circuitry to minimize the effect of nonlinear thermopile response to ambient temperature. Conduction losses are minimized by isolating the thermopile from the other components of the pyranometer.

From Kirchhoff's law, as discussed in Chapter 2, a good absorber of radiation is also a good emitter of radiation, and a good reflecting surface is also a poor emitting surface. At thermal equilibrium, absorption is equal to emission. As a first guess, one might think that the highly reflective white surface would also be a good reflector (poor emitter) at all wavelengths, including infrared. This is not the case because the emittance and reflectance characteristics are dependent on wavelength and the highly reflective white paint used in the black-and-white type pyranometer becomes as good an emitter at the mid-IR wavelengths as are the black surfaces. Thermal losses result from the pyranometer dome radiating to the sky. As the dome cools, the receiver surface cools by radiating to the cooler dome. At ambient temperatures, almost all of this radiation is in the mid-IR. Therefore both surfaces emit nearly the same amount of mid-infrared radiation, resulting in a minimal differential thermal offset.

As discussed in Section 5.2.1, black-disk thermopile pyranometers can have offsets as large as -20 to -40 Wm^{-2}. This can lead to intolerable errors if using these instruments to measure diffuse irradiance that can be on the order of 100 Wm^{-2} or less on clear days. This radiative loss suggests that black-and-white type pyranometers are superior to black-disk thermopile pyranometers for diffuse measurements unless the latter are corrected precisely for thermal offsets. This issue will be discussed more thoroughly in Chapter 6 on high-quality diffuse measurements.

5.4 PHOTODIODE-BASED PYRANOMETERS

Major features of photodiode pyranometers are as follows:

- Fast response
- Based on short-circuit current from solar cells (silicon photodiode)
- Output signal can be monitored either in a current or voltage mode
- Responsivity has a temperature dependence
- Responsivity depends on the spectral characteristics of the incident radiation
- Responsivity not affected by tilt of instrument
- Tilted measurements affected by spectral characteristics of ground-reflected irradiance

Understanding the characteristics of the silicon photodiode is helpful in understanding the rationale for methods used to ameliorate the problems faced when using a photodiode pyranometer. The cell used in a photodiode is often enhanced to have a better response to the blue portion of the solar spectrum than a typical solar cell.

Photodiode-based pyranometers use silicon photodiodes similar in design to photovoltaic or solar cells. Photodiodes are used to measure energy flux, and those packaged in a pyranometer are specifically designed to monitor incident solar radiation. Photodiodes measure the incident flux by monitoring the voltage across a low-temperature-coefficient (precision) resistor through which the photocurrent flows. This is equivalent to measuring the short-circuit current of a solar cell. As opposed to photodiodes, the much larger area solar cells are used to convert the incident solar energy into electrical power. Therefore, solar cells are operated differently than photodiodes. While solar cell output is related to incident solar energy, the photocurrent from a photodiode pyranometer is a much more accurate measure of the incident irradiance. The measurement accuracy of a photodiode pyranometer can be enhanced by using a foreoptic to improve the angular response of the instrument.

At first glance it would seem relatively easy to construct a pyranometer using a photodiode. However, the external quantum efficiency of photodiodes that generate the current is dependent on the wavelength of the incident light and photodiodes using silicon-based solar cells respond well to wavelengths between 450 nm and 900 nm but do not respond efficiently to wavelengths shorter than ~300 nm and longer than ~1100 nm (see Figure 5.19). The spectral response of silicon is another way of examining the applicability of photodiodes as pyranometers. The spectral response is the amount of current that is produced divided by the energy incident

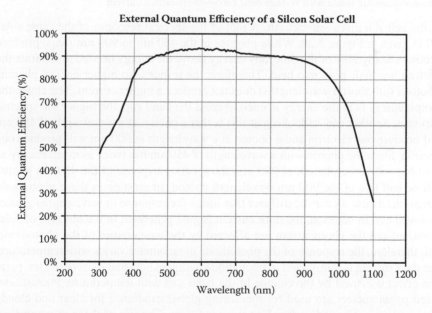

FIGURE 5.19 External quantum efficiency of a solar cell as a function of wavelength. Quantum efficiency is the percentage of electrons produced divided by the number of photons incident on the solar cell. A 90% quantum efficiency means that 90% of the photons of the particular wavelength incident on the material produce an electron. Measurements are made with voltage near zero—short-circuit current.

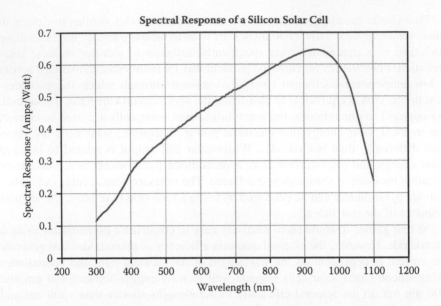

FIGURE 5.20 A plot of the spectral response of a typical silicon solar cell. Data comes from the same silicon cell used in Figure 5.19. The spectral response is the amount of current that is produced per incident irradiance on the cell. The units are normalized amps/watt. Measurements are made with voltage near zero—short-circuit current.

on the cell at a given wavelength. A plot of the spectral response of the same solar cell is given in Figure 5.20. While photons in the 450 nm to 900 nm range produce electrons nearly 90% of the time, any energy above the energy needed to separate the electron hole pair goes into heat. Therefore, the response to higher-energy photons (photons with shorter wavelengths) does not produce a higher current. The closer the photon energy is to the energy needed to cross the band gap, the higher the spectral response. Another way of looking at this is that a photon at a wavelength of 450 nm will produce one electron and a photon at a wavelength at 900 nm will produce one electron. Since the photon with a wavelength of 450 nm has twice as much energy as a photon with a wavelength of 900 nm, the spectral response of the 450 nm photon will be half that of the 900 nm wavelength photon. In most cases these photodiode pyranometers use an acrylic diffuser that limits the response to wavelengths greater than 400 nm. In addition, the dark current (signal generated in the absence of solar radiation) and the photocurrent are affected by the temperature of the photodiode, and, therefore, the response of the photodiode pyranometer varies with temperature. The dark current is minimized in photodiode pyranometers and most of the temperature effect is caused by the change in the band gap with temperature. Photodiode-based pyranometers are used for measuring global irradiance for clear and cloudy skies and are calibrated under clear-sky conditions. The use of these pyranometers in measurement situations that have a different spectral distribution is discouraged. The LI-COR model LI-200 user's manual states that the pyranometer should not be used under vegetation or artificial lights, in a greenhouse, or for reflected solar radiation. Photodiode pyranometers should not be used for diffuse or direct normal solar measurements without adjustments correcting the responsivity's spectral dependence (see Chapter 7).

The peak spectral response of a typical single crystalline silicon solar cell is around 900 nm, and the silicon solar cell doesn't respond to wavelengths longer than about 1100 nm. The quantum efficiency of the solar cells is highest between 450 and 900 nm but falls off rapidly below 450 nm and above 1000 nm (King and Myers, 1997) (see Figure 5.19).

At higher temperatures, the photodiode is more responsive. The temperature response is actually wavelength dependent with a small, negative temperature dependence below 900 nm and a large positive temperature dependence above 900 nm resulting in a net positive temperature response.

The temperature response of a photodiode is different than the temperature response of solar cells. Photodiodes operate at near short-circuit currents with very little voltage drop in the circuit. Solar panels that are used to produce electricity operate at a max power point that is nearer the open-circuit voltage than the zero voltage of a genuine short-circuit voltage. The open-circuit voltage decreases significantly with temperature, and this causes solar panel output to drop with temperature. The temperature dependence of the maximum power point output of a solar panel is not directly related to the slight increase in short-circuit current as the temperature increases.

PV reference cells are used to compare actual PV system performance with that expected from reference conditions, such as ASTM standard spectra for 1000 Wm^{-2} and 25°C (ASTM E913-97, 1997). Used properly, PV reference cells should be made of the same material as the PV system being tested. Reference cells are made to test whether system performance is as expected and not to measure the precise irradiance incident on the PV system. Metallic current collection grids on the surface of the cell and antireflective coatings on the protective glazing of the PV cell also affect the cell's output depending on the angle of incidence of the incoming irradiance. PV reference cells do not make good pyranometers since they have a poor cosine response.

5.4.1 Characterizing a Photodiode Pyranometer

The factory specifications for the LI-COR model LI-200 (Figure 5.21) and the Kipp & Zonen SP Lite (Figure 5.22) pyranometers are given in Tables 5.6 and 5.7, respectively. Photodiode pyranometers do not meet the ISO and WMO specifications as second-class pyranometers because of their limited spectral response. However, these pyranometers may perform as well as many second-class pyranometers in certain applications. In fact, many of the specifications of these photodiode pyranometers are significantly better than second-class specifications and even meet or surpass some of the first-class specifications. For example, the response time of a LI-COR LI-200 is extremely fast (approximately 10 μsec), about five orders of magnitude faster than a thermopile pyranometer. Therefore, photodiode pyranometers are often used when fast response times are required (see rotating shadowband radiometers, Chapter 7).

Pyranometer calibration plots of an Eppley PSP, a LI-COR LI-200, and a Schenk Star are given in Figure 5.23 and summarized in Table 5.8. The data in Figure 5.23 were taken on July 6, 2010, and show the typical cosine response characteristics of each instrument. A calibration ratio is created by dividing the measured GHI of each instrument, using previous responsivities, by the reference GHI calculated from the DNI and DHI. The

FIGURE 5.21   LI-COR model LI-200 pyranometer.

DNI was measured using an AHF absolute cavity radiometer, and the DHI was obtained from an Eppley PSP shaded by a shade ball. The thermal offset from the Eppley PSP was corrected by use of pyrgeometer and relative humidity measurements (Younkin and Long, 2003). Normally a Schenk Star or an Eppley model 8-48 pyranometer are used for the DHI measurements to minimize the thermal offset issue, but in this case the only

FIGURE 5.22   Kipp & Zonen model SP Lite pyranometer.

TABLE 5.6
LI-COR 200 Pyranometer Specifications

Characteristic	Specification
Sensitivity	90 μA per 1000 Wm^{-2}
Stability	< ±2% change over a 1-year period
Response time	10 μsec
Temperature dependence	< 0.15%/°C
Cosine correction	Corrected up to an 80° angle of incidence
Azimuth	< ±1% error over 360° at 45° elevation
Tilt	No error induced from orientation
Operating temperature	−40 to 65°C
Relative humidity	0 to 100%
Calibration	Calibrated against an Eppley PSP under natural daylight conditions, typical uncertainty is ±5%
Detector	High-stability silicon photovoltaic detector (blue enhanced)
Sensor housing	Weatherproof anodized aluminum case with acrylic diffuser and stainless steel hardware

available Schenk Star type pyranometer was being calibrated. The responsivities have been normalized to give the calibration ratio of 1 at a 45° solar zenith angle. It is standard practice to specify the pyranometer responsivity at a 45° solar zenith angle (Myers, Stoffel, Reda, Wilcox, and Andreas, 2002; Myers, Reda, Wilcox, and Andreas, 2004). Because the responsivity changes with solar zenith angle, only data taken within +/–1° of 45° were used to determine the instrument's responsivity at 45°. The data at 44.54° were used to determine the responsivity of the instruments. Typically the variance in the PSP data at each solar zenith angle is less than the LI-200 and Schenk Star pyranometer because the PSP has a large thermal mass and hence is less influenced by rapid changes in the environment. The Schenk Star pyranometer responsivity varies more than the PSP

TABLE 5.7
Kipp & Zonen SP Lite Pyranometer Specifications

Characteristic	Specification
Spectral range	400 to 1100 nm
Sensitivity	60 to 100 (option, 10 ± 0.5) μV/Wm^{-2}
Sensitivity change per year	< 2%
Response time (95%)	<<1 s
Directional error up to 80°	<5%
Temperature dependence	0.15%/°C
Operating temperature range	−30°C to +70°C
Nonlinearity 0 to 1000 W m^{-2}	< 1%
Maximum solar irradiance	2000 Wm^{-2}
Field of view	180°

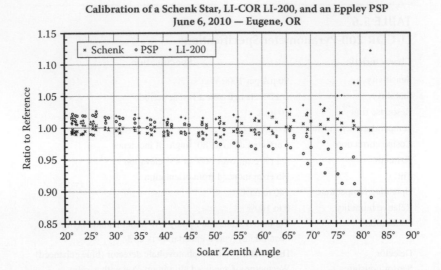

FIGURE 5.23 Calibration ratio data for an Eppley PSP pyranometer, a Schenk Star pyranometer, and a LI-COR model LI-200 pyranometer plotted against solar zenith angle. These are 1-minute data. The reference data are the DNI from an Eppley AHF absolute cavity radiometer projected onto a horizontal surface plus the diffuse irradiance from an Eppley PSP that has been corrected for thermal offsets.

TABLE 5.8

Relative Responsivity of Three Pyranometers Normalized to 1 at *sza* of 45°

Zenith Angle	Std Dev	Schenk Star	Std Dev	Eppley PSP	Std Dev	LI-COR LI-200	Std Dev
22.53	0.99	0.9926	0.0024	1.0154	0.0046	1.0078	0.0030
26.90	1.39	0.9954	0.0055	1.0146	0.0063	1.0082	0.0087
32.15	1.64	0.9969	0.0030	1.0116	0.0046	1.0082	0.0110
36.96	1.19	0.9958	0.0069	1.0042	0.0065	1.0072	0.0101
42.19	1.79	0.9956	0.0077	0.9985	0.0078	1.0055	0.0112
44.54	*0.47*	*1.0000*	*0.0080*	*1.0000*	*0.0089*	*1.0000*	*0.0129*
47.26	1.35	1.0002	0.0081	0.9957	0.0103	1.0012	0.0137
52.41	1.71	1.0003	0.0125	0.9880	0.0141	1.0042	0.0125
57.49	1.72	1.0012	0.0095	0.9816	0.0133	1.0037	0.0157
62.68	1.36	1.0031	0.0097	0.9783	0.0118	1.0095	0.0188
67.81	1.82	1.0095	0.0089	0.9728	0.0184	1.0055	0.0228
72.70	0.64	1.0105	0.0124	0.9536	0.0221	1.0136	0.0231
77.42	1.75	1.0088	0.0090	0.9277	0.0274	1.0390	0.0339
82.04	na	0.9956	na	0.8899	na	1.1214	na

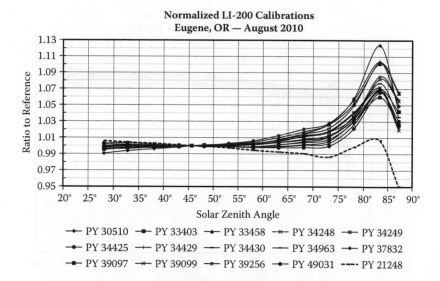

FIGURE 5.24 Average calibration ratio plots of 15 LI-200 pyranometers that have been normalized so that the ratio is 1 at a zenith angle of 45°. The oldest pyranometer show the best cosine fit at large angles.

because it has a smaller thermal mass and only one glass dome. However, this instrument has a better cosine response than this PSP at the station. The LI-COR LI-200 pyranometer exhibits more variance because the spectral composition of the irradiance changes, but the average irradiance measured is a good representation of the reference values. The cosine response of the LI-200 pyranometer starts to significantly deviate from true cosine response once the zenith angle is greater than 75°.

Figure 5.24 shows a calibration ratio plot of 15 LI-200 pyranometers. Some instruments representing the same model do perform slightly better than others, and while the shapes of the cosine response curves are similar among the same model of instrument they do vary. This is especially true at large angles of incidence. The large deviation around 85° is related to the shape of the diffuser and the artificial horizon on the LI-200 pyranometer.

5.4.2 Corrections Made to Photodiode Pyranometers

The designers of photodiode pyranometers consider two factors in their construction (Kerr, Thurtell, and Tanner, 1967):

1. The diffuser of the pyranometer must be designed to produce, as close as possible, a Lambertian response to the incident solar radiation. With a Lambertian response, the incident energy from a point source is proportional to the intensity of the source times the cosine of the incident angle.
2. The solar cell chosen needs to have a similar response throughout most of the day to give a better measurement of the incident solar irradiance.

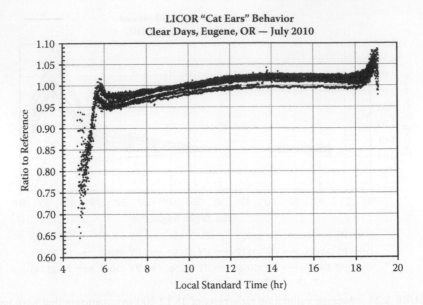

FIGURE 5.25 Plot of the ratio of a LI-COR LI-200 pyranometer divided by GHI calculated from an Eppley NIP for DNI and a Schenk Star pyranometer for the DHI against time. Values are 1-minute averaged data. Clear days in July 2010 were selected for the plot.

The typical responsivity of the LI-COR LI-200 pyranometer is fairly flat across the day and shows sharp spikes in early morning and late afternoon hours (see Figure 5.25). Only clear days were selected for the plot. For LI-200 pyranometers, two peaks in Figure 5.25 are observed. These angles correspond to high solar zenith angles around 85° in the morning and the afternoon. The plot reminded Augustyn et al. (2002) of "cat ears" when he was developing a model to correct the cosine response of the LI-200 pyranometer. The cause of the *cat ears effect* is the shape of the diffuser that improves the uniformity of the responsivity over much of the day. As with all pyranometers, modifications to correct for one limitation often lead to another limitation in a different situation. Therefore, it is always useful to know how the instrument performs under a wide variety of circumstances.

Notice that there is a distinct slope to the responsivity over the day with the ratio lower in the morning than in the afternoon. Temperature dependence of the pyranometer is likely to be partially responsible for some of this apparent slope. By multiplying the LI-200 pyranometer GHI values by

$$1 - 0.0035 \cdot (temperature - 25°C) \qquad (5.2)$$

the shape of the responsivity ratio is more symmetric (see Figure 5.26). The temperature used in the equation is the ambient air temperature in Celsius. An ambient temperature of 25°C was chosen as the reference temperature because it is the standard temperature used when specifying solar cell performance. The temperature data in Figure 5.26 range from 10°C to 35°C.

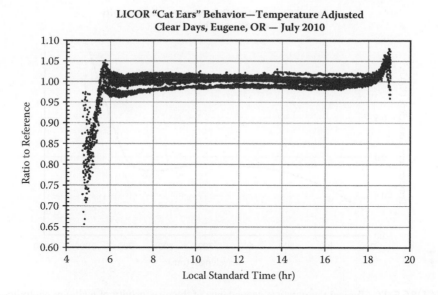

FIGURE 5.26 Plot of the ratio of a LI-COR LI-200 pyranometer divided by GHI calculated from an Eppley NIP for DNI and a Schenk Star pyranometer for the DHI against time. The LI-200 GHI values have been multiplied by $1 - 0.0035*(\text{Temperature} - 25°\text{C})$. Values are 1-minute averaged data. Clear days in July 2010 were selected for the plot.

These corrections are two to four times more than expected for silicon solar cells, indicating that other factors may mimic this assumed temperature dependence (Kerr et al., 1967). For example, the optical alignment of the detector could be partially responsible for the asymmetry seen in Figure 5.25 (Michalsky et al., 1995).

Many other factors can contribute to the shape and distribution seen in Figure 5.25, and care should be taken to untangle spectral change effects from cosine and temperature response effects. It is important to theoretically and experimentally identify the magnitude and direction of such effects on measurement performance before claiming that results show the physical effects of a specific variable. Before applying a correction with confidence, it has to be verified for a wide variety of measurement situations and at a number of different geographic locations.

For solar cells, the temperature coefficient introduces a bias of a few percent in measurements made in the 0 to 40°C range. The temperature coefficient for short-circuit current from silicon solar cells varies from −0.0004/°C to around 0.0015/°C when measured at different wavelengths (see Figure 5.27). A more detailed examination shows that the temperature coefficient for a silicon photodiode is positive in the wavelengths between 850 to 1100 nm and zero or slightly negative for wavelengths below 850 nm (Kerr et al., 1967). To minimize the uncertainty associated with temperature dependence, some photodiode pyranometers are heated to maintain a constant (higher than ambient) temperature, while others measure the pyranometer temperature and apply a correction factor to account for the change in responsivity with temperature. Some of the least expensive photodiode pyranometers ignore the instrument calibration change with temperature.

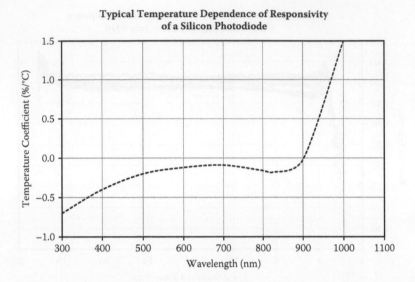

Typical Temperature Dependence of Responsivity of a Silicon Photodiode

FIGURE 5.27   Typical temperature dependence of the responsivity of a silicon photodiode versus wavelength. The spectral response will vary depending on the processing; however, the general features are driven by the temperature effect on the band gap. (From Myers, D., *Energy*, *30*, 2005. With permission.)

Pyranometers are calibrated during clear days and produce results such as shown in Figure 5.23. The measured irradiance under cloudy conditions compares favorably to measurements made with higher precision pyranometers. Clouds scatter all solar wavelengths nearly equally since the cloud particles are large compared with the solar wavelengths (geometric scattering limit).

Black-disk thermopile pyranometers such as an Eppley PSP have a fairly uniform spectral response over the wavelengths of interest. However, photodiode pyranometers, such as the LI-COR LI-200 and Kipp & Zohen SP Lite have a marked spectral response to the incident solar radiation. As shown in Figure 5.28, the LI-200 has the maximum sensitivity to wavelengths near 900 nm and does not respond to solar radiation with wavelengths over 1100 nm or under 400 nm. The effects of the acrylic diffuser on the LI-200 responsivity can be seen by comparing Figure 5.28 with Figure 5.20, which shows the responsivity for a silicon solar cell without an acrylic diffuser.

King, Kratochvil, and Boyson (1997) provide a good study of the spectral responsivity of the LI-200 pyranometer. They examine the responsivity on a clear day when approximately 90% of the global irradiance came from DNI. The responsivity of the photodiode-based pyranometer to clear-day DHI is significantly different from the responsivity for the DNI (Figure 5.29) (Vignola, 1999). On a clear day, the responsivity of the photodiode to the blue sky dome is about 70% of the responsivity determined for full-spectrum GHI or DNI. The photodiode pyranometers do not respond as well to the blue portion of the solar spectrum as seen in Figures 5.20 and 5.28, and when the diffuse spectrum consists largely of blue light, such as on a clear day, the photodiode will underreport the DHI. The DHI responsivity also changes as the solar zenith angle increases and the diffuse spectrum shifts toward red. The DHI

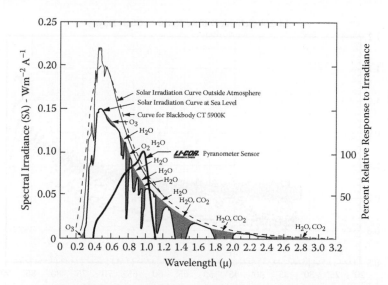

FIGURE 5.28 Percent relative response of a LI-COR LI-200 pyranometer plotted against wavelength from the LI-200 pyranometer brochure from LI-COR. The percent scale is on the right-hand axis. Also plotted are the extraterrestrial solar radiation curve, solar radiation curve at sea level, and a curve for black-body radiation at a temperature of 5900 K. Peak responsivity of the LI-200 pyranometer is to wavelengths between 950 nm and 1000 nm.

responsivity of the LI-200 increases as the DHI reddens, and at a solar zenith angle of 85° it is about 87% of the responsivity when measuring GHI at 45°. Some of the spread in the responsivity ratios in Figures 5.25 and 5.26 results from different atmospheric composition and hence differences in the spectral characteristics of the incident radiation. The haziest day (as determined by the lowest DNI values and highest DHI values) also has the lowest responsivity ratio values in Figure 5.26.

On clear days, the spectral characteristics of both the DNI and DHI are different and their spectral distribution changes over the day. The DNI spectrum shifts toward the IR as the path length through the atmosphere increases and more of the blue portion of the spectrum is scattered from the DNI. The DHI on clear days contains a considerable amount of blue light caused by Rayleigh scattering. Near sunrise or sunset, both DNI and DHI are redder as the blue light has been preferentially removed by Rayleigh scattering. This results in the reddish color of the sun and sky in the morning and evening. This red shift in the morning and evening hours occurs for DHI on cloudy as well as clear days because clouds are neutral attenuators.

On clear days it is difficult to separate angular response problems from spectral response changes. This can be seen in Figure 5.25 where the *cat ears* effects dominate the change in responsivity at larger solar zenith angles. On cloudy days, the scattering by clouds provides a more uniform distribution of irradiance, and the angular response does not change over the day. Figure 5.30 plots the ratio of DHI from a LI-COR LI-200 pyranometer divided by the DHI from the Schenk pyranometer on cloudy days. This ratio isolates the effects of changes in spectral distribution on the LI-200 responsivity. The angular responses of both instruments affect the ratio,

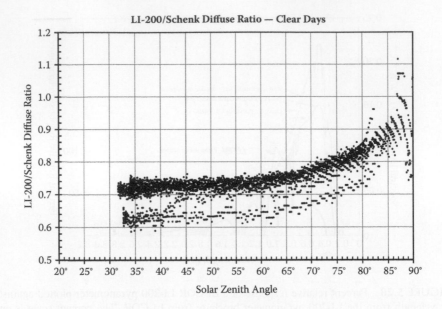

FIGURE 5.29 Comparison of diffuse measurements from LI-200 pyranometer in a rotating shadowband pyranometer and a Schenk Star pyranometer with a shade ball. Five clear days in August were chosen for the plot. The lowest values occurred on the clearest day. The LI-200 diffuse measurements are around 30% below the measurements of broadband thermopile pyranometers. From Figure 5.22, it can be seen that the photodiode pyranometer responds better to the redder portion of the solar spectrum.

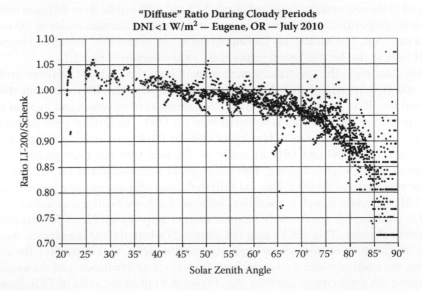

FIGURE 5.30 Plot of the responsivity ratio of a LI-200 pyranometer as a function of zenith angle for periods in July 2010 when the DNI less than or equal to 1 Wm^{-2}. Data are 1-minute values.

but the effect remains approximately constant over the day because the radiation is coming from all parts of the sky. Note that the responsivities of the instruments for cloudy periods are about the same during most of the day, because the DHI spectrum that the LI-COR LI-200 pyranometer sees is much like the GHI spectrum because the clouds act as neutral filters scattering all wavelengths equally. However, as the solar zenith angle increases, the total spectrum shifts toward red and becomes more like the spectrum that the LI-200 sees on clear days near sunrise and sunset; therefore, the responsivity of the LI-200 decreases to about 85% at 85°, the same value as in Figure 5.29. If the photodiode pyranometer is to be used for DHI measurements, large systematic errors will occur unless corrections are made to account for the changes in the spectral distribution and the resulting changes in pyranometer responsivity.

The response of the photodiode pyranometer to clear-sky diffuse is about 30% lower than a thermopile pyranometer. Under cloudy skies, the spectral distribution is similar to the GHI solar spectrum because the clouds are neutral scatters. Therefore, the photodiode pyranometer measurement of cloudy sky diffuse is comparable to thermopile pyranometer measurements.

The shade–unshade and summation calibration techniques will be discussed in detail in Section 5.5, but there is a systematic difference in responsivities determined for photodiode pyranometers using the two techniques that result from the different DNI and DHI spectra. The calibration of the photodiode pyranometer by the shade–unshade versus the summation technique is minimally affected by the lower diffuse responsivity since the diffuse component may make up only about 10% of the total irradiance on a clear day. A 30% error in a 10% component is only 3%, which is less than the typical uncertainty of ±5% associated with the calibration of a photodiode pyranometer. The poorer response of the photodiode pyranometer to clear-day DHI results in the responsivity determined by the summation method being about 3% lower than the responsivity obtained by a shade–unshade calibration.

It is recommended that a photodiode pyranometer be calibrated by the summation technique as this is similar to the way the pyranometers are used in the field for GHI measurements. For calibrations where it is not possible to use the summation technique, a side-by-side comparison can be made near a solar zenith angle of 45° with a reference photodiode pyranometer of the same model that has been calibrated by the summation technique. The estimated measurement uncertainties associated with side-by-side pyranometer comparisons are larger than methods referenced directly to pyrheliometer measurements.

Pyranometers operate under all weather conditions, unlike the clear-sky conditions used to calibrate the pyranometers. The absolute uncertainty of a photodiode pyranometer is often quoted as ±5%. In the field, the measurement uncertainty of these devices depends on many conditions, such as air temperature, time of day (solar zenith and azimuth angles), irradiance levels, spectral composition, and soiling or moisture on the diffuser. The temperature response of the LI-200 pyranometer under all sky conditions is illustrated in Figure 5.31. The responsivity of the LI-200 pyranometer in this example appears to increase by 0.3% per °C; however, the temperature dependence of the reference instruments has not been addressed. Other environmental factors may have also influenced the distribution of the data in Figure 5.31. If the temperature dependence of the pyranometer has been accurately

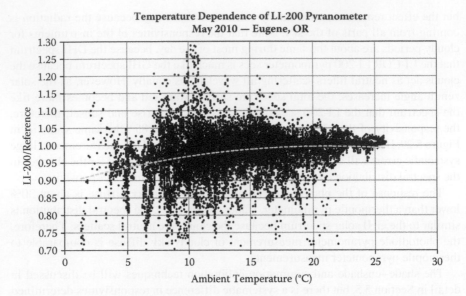

FIGURE 5.31 Ratio of LI-200 pyranometer readings divided by reference global values plotted against temperature. The ratio increases at a rate of approximate 0.3% per degree Celsius.

determined, these systematic errors may be removed. A multivariate correlation is needed to analyze the true dependence of the responsivity on a variable because some factors can mimic changes seen in other variables. In this case, the temperature dependence of the reference instruments may not be the only contributor.

Figure 5.32 contains a plot of global irradiance measured by a LI-COR LI-200 pyranometer divided by the reference global irradiance plotted against the *clearness index* for May 2010 in Eugene, Oregon. The clearness index (k_t) is computed as the ratio of measured GHI to the corresponding extraterrestrial value. The points with the low response ratios around clearness index of 0.2 occur mostly at zenith angles above 70° and probably represent the same spectral phenomena as shown in Figure 5.30.

5.4.3 Reference Solar Cells

Reference solar cells are sometimes used to check the relative performance of PV systems. Reference solar cells are specially packaged and characterized according to standard practices (ASTM, 2008, 2009, 2010; IEC, 2007), have a high-quality calibration history, and are used to check performance of PV modules made of similar materials when tested in solar simulators. In addition, reference cells are commonly used to adjust and monitor the intensity of solar simulators. The reference solar cells are designed to mimic the performance of PV modules on clear days when the incident angles are not extreme and are not designed to be pyranometers. Reference cells are useful for checking relative performance of PV modules and evaluating the degradation of performance over time, as long as periodic calibrations of the reference cell are maintained.

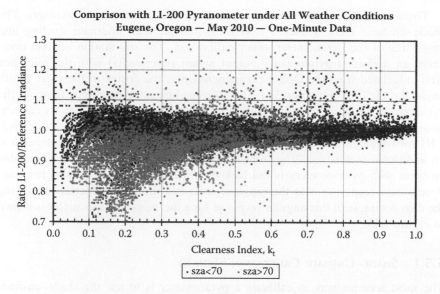

FIGURE 5.32 Plot of ratio of LI-200 pyranometer data to reference global irradiance plotted against zenith angle. One-minute data from May 2010 were used in the plot. May was a particularly cloudy month. Note that under cloudy conditions (clearness index < 0.4) and zenith angles greater than 70° the LI-200 pyranometer deviates by more than 10% from reference measurements. Reflection off clouds accounts for the high number of data points with a clearness index above 0.8.

Reference cells are normalized to standard operating conditions (an ambient temperature of 25°C, global irradiance of 1000 Wm⁻², air mass of 1.5, and no wind). The output of one reference cell is not comparable to a reference cell made with a different cell type, although they should compare well to identical solar cells in identical packaging. Because reference solar cells are behind a flat glazing, they suffer from a non-Lambertian response similar to solar modules. Pyranometers are designed to measure the total solar spectrum from 300 to 3000 nm under all weather conditions. Therefore, a reference solar cell is not a pyranometer and should not be treated as such.

5.5 CALIBRATION OF PYRANOMETERS

Proper calibrations depend on traceability to an internationally accepted standard. A critical aspect of pyranometer calibrations is its traceability to the World Radiometric Reference (WRR). The WRR is the internationally recognized measurement reference for solar irradiance (Fröhlich, 2006). A detector-based reference, the WRR is dependent on measurements from the World Standard Group (WSG) of electrically self-calibrated absolute cavity radiometers limited to DNI above 700 Wm⁻² under clear-sky conditions. Another important consideration is to calibrate an instrument in the same environment it will be used for routine measurements. Although this concept may not always lead to practical calibration methods, the estimated measurement uncertainties are generally more applicable.

There are two accurate outdoor methods used to calibrate pyranometers. The shade–unshade method uses a DNI measurement with alternate shading and unshading of the pyranometer under calibration. The summation method compares an unshaded pyranometer output signal (microvolts) with the reference GHI calculated from the DNI measurement times the cosine of the solar zenith angle plus the DHI (Equation 5.1). The estimated measurement uncertainty in the pyranometer calibrations is larger than the uncertainty in pyrheliometer calibrations because there is no absolute pyranometer available to obtain a reference GHI measurement. In addition, the responsivity of pyranometers is dependent on additional factors resulting from the 180° field of view. The most noticeable problem with pyranometers is the lack of a true Lambertian angular response (true cosine response) over the range of solar zenith angles experienced during the day. A sample of this angular response for a thermopile pyranometer is shown in Figure 5.23.

5.5.1 SHADE–UNSHADE CALIBRATION METHOD

The most accurate way to calibrate a pyranometer is to use the shade–unshade method (ASTM E913-97, 1997). This is the preferred method to establish reference pyranometers, especially pyranometers used for diffuse measurements. The responsivity ($\mu V/Wm^{-2}$) of the pyranometer is determined by subtracting the shaded voltage reading from the unshaded voltage reading and dividing this difference by the reference DNI projected onto the horizontal surface.

$$R = (UnshadedVoltage - ShadedVoltage) / (DNI \cdot \cos(sza)) \qquad (5.3)$$

The accuracy of the shade–unshade method is dependent on the accuracy to which the DNI is measured. One can use either a high-quality thermopile pyrheliometer or, preferably, an absolute cavity radiometer for the DNI value. A cavity radiometer can measure the DNI to an accuracy of better than 0.5%, whereas a thermopile pyrheliometer has an absolute accuracy of between 0.7% and 1.7% (Michalsky et al., 2011). The field of view of a cavity radiometer is 5°, while that of some thermopile pyrheliometers is 5.7°. A shade disk or sphere has to be built that will shade the pyranometer such that the field of view blocked by the disk is the same as the field of view of the instrument used to obtain the DNI value. For example, a disk with a 4.97 cm radius would be equivalent to the 5.7° field of view for the disk if it was 1 meter away. If the disk was only 50 cm away, the radius needed would be 2.49 cm. For a 5.0° field of view the radius for a distance of 1 meter would be 4.36 cm, and for a distance of 50 cm the radius would be 2.18 cm. The size of the disk should be such that it shades the dome of the pyranometer, and, hence, prevents reflections from the dome impinging on the receiver. The disk should be painted black. Use an ultra-flat black paint such as Rust-Oleum Black to coat the disk and the supports. For information on standards for the shade–unshade calibrations, see ASTM E913-97 (1997) and McArthur (2004).

When performing a shade–unshade calibration, it is necessary to wait until the pyranometer reaches equilibrium when the instrument is shaded and when it is unshaded. For photodiode pyranometers, the response time is ~10 μseconds; there-fore, the shade–unshade intervals are very short. For a thermopile pyranometer, the response time may be several seconds depending on the design. Response time is the amount of time it takes for the instrument to respond to a step change in irradiance. In this chapter, response time is the time it takes to be within $1/e^{th}$ (~37%) of the new value. Ideally, the reading should be the true value or the equilibrium read-ing. During a shade–unshade calibration under the required clear-sky conditions, the incident irradiance may change by an order of magnitude from GHI to DHI. It takes between 20 and 60 times the $1/e$ response time for consistent readings (Reda, Stoffel, and Myers, 2003). It is important to note that it takes three times longer to achieve an equilibrium state when the shaded pyranometer is unshaded and exposed to full sun than when the unshaded pyranometer is shaded for the calibration. At full sun (e.g., DNI = 1000 Wm⁻²), a reading taken after 60 times the $1/e$ response time is representative of the true GHI. During the shaded period, it takes 20 time constants for the readings to reach a consistent DHI level (Figure 5.33). The response time is related to the thermal mass, thermal conductivity, and other design elements of the pyranometer. As in Figure 5.33, the best way to determine when the signal has

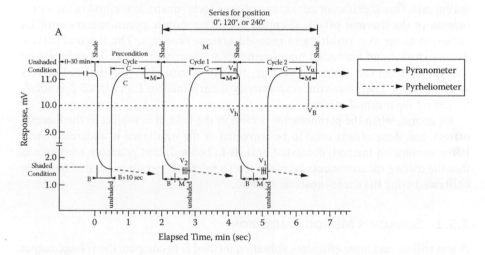

FIGURE 5.33 Schematic for the shade–unshade calibration. The timeline is the bottom axis, and the voltage reading in mV is the vertical axis. The timeline for the shade–unshade method starts 30 minutes after the pyranometer was set outside to reach thermal equilib-rium. Three time scales are indicated on the plot, *B*, *C*, and *M*. *B* is equal to 20 or 30 times the instrument $1/e$ response time, *C* is 60 times the instrument response time, and *M* is a 30 second period when data are collected. The dashed line is the DNI irradiance divided by 100 to be on the same scale as the voltage measurement of the pyranometer being calibrated. The responsivity of the pyranometer is determined by the ratio of the voltage reading at full sun minus the voltage reading when shaded divided by the beam irradiance projected onto a horizontal surface. In this example, the responsivity is determined when the pyranometer is rotated 0, 120, and 240 degrees. This was done to check the azimuthal response of the instru-ment. (From Reda, I., T. Stoffel, and D. Myers, *Solar Energy*, *74*, 103–112, 2003, and Myers, D., *Energy*, *30*, 1517–1531, 2005. With permission.)

stabilized is to plot the output voltage from the pyranometer when it is shaded and unshaded. Identifying this stable period requires sound judgment and experience to separate voltage creep (or overshoot) due to true irradiance change from instrument stabilization. The data used in determining the responsivity are taken when it is judged that equilibrium has been attained.

When performing a shade–unshade calibration, the timing of the measurements is important to accurately compute the solar zenith angle. One should also measure the ambient air temperature because thermopile- and photodiode-based pyranometer readings have a known dependence on temperature. Shade–unshade calibrations should be made only when the skies are clear of clouds and the DNI readings are stable (e.g., changing by less than 3 Wm^{-2} during a 1-minute interval). Ideally, the measurements for establishing a reference pyranometer calibration should be repeated over multiple days throughout a year to evaluate the pyranometer's change in responsivity with different combinations of solar zenith and azimuth angles. In lieu of this challenging requirement it is often better to fit a curve (responsivity as a function of solar zenith angle or air mass) over a few days of calibration data that average the random variations, which makes it easier to detect and account for biases in the pyranometer measurements. One significant advantage of the shade–unshade method is the elimination of the thermal offsets (Section 5.2). Thermopile pyranometers emit IR radiation to the sky, resulting in reduced voltage readings. The thermal offsets during shaded and unshaded measurements are nearly identical. By subtracting the shaded from the unshaded voltage, the offset caused by the thermal offsets is eliminated. Therefore, the responsivity determined in Equation 5.3 is unaffected by the thermal offset.

Of course, when the pyranometer is used in the field, it is subject to the thermal offsets, and these offsets need to be corrected or the irradiance is underestimated. If the summation method, discussed next, is to be used, best practices recommend that the diffuse measurements be made with a reference pyranometer that has been calibrated using the shade–unshade method.

5.5.2 Summation Method Calibration

A less tedious and more efficient calibration method is to compare the voltage output from the pyranometers under test with a reference global irradiance determined by a DNI measurement made with an absolute cavity radiometer or a first-class thermopile pyrheliometer and a DHI measurement made with a shade–unshade-calibrated and offset-corrected pyranometer mounted under a shade disk. The DNI value is projected onto a horizontal surface by multiplying this irradiance by the cosine of the solar zenith angle, and the reference GHI value is obtained by adding the DHI value to the direct horizontal value (Myers el al., 2002, 2004; Reda et al., 2003; Wilcox, Andreas, Reda, and Myers, 2002). This method can be automated for the simultaneous calibration of multiple pyranometers with responsivity values obtained throughout the day. Ideally, this should be performed near the summer solstice to allow for the widest range of solar zenith angles, but calibrations can be obtained as long as the solar zenith angle at solar noon is less than 45°. The key to the accuracy of the summation calibration method is to determine the reference GHI value using a DHI value from a pyranometer that has minimal or correctable thermal offset. By design, black-and-white type thermopile pyranometers have minimal thermal offsets. There are also single black thermopile pyranometers that exhibit minimal or correctable thermal offsets.

For consistent responsivity results, pyranometer calibrations are performed on clear days with minimal variations in atmospheric transmission (i.e., aerosol optical depth). There are many ways to specify a stable atmosphere. Instruments with fast response times, like photodiode pyranometers, can be used to check for rapid variations in the GHI due to variations in atmospheric water vapor, aerosols, and thin (subvisual) clouds, or the variability of the DNI and DHI irradiance over sub 1-minute intervals can be used to select the best calibration periods (Reda et al., 2003; Wilcox et al., 2002).

For a comprehensive determination of responsivity, the calibrations should be made over a full day or parts of 2 days and include a wide range of solar zenith and azimuth angles. To understand the change in responsivity of a pyranometer over time, it is important to measure the environmental parameters that influence the instrument responsivity. (See the BORCAL certificate in Appendix G for examples of these parameters.) Specifically, in addition to the irradiance measurements, concurrent ambient temperature, relative humidity, wind speed and direction, and barometric pressure should be recorded. If an initial responsivity is obtained when the ambient temperature is 10°C and an additional responsivity is obtained when the ambient temperature is 25°C, some of the difference between the two measured responsivities may be related to the change in temperature. If it is known how the responsivity varies with temperature, the change in responsivity over time can be better estimated by adjusting the responsivities to a standard temperature. Influence of wind speed, downwelling infrared, spectral distribution of the incident radiation, and other factors influence the responsivity of the pyranometer. It is not straightforward to separate the influence of these factors on the responsivity because the effects are often small, and they may be inseparable. For example, wind speed affects the temperature difference between the dome and body of the pyranometer, and relative humidity affects the thermal offsets and the advective efficiency of wind on heat transfer. Therefore, calibrations should be performed over a wide variety of meteorological conditions.

A single responsivity can be assigned to a pyranometer for all measurements, or, for more accuracy, a responsivity that is a function of the primary variables that affect the responsivity can be used. A prime example of the latter is a pyranometer responsivity that varies with solar zenith angle. Generally, a single responsivity of the instrument, near 45°, is used to specify the calibration for irradiance measurements. An example of a pyranometer responsivity changing over a day is given in Figure 5.34, and the responsivity plotted against the solar zenith angle is illustrated in Figure 5.35.

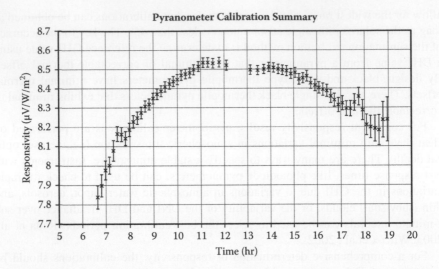

FIGURE 5.34 Plot of 1-minute responsivity data against time of day. An ideal pyranometer would have a constant responsivity over the day. Often the deviation from a flat responsivity is attributed to imperfect Lamberian (or cosine) response. Note: thermal offsets and other factors also play a role in the deviation from a true cosine response.

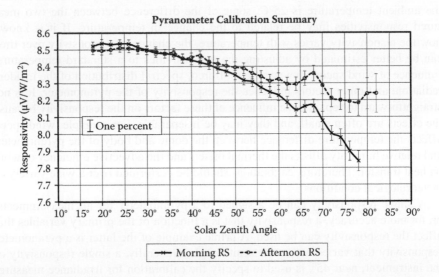

FIGURE 5.35 Same data as in Figure 5.28 with the responsivity plotted against zenith angle. A true cosine response would yield a straight line. Note that there is an asymmetry between the morning values and the afternoon values. Leveling and other factors such as changes in the azimuth response of the pyranometer and temperature effects on the responsivity can contribute to this asymmetry.

5.6 PYRANOMETER CALIBRATION UNCERTAINTIES

As discussed in Chapter 4, the *Guide to the Expression of Uncertainty in Measurements* (*GUM*; JCGM, 2008) presents the *GUM* protocols that specify a uniform manner for obtaining uncertainties. With a set of measurements, one can determine the mean and standard deviation of the data set with respect to a reference value. This information provides only a subset of the uncertainties. Many other factors contribute to the uncertainty such as the calibration of reference instruments and accuracy of the measured radiometer output voltages, for example. The *GUM* protocols help classify the uncertainties and prescribe a methodology for combining the uncertainties to give an overall uncertainty of the measured values. Estimated measurement uncertainty is a very important part of understanding and applying radiometer data. Uncertainties as applied to the calibration and use of pyranometers are covered in this section (the *GUM* procedure was applied in Chapter 4 to pyrheliometer measurements). Uncertainties as applied to pyranometers are more complex and will be approached from a slightly different perspective than covered in Chapter 4. A brief review of the *GUM* procedure will be followed by examples of uncertainties calculations taken from the National Renewable Energy Laboratory (NREL) calibration procedures for broadband radiometer calibration (Myers et al., 2004).

The *GUM* procedure can be summarized in five steps:

1. Determine the measurement equation.
2. Calculate or estimate the standard uncertainty for each variable in the measurement equation.
3. Calculate the sensitivity coefficients.
4. Calculate the combined standard uncertainty (u_C).
5. Calculate the expanded uncertainty (U_E) by multiplying the combined standard uncertainty by the coverage factor (k), typically by applying student's t-distribution factors for 95% or 99% confidence.

There are two types of uncertainties in a measurement. Type A uncertainties are those that can be calculated from a series of observations (labeled u_A). Type A analysis is statistical in nature and determines the random errors associated with any measurement. Type B uncertainties are all of the other uncertainties that are not subject to the statistical analysis of the observations (labeled u_B). Type B uncertainties are associated with a variety of effects that can affect the measurements and add uncertainty. This type of uncertainty includes the meteorological and situational influences that affect the measurement, systematic errors caused by the data logger used for the voltage measurement, and uncertainties in the calibration system used to calibrate the pyranometer.

GUM (JCGM, 2008) treats the two uncertainty types separately and then determines a combined uncertainty (u_C). Uncertainties with a small "u" refer to uncertainties that are within one standard deviation (a 68% confidence level), while uncertainties labeled with a capital "U" refer to uncertainties at the 95% level of confidence. For uncertainties that are from a Gaussian (or "normal") distribution with a

TABLE 5.9

Standard Deviation and Confidence Level for a Gaussian Distribution

Number of Standard Deviations	Confidence Level (%)
1 σ	68.27
1.96 σ	95.0
2 σ	95.45
3 σ	99.73
4 σ	99.99

large number of data points, 68% of the data points lie within 1 standard deviation of the mean, while 95% of the data points lie within 1.96 times the standard deviation (see Table 5.9 for the relationship between multiples of the standard deviation and the confidence levels for a Gaussian distribution). The u_{std} is the estimate of an equivalent standard deviation (of a specified distribution) of the error source (Stoffel et al., 2010). The combined uncertainty (u_C) is computed from the Type A and Type B standard uncertainties summed in quadrature. When terms are added in quadrature, the sum of the square of the terms is calculated, and the square root of the total is the combined standard uncertainty of all terms. It is important to use the same confidence level for all the uncertainty terms. If one term is assigned at 68% confidence level (1 standard deviation), then all terms should be referenced to the 68% confidence level. When the square root of the sum of the squares is determined, the confidence level can be expanded. In the *GUM* methodology, the expanded uncertainty, U_E, is determined from the combined uncertainty u_C determined at the 68% level of confidence and multiplied by a coverage factor k to obtain the desired level of confidence:

$$U_E = k \cdot \left[\Sigma u_B^2 + \Sigma u_A^2 \right]^{1/2} \tag{5.4}$$

For a large enough statistical sample, a coverage factor of 1.96 is used to obtain a 95% level of confidence or a coverage factor of 2.58 is used to obtain a 99% confidence level. This all-inclusive uncertainty is called the expanded uncertainty (U_E).

To be more precise, the combined standard uncertainty for a measurement is obtained from the square root of the sum of the variances of the variables that affect the measurement, weighted according to how the measurement results change with changes in these quantities; that is, by sensitivity coefficients. These sensitivity coefficients are explained in the example uncertainty analysis given in the next section.

5.6.1 Uncertainty Analysis Applied to Pyranometer Calibration

The procedure for determining the uncertainty in calibrating a pyranometer based upon the summation technique will be illustrated. This example is taken from Reda et al. (2008). For calibrating pyranometers with a thermal offset, the equation for the responsivity is

$$R = \frac{V - offset}{DNI \cdot \cos(sza) + DHI} \qquad (5.5)$$

where

R = responsivity in [$\mu V/(Wm^{-2})$]

V = thermopile output voltage in (μV)

offset = the offset voltage determined from nighttime measurements of the pyranometer offset as a function of the net radiation measured by an infrared radiometer (pyrgeometer); the pyranometer responsivity to net IR irradiance R_{net}, thus established, can be used during the daytime with net infrared measurements W_{net} to derive the offset, which will equal $R_{net} \cdot W_{net}$

R_{net} = longwave net responsivity of the pyranometer in [$\mu V/(Wm^{-2})$]

W_{net} = infrared net irradiance measured by collocated pyrgeometer in (Wm^{-2})

DNI = direct normal irradiance measured by the reference pyrheliometer in (Wm^{-2})

sza = solar zenith angle

DHI = diffuse horizontal irradiance measurement (Wm^{-2})

Thermopile pyranometers are subject to an IR radiative loss or thermal offset. As discussed in Section 5.3.2, black-and-white pyranometers have a minimal thermal offset. Sometimes it is possible to determine both the broadband solar and *IR* responsivities (R_{net}) for the pyranometer. The instrument-specific R_{net} can be measured in a black-body chamber (Reda et al., 2005). Generic offsets have been determined using a black-body chamber for Eppley and Kipp & Zonen pyranometers by model, but there are still some variations within model types. The R_{net} is typically 10% or smaller than the broadband solar responsivity (R). Having R_{net} is useful only when pyrgeometer data are available to determine net IR flux and compute the thermal offset for a specific or generic pyranometer. If a pyrgeometer is not present, one can average the nighttime thermal offsets from the night before and night after the daytime measurements to estimate an offset correction. This will, of course, carry a larger uncertainty than the measured offset estimate (Dutton et al., 2001; Vignola et al., 2007).

The sensitivities of the coefficients in Equation 5.5 to changes in the variables are found by calculating partial derivatives of the responsivity (R) with respect to each of the variables:

$$c_V = \frac{\partial R}{\partial V} = \frac{1}{[DNI \cdot \cos(sza) + DHI]} \qquad (5.6)$$

$$c_{R_{net}} = \frac{\partial R}{\partial R_{net}} = \frac{-W_{net}}{[DNI \cdot \cos(sza) + DHI]} \qquad (5.7)$$

$$c_{W_{net}} = \frac{\partial R}{\partial W_{net}} = \frac{-R_{net}}{[DNI \cdot \cos(sza) + DHI]} \qquad (5.8)$$

$$c_{DNI} = \frac{\partial R}{\partial DNI} = \frac{-(V - R_{net} \cdot W_{net}) \cdot \cos(azm)}{[DNI \cdot \cos(sza) + DHI]^2} \tag{5.9}$$

$$c_Z = \frac{\partial R}{\partial Z} = \frac{DNI \cdot \sin(azm) \cdot (V - R_{net} \cdot W_{net})}{[DNI \cdot \cos(sza) + DHI]^2} \tag{5.10}$$

$$c_{DHI} = \frac{\partial R}{\partial DHI} = \frac{-(V - R_{net} \cdot W_{net})}{[DNI \cdot \cos(sza) + DHI]^2} \tag{5.11}$$

Table 5.10 contains the Type B standard uncertainties for the variables in Equation 5.5. All of these uncertainties are assumed to come from rectangular distributions implying that any value between plus or minus the u-value in column three of Table 5.10 is equally likely. For this type of distribution the standard uncertainty is calculated as $u/\sqrt{3}$.

The pyranometer responsivity using the data in Table 5.10 and Equation 5.5 is 8.0736 $\mu V/Wm^{-2}$. The *combined* uncertainty of the calibration is determined by combining the standard uncertainties with sensitivity coefficients applied in quadrature

$$u_B^2 = c_V^2 \Delta V^2 + c_{R_{net}}^2 \Delta R_{net}^2 + c_{W_{net}}^2 \Delta W_{net}^2 + c_{DNI}^2 \Delta DNI^2 + c_Z^2 \Delta Z^2 + c_{DHI}^2 \Delta DHI^2 \tag{5.12}$$

Using the data in Table 5.10, u_B is calculated to be 0.022 $\mu V/Wm^{-2}$. The uncertainties of Type A arise from statistical sampling of the measured voltage V and the measured irradiances DNI and DHI used in the determination of R. Therefore, an equation of the form

$$u_A^2 = c_V^2 \Delta V^2 + c_{DNI}^2 \Delta DNI^2 + c_{DHI}^2 \Delta DHI^2 \tag{5.13}$$

determines the *combined* uncertainties of the Type A variables. The total *combined* uncertainty is computed from the Type A and Type B standard uncertainties by adding them in quadrature. A coverage factor k is applied to the combined uncertainty

---

TABLE 5.10
Estimated Standard Uncertainties of Type B

Variable	Value	u	Offset	a = u + offset	Distribution	DF	u_{std}
V	7930.3 μV	0.079 μV	1 μV	1.079 μV	Rectangular	1000	0.62
R_{net}	0.4 $\mu V/(Wm^{-2})$	0.04 $\mu V/(Wm^{-2})$	–	0.4 $\mu V/(Wm^{-2})$	Rectangular	1000	0.02
W_{net}	−150 Wm^{-2}	7.5 Wm^{-2}	–	7.5 Wm^{-2}	Rectangular	1000	4.33
DNI	1000 Wm^{-2}	4 Wm^{-2}	–	4 Wm^{-2}	Rectangular	1000	2.31
Z^*	20°	0.00002	–	0.00002	Rectangular	1000	0.00001
DHI	50 Wm^{-2}	1.5 Wm^{-2}	1 Wm^{-2}	2.5 Wm^{-2}	Rectangular	1000	1.44

* Since the uncertainty in Z is 0.003°, at 20° the uncertainty is cos(Z + 0.003) − cos(Z).

to determine an *expanded* uncertainty, usually, a 95% confidence interval or a 99% confidence interval

$$U_E = k \cdot \left[u_A^2 + u_B^2 \right]^{1/2} \tag{5.14}$$

The expanded uncertainty (U_E) is defined as the interval that is expected to contain 95% ($k \cong 2$) or 99% ($k \cong 3$) of the values from measurements. Note that the $+U_E$ and $-U_E$ can be different values as some errors do not have symmetrical distributions. Understanding the *GUM* procedure requires some statistics background, especially for determining the coverage factor. For example, a student's t-distribution should be used to determine k when the number of samples is small ($n < 30$) and the distribution is nearly normal. For a large number of samples, the student's t-distribution yields a coverage factor of 1.96 for 95% uncertainty as it would be for a Gaussian distribution of an infinite number of samples.

5.6.2 An Example of the *GUM* Procedure Applied to the Calibration Uncertainties of a Pyranometer

The responsivity of pyranometers changes as a function of solar zenith angle. The responsivity behavior is similar among pyranometers of the same model. Because of this change with solar zenith angle, it is necessary to be specific when stating a single calibration *constant* for a pyranometer. The NREL has a calibration procedure based on the summation method called Broadband Outdoor Radiometer CALibration (BORCAL) (Myers et al., 2002). Pyranometers calibrated with the BORCAL method are assigned the responsivity determined when the solar zenith angle is 45° with the expanded uncertainty based on the calibration results for solar zenith angles between 30° and 60°. Responsivities are also reported at 2° solar zenith angle increments, and a composite responsivity based on all the solar zenith angles in the calibration is reported (see Appendix G for a sample BORCAL report). The responsivity and the uncertainty associated with each method of calculating the responsivity are different because the responsivity is characterized over different solar zenith angle ranges.

In Section 5.6.1 the Type B uncertainties were discussed in detail. Examples of the Type A uncertainties will be discussed here for two different methods of defining the pyranometer responsivity.

As shown in Figures 5.34 and 5.35, the responsivity of a pyranometer changes over the day with the solar zenith angle. For the u_A uncertainty for the responsivity at 45°, a simple standard deviation is determined for the data used to obtain the responsivity. The combined uncertainty is obtained by adding in quadrature the u_B determined in Section 5.6.1 with the u_A determined by the statistical analysis. The expanded uncertainty (U_E) is obtained by multiplying the combined uncertainty by the coverage factor.

The Type A uncertainties for the responsivity data binned by solar zenith angle are determined in a similar manner to the Type A uncertainty determined for the responsivity at 45°. These u_A uncertainties are the statistical uncertainties of the data used to obtain the responsivity for the given 2° solar zenith angle bin.

The complexity starts to enter when looking at the combined uncertainty over the full range of solar zenith angles or over a given range of solar zenith angles. The first task is to create a spline fit through the data points

$$R_{i,AM}(Z) = \sum_{j=0}^{3} a_j (sza - sza_i)^j \qquad (5.15)$$

$$R_{i,PM}(Z) = \sum_{j=0}^{3} a_j (sza - sza_i)^j \qquad (5.16)$$

This is a third-degree polynomial through four data points. A natural cubic spline interpolation is a piecewise cubic fit that is twice differentiable and the second differential is zero at the end points. The a_j are determined by the spline fit to the nearest data points. With the spline responsivities determined for each measurement, the root mean square error can be calculated for all data. The root mean square error is the difference between the measured responsivity and the responsivity determined for each bin by the piecewise fit.

$$r^2 = \left[\frac{\sum_{i=i}^{m} (R_{i,meas} - R_{i,AM})^2}{m} + \frac{\sum_{i=i}^{n} (R_{i,meas} - R_{i,PM})^2}{n} \right] \Big/ 2$$

$$= \left[\frac{\sum_{i=i}^{m} r_{i,AM}^2}{m} + \frac{\sum_{i=i}^{n} r_{i,PM}^2}{n} \right] \Big/ 2 \qquad (5.17)$$

where r^2 represents the sum of the squares of the AM and PM residuals with respect to the spline fit divided by the total number of data points. The number of responsivity data points in the AM is m and the number of points in the PM is n.

Next, the task is to calculate the standard deviation of the residuals from the spline fit to determine the average residual. Since there are separate spline fits for the AM and PM, a standard deviation for the morning and afternoon hours can be determined.

$$\sigma_{AM} = \sqrt{\frac{\sum_{i=1}^{m} (r_{AM} - r_{i,AM})^2}{m-1}} \qquad (5.18)$$

where r_{AM} in this equation represents the average residual for the spline fit in the morning and $r_{i,AM}$ is the individual residual of each data point. The afternoon standard deviation is the same format except the AM values are replaced by the PM values. The standard uncertainty (u_{int}) from the Type A analysis is then

$$u_{int} = \sqrt{(r_{AM}^2 + r_{PM}^2)/2 + (\sigma_{AM}^2 + \sigma_{PM}^2)/2} \qquad (5.19)$$

and the degrees of freedom are $m + n - 2$. For this example, the uncertainty for the interpolated function (spline fit in this example) is ±0.21%, and the Type B uncertainty (u_B) is ±0.61%. Combining their uncertainty in quadrature gives the combined

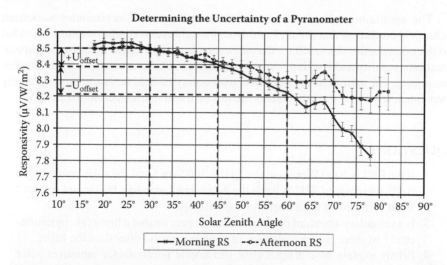

FIGURE 5.36 Responsivity plotted against zenith angle as in Figure 5.29. The uncertainty offsets U_+(offset) and U_-(offset) are shown on the plot.

uncertainty (u_C) of ±0.65%. The coverage factor for this data set is 1.96, which yields an expanded uncertainty at a 95% level of ±1.27%.

In the NREL BORCAL reports, the assigned responsivity is specified at 45° solar zenith angle. This responsivity is accompanied by + and − range of uncertainty at the 95% level of confidence for measurements made when the solar zenith angle is between 30° and 60°. This confidence level is determined in the following manner. First Type A and Type B uncertainties for the 45° responsivity are determined. The percent expanded uncertainty for the 45° responsivity is then calculated. Next, the minimum and maximum responsivities in the range from 30° to 60° are determined from the calibration measurements (see Figure 5.36). These plus and minus differences are obtained, and then the percent uncertainty in the responsivity is calculated from

$$\%U(differences)_\pm = (R_{max}/R_{min} - R_{45})_\pm / R_{45} \tag{5.20}$$

The expanded uncertainty is calculated by

$$U_{E\pm} = U_{B-45\pm} + U_{differences\pm} \tag{5.21}$$

In this example, the expanded uncertainty (U_E) for the responsivity at 45° is ±0.64%. The difference between the responsivity at 45° and the maximum and minimum responsivities in the 30° to 60° range provides a $+U_{differences}$ and a $-U_{differences}$. The differences shown in Figure 5.36 are used in Equation 5.21 to calculate the percent differences $+U_{differences} = 1.54\%$ and $-U_{differences} = -3.01\%$. Therefore, the expanded uncertainty for the plus range would be 2.18% and −3.64% for the negative range. Note that the offset uncertainties are at the 95% level and are the values from simple addition—not added in quadrature. Uncertainties for other zenith angle ranges can be calculated in a similar manner.

The uncertainties discussed so far are for calibrations. Uncertainties associated with field conditions that are different from the calibration conditions must be added to the uncertainties discussed for the calibration (e.g., data logger measurement specifications, stability and leveling of mounting, degree of solar access, electrical interference, etc.). The estimated measurement uncertainty of field measurements is largely dependent upon how the instruments are installed and maintained in the field.

QUESTIONS

1. If the DNI is 1000 Wm^{-2} and the DHI is 100 Wm^{-2} and the solar zenith angle is 45°, what is the GHI? What is the direct irradiance on a horizontal surface?
2. What is meant by a Lambertian response?
3. Is a secondary-standard pyranometer more precise that a first-class pyranometer? List three reasons one type of pyranometer is better than the other.
4. Briefly explain how a black-disk thermopile pyranometer measures solar energy.
5. Why do many first-class pyranometers have two quartz domes?
6. What is meant by a thermal offset in a pyranometer? What is one way to estimate the thermal offset of a pyranometer?
7. What is a ventilator, and why is it used?
8. If a pyranometer has a responsivity of 8.05 μV/Wm^{-2} at a solar intensity of 1000 Wm^{-2}, a responsivity of 8.00 μV/Wm^{-2} at 500 Wm^{-2}, and 7.95 μV/Wm^{-2} at 100 Wm^{-2}, what is the deviation from linearity?
9. Why should a pyranometer be calibrated on a yearly basis?
10. What is meant by advective heat loss?
11. Describe how a black-and-white pyranometer works.
12. In equilibrium, a black surface has an emissivity of 95%; what is its absorptivity?
13. Can a white surface be highly reflective in the visible portion of the spectrum and highly absorptive in the infrared portion of the spectrum?
14. Explain why black-and-white pyranometers have smaller thermal offsets than first-class pyranometers.
15. List a number of reasons photodiodes are better at measuring solar radiation than a photovoltaic module.
16. Will a silicon-based photodiode pyranometer respond to sunlight with a wavelength of 1400 nm? Explain why.
17. Why does the LI-COR LI-200 photodiode pyranometer exhibit its *cat ears* response?
18. If the ambient temperature is 25°C, what should be the change made to the Eppley Model 8-48 pyranometer's responsivity? If the temperature is 0°C, what should be the change made to the pyranometer's responsivity?
19. On a clear day, a photodiode pyranometer responds differently to DHI than to DNI. Explain why? During the middle of the day, is there a large effect on the GHI measurements?
20. Explain the shade–unshade method of calibration.

21. Explain the summation calibration technique.
22. Name several causes of Type A uncertainties.
23. Name several causes of Type B uncertainties.
24. If the Type A uncertainties are 3% and the Type B uncertainties are 4%, what is the combined uncertainty? What is the expanded uncertainty at a 95% level of confidence?

REFERENCES

Abbot, C. G. and L. B. Aldrich. 1916. The pyranometer—An instrument for measuring sky radiation. *Smithsonian Miscellaneous Collection* 66(7).

ASTM E913-97. 1997. *Standard method for calibration of reference pyranometers with axis vertical by the shading method.* West Conshohocken, PA: American Society for Testing and Materials.

ASTM Standard E1036-08. 2008. *Standard test methods for electrical performance of non-concentrator terrestrial photovoltaic modules and arrays using reference cells.* West Conshohocken, PA: American Society for Testing and Materials International.

ASTM Standard E948-09. 2009. *Standard test method for electrical performance of photovoltaic cells using reference cells under simulated sunlight.* West Conshohocken, PA: American Society for Testing and Materials International.

ASTM Standard E1362-10. 2010. *Standard test method for calibration of non-concentrator photovoltaic secondary reference cells.* West Conshohocken, PA: American Society for Testing and Materials International.

Augustyn, J., T. Geer, T. Stoffel, R. Kessler, E. Kern, R. Little, and F. Vignola. 2002. Improving the accuracy of low cost measurement of direct normal solar irradiance. In Conference Proceedings of the 2002 American Solar Energy Society, Solar 2002, Reno, NV: American Solar Energy Society, Boulder, CO.

Augustyn, J., T. Geer, T. Stoffel, R. Kessler, E. Kern, R. Little, F. Vignola, and B. Boyson. 2004. Update of algorithm to correct direct normal irradiance measurements made with a rotating shadow band pyranometer. In Conference Proceedings of the 2004 American Solar Energy Society, Solar 2004, Portland, OR: American Solar Energy Society, Boulder, CO.

Bunch, B. and A. Hellemans. 2004. *The history of science and technology: A browser's guide to the great discoveries, inventions, and the people who made them, from the dawn of time to today.* Boston: Houghton Mifflin.

Dutton, E. G., J. J. Michalsky, T. Stoffel, B. W. Forgan, J. Hickey, D. W. Nelson, T. L. Alberta, and I. Reda. 2001. Measurement of broadband diffuse solar irradiance using current commercial instrumentation with a correction for thermal offset errors. *Journal of Atmospheric Oceanic Technology* 18:297–314.

Gulbrandsen, A. 1978. On the use of pyranometers in the study of spectral solar radiation and atmospheric aerosols. *Journal of Applied Meteorology* 17:899–905.

International Electrotechnical Commission (IEC). 2007. *IEC 60904-2: Photovoltaic device—Part 2: requirements for reference solar cells.* International Electrotechnical Commission, 3, rue de Varembé, PO Box 131, CH 1211, Geneva 20, Switzerland.

Joint Committee for Guides in Metrology. (JCGM). 2008. *Evaluation of measurement data— Guide to the expression of uncertainty in measurement. GUM 1995 with minor revisions.* Bureau International des Poids et Mesures. Available at: http://www.bipm.org/en/publications/guides/gum.html

Kerr, J. P., G. W. Thurtell, and C. B. Tanner. 1967. An integrating pyranometer for climatological observer stations and mesoscale networks. *Journal of Applied Meteorology* 6:688–694.

King, D. L., W. E. Boyson, B. R. Hansen, and W. I. Bower. 1998. *Improved accuracy for low-cost solar irradiance sensors.* Paper presented at the 2nd world conference and exhibition on photovoltaic solar energy conversion, July 6–10, Vienna, Austria.

King, D. L., J. A. Kratochvil, and W. E. Boyson. 1997. *Measuring solar spectral and angle-of-incidence effects on photovoltaic modules and solar irradiance sensors.* Paper presented at the 26th IEEE photovoltaic specialists conference, September 29–October 3, Anaheim, CA.

King, D. L. and D. R. Myers. 1997. *Silicon-photodiode pyranometers: Operational characteristics, historical experiences, and new calibration procedures.* Paper presented at the 26th IEEE photovoltaic specialists conference, September 29–October 3, Anaheim, CA.

Long, C. N., K. Younkin, and D. M. Powell. 2001. Analysis of the Dutton et al. IR loss correction technique applied to ARM diffuse SW measurements. In: *Proceedings of the eleventh atmospheric radiation measurement (ARM) science team meeting,* ARM-CONF-2001. Washington, DC: United States Department of Energy.

McArthur, L. J. B. 2004. *Baseline Surface Radiation Network* (BSRN). *Operations Manual.* WMO/TD-No. 879, World Climate Research Programme/World Meteorological Organization.

Michalsky, J., E. G. Dutton, D. Nelson, J. Wendell, S. Wilcox, A. Andreas, P. Gotseff, D. Myers, I. Reda, T. Stoffel, K. Behrens, T. Carlund, W. Finsterle, and D. Halliwell. 2011. An extensive comparison of commercial pyrheliometers under a wide range of routine observing conditions. *Journal Atmospheric Oceanic Technology* 28:752–766.

Michalsky, J, J. Liljegren, and L. Harrison. 1995. A comparison of sun photometer derivations of total column water vapor and ozone to standard measures of same at the Southern Great Plains Atmospheric Radiation Measurement site. *Journal of Geophysical Research* 100(D12):25995–26003.

Myers, D. 2005. Solar radiation modeling and measurements for renewable energy applications: Data and model quality. *Energy* 30:1517–1531.

Myers, D. R., I. Reda, S. Wilcox, and A. Andreas. 2004. *Optical radiation measurements for photovoltaic applications: Instrument uncertainty and performance.* National Renewable Energy Laboratory (NREL) conference paper NREL/CP-560-36320.

Myers, D. R., T. L. Stoffel, I. Reda, S. M. Wilcox, and A. Andreas. 2002. Recent progress in reducing the uncertainty in and improving pyranometer calibrations. *Transactions of the American Society of Mechanical Engineers* 124:44–49.

Pulfrey, D. L. 1978. *Photovoltaic power generation.* New York: Van Nostrand Reinhold.

Reda, I., J. Hickey, C. Long, D. Myers, T. Stoffel, S. Wilcox, J. J. Michalsky, E. G. Dutton, and D. Nelson. 2005. Using a blackbody to calculate net longwave responsivity of shortwave solar pyranometers to correct for their thermal offset error during outdoor calibration using the component sum method. *Journal of Atmospheric and Oceanic Technology* 22(10):1531–1540.

Reda, I., D. Myers, and T. Stoffel. 2008. Uncertainty estimate for the outdoor calibration of solar pyranometers: A meteorologist perspective. *National Conference of Standards International (NCSLI) Journal of Measurement Science* 3(4):58–66.

Reda, I., T. Stoffel, and D. Myers. 2003. A method to calibrate a solar pyranometer from measuring reference diffuse irradiance. *Solar Energy* 74:103–112.

Riihimaki, L. and F. Vignola. 2008. Establishing a consistent calibration record for Eppley PSPs. Paper presented at the Proceedings of the 37th ASES Annual Conference, San Diego, CA.

Stoffel T. L. and D. R. Myers. *Accuracy of silicon versus thermopile radiometers for daily and monthly integrated total hemispheric solar radiation, BSRN instrument book.* Available at: http://www.bsrn.awi.de/fileadmin/user_upload/Home/Publications/McArthur.pdf

Stoffel, T., D. Renné, D. Myers, S. Wilcox, M. Sengupta, R. George, and C. Turchi. 2010. *Concentrating solar power: Best practices handbook for the collection and use of solar resource data.* Technical Report NREL/TP-550-47465, National Renewable Energy Laboratory, Golden, CO.

Velmre, E. 2007. Thomas Johann Seebeck (1770–1831). *Proceedings of the Estonian Academy of Sciences, Engineering.* 13(4):276–282.

Vignola, F. 1999. Solar cell based pyranometers: Evaluation of the diffuse response. *Proceedings of the 1999 Annual Conference American Solar Energy Society,* 260.

Vignola, F., C. Long, and I. Reda. 2007. *Evaluation of methods to correct for IR loss in Eppley PSP diffuse measurements.* Paper presented at the SPIE conference, San Diego, CA.

Vignola, F., C. Long, and I. Reda. 2008. *Modeling IR radiative loss from Eppley PSP pyranometers.* Paper presented at the SPIE conference, San Diego, CA.

Vignola, F., C. Long, and I. Reda. 2009. Testing a model of IR radiative losses. Paper presented at the SPIE conference, San Diego, CA.

Walker, J. H., R. D. Saunders, J. K. Jackson, and D. A. McSparron. 1987. *NBS measurement services: Spectral irradiance calibrations.* Washington, DC: U.S. Department of Commerce, National Bureau of Standards.

Wilcox, S., A. Andreas, I. Reda, and D. Myers. 2002. *Improved methods for broadband outdoor radiometer calibration (BORCAL).* Paper presented at the 12th Atmospheric Radiation Measurement (ARM) science team meeting proceedings, St. Petersburg, FL.

Wilcox, S., D. Myers, N. Al-Abbadi, and M. Bin Mahfoodh. 2001. Using irradiance and temperature to determine the need for radiometer calibrations. Paper presented at the forum on solar energy, the power to choose, Washington, DC.

Wilcox, S. and T. Stoffel. 2009. *Solar Resource and Meteorological Assessment Project (SOLRMAP) solar and meteorological station options: Configurations and specifications.* National Renewable Energy Laboratory, Golden, CO.

WMO. 2008. *WMO guide to meteorological instruments and methods of observation,* WMO-No. 8 (7th ed). World Meteorological Organization, Geneva, Switzerland.

Younkin, K. and C. N. Long. 2003. *Improved correction of IR loss in diffuse shortwave measurement: AN ARM values-added product.* ARM TR-009. Pacific Northwest National Laboratory, 47pp.

USEFUL LINKS

WMO guide to meteorological instruments and methods of observation, http://www.wmo.int/pages/prog/www/IMOP/publications/CIMO-Guide/CIMO_Guide-7th_Edition-2008.html.

BSRN operator's manual, http://www.bsrn.awi.de/fileadmin/user_upload/Home/Publications/WCRP21_TD1274_BSRN.pdf.

Seebeck values, http://www.omega.com/temperature/z/pdf/z021-032.pdf.

Standards, http://www.ictinternational.com.au/brochures/pyranometer-classification.pdf.

Standards, http://www.middletonsolar.com/documents/PyrClass.pdf.

Solar cell temperature dependence, http://pvcdrom.pveducation.org/CELLOPER/TEMP.HTM.

Transmission of wg295 glass, http://www.optical-filters.com/wg295.html.

6 Diffuse Irradiance

A little later in the morning, if one hides the solar disk by holding the tip of a single finger at arm's length before his eyes, the deep blue sky color holds right up to the edge of the sun itself.

Charles Greeley Abbot
(1872–1973)

6.1 INTRODUCTION

Diffuse horizontal irradiance (DHI) is defined as the solar irradiance that has been scattered by molecules, aerosols, and clouds in the atmosphere and received on a horizontal surface (refer to Figure 6.1). Another perspective on diffuse irradiance is that DHI is the *skylight* portion of the global horizontal irradiance (GHI) remaining after removing the direct normal irradiance (DNI). DHI is measured by a properly shaded pyranometer, or it can be computed from measurements of DNI and GHI using Equation 4.1. Light reflected by objects above the surface contributes to the DHI, but light reflected from the horizontal surface is not included. However, the radiation reflected from the surface that is subsequently reflected back to the surface by the atmosphere is included; that is, multiple scattering components are included in DHI.

A very clear atmosphere gives the sky a deep blue color because of Rayleigh scattering of solar radiation by air molecules. (Scattering and absorption of solar radiation will be discussed in more detail in Chapter 12.) This blue color is produced because air molecules scatter radiation nearly inversely with the fourth power of the wavelength, that is, *scattering* $\propto \lambda^{-4}$. The sun appears yellow or red near the horizon since Rayleigh scattering preferentially removes the bluest light from the transmitted DNI.

Aerosol-scattered light depends on the aerosol particle size. For "typical" continental aerosols, the scattering is approximately $\propto \lambda^{-1.3}$. Heavily polluted skies typically have an even larger negative exponent, and they scatter blue light preferentially, but not as strongly, as in Rayleigh scattering. This wavelength dependence for aerosols is most noticeable in extremely polluted conditions where the solar disk appears red even with the sun high in the sky, when the amount of Rayleigh scattering would ordinarily be insufficient to cause noticeable reddening. Although the sun's disk appears red for these hazy conditions, skylight appears not blue but milky white, especially near the horizon because all photons are scattered multiple times during these heavy pollution episodes. The range of sky color from dark blue to milky blue to milky white is a rough gauge of the amount of aerosol in the atmosphere.

Cloud particles have large sizes relative to visible wavelengths, and their scattering wavelength dependence is nearly neutral; therefore, they appear to have the white color of the sun. The spatial distribution of skylight is complex and unpredictable for partly cloudy conditions but can be described mathematically for some conditions. Moon and Spencer (1942) developed a simple and useful approximation of the spatial distribution

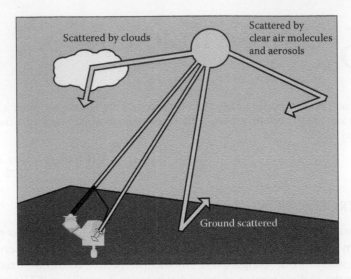

FIGURE 6.1 This figure demonstrates the source of diffuse solar radiation that falls on a horizontal surface. The black ball on the tracker blocks direct beam radiation from the sun but allows sunlight scatted by air molecules, aerosols, and clouds to reach the pyranometer. Ground-scattered solar radiation, beneath the pyranometer horizon, does not reach the pyranometer detector unless it is further scattered by the atmosphere. (Graphics by Bobby Hart.)

of skylight for thick uniform clouds. This sky is, basically, three times brighter in the zenith than at the horizon and varies as the cosine of the zenith angle (the angle from the zenith position, not the solar zenith angle). For very clear skies, where Rayleigh scattering dominates, the skylight is roughly half as bright 90° from the sun as it is near the sun. This is not quite correct because of multiple scattering (photons scattered more than once) that occurs in the real atmosphere. Cloudless but aerosol-laden skies are an intermediate case with respect to the complexity of mathematically describing the spatial distribution of skylight. Figure 6.2 is a low-resolution image of the skylight distribution for clear conditions with moderate aerosol. A disk normally blocks the area of the sky surrounding the sun, but in spite of this there is an enhanced bright aureole around the sun, called circumsolar radiation. A bright horizon caused by multiple scattering skirts the image. As for Rayleigh skies, there is minimum sky brightness roughly 90° from the sun. Forward-scattered radiation (from the sun direction to 90°) is higher than backscattered radiation (90–180° from the sun) for cloudless, aerosol-laden skies. This enhancement in the forward direction depends mainly on the size of the aerosols with larger aerosols producing more scattering in the forward hemisphere.

6.2 MEASUREMENT OF DIFFUSE IRRADIANCE

Diffuse irradiance measurements have become more sophisticated with the development of automated tracking devices that dependably keep the direct solar radiation from reaching the detector of the pyranometer. The first part of this section describes the use of the fixed shadowband, which is still in use, and the attempts to correct the nagging problem of excessive skylight blockage by the band. Most of the correction

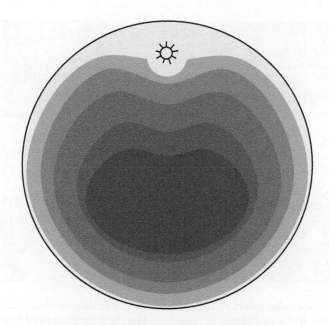

FIGURE 6.2 An illustration of an all-sky brightness distribution on a clear but hazy day. This demonstrates the bright aureole about the sun. There is a minimum brightness near an angle of 90° from the sun. A bright horizon is produced from light scattering multiple times. (Graphics by Bobby Hart.)

schemes for this skylight blockage use as their "best estimate" the diffuse calculated by taking the difference between the total horizontal irradiance and the horizontal beam irradiance. Diffuse values obtained using this subtraction method are prone to large systematic errors that make comparisons between different sites unreliable and create models that contain systematic errors in the calculated diffuse values. The tracking disk is the preferred method for measuring diffuse irradiance, and this method will be discussed in Section 6.2.2. A discussion of the rotating shadowband radiometer measurements of diffuse horizontal irradiance completes this section.

6.2.1 Fixed Shadowband Measurements of Diffuse Irradiance

Before the development of automatic solar trackers capable of dependably following the sun in two dimensions (solar elevation and azimuth or hour angle and declination), most diffuse measurements were performed with an adjustable fixed shadow band or shade ring that blocked the sun from sunrise to sunset. This band blocked a large part of the sky and had to be adjusted as the sun's declination changed: more often when the sun is near the equinoxes when the solar declination changes rapidly (~ 0.4° per day) and less often when the sun is near the solstices when the solar declination changes slowly (< 0.1° per day). The shadowband should be aligned to keep the pyranometer detector within 1.5° of the center of the shadow cast by direct sunlight.

To align the fixed shade, the plane of the arc formed by the shadowband or the shade ring should be perpendicular to the polar axis. The plane of the irradiated surface of

FIGURE 6.3   (See color insert.) The Eppley model SBS shadowband blocks direct sunlight on a clear summer day in Eugene, Oregon. Note that the detector surface is well centered beneath the shadowband. The shadowband blocks a significant part of the brightest region of the clear-day skylight.

the pyranometer is adjusted to intersect the polar axis. This latter adjustment allows the shade to block the pyranometer surface throughout the day once the band is adjusted on the polar axis according to the declination of the sun. Adjustment details are explained in the installation instructions from the manufacturers. Kipp & Zonen B.V. and Eppley Laboratory, Inc. continue to offer fixed shading devices. Figure 6.3 illustrates the alignment of the Eppley fixed shadowband for a clear summer day in Eugene, Oregon.

Because the fixed shadowband or shade ring can block over 20% of the diffuse irradiance, a correction is needed for the excluded diffuse irradiance. This correction factor for the fixed shadowband or shade ring is a frequent topic in the solar radiation literature. Drummond (1956) developed a geometrical correction for the blocked skylight, assuming an isotropic (uniform) distribution of skylight. This method is applicable for any location on Earth. However, as previously described, diffuse radiation is not uniformly distributed. Papers since Drummond have developed corrections or evaluated correction schemes for the true distribution of skylight for different conditions from overcast to clear skies. Corrections derived for one climate are often not applicable to another because different sky conditions prevail at dissimilar locations.

Kudish and Evseev (2008) recently evaluated four of these correction methods: Drummond (1956), LeBaron, Michalsky, and Perez (1990), Battles, Alados-Arbodelas, and Olmo (1995), and Muneer and Zhang (2002). Gueymard and Myers (2009) pointed out, however, that this evaluation is invalid because the data used for establishing the "true diffuse values" employed the subtraction method to calculate diffuse values from GHI and direct horizontal values. Diffuse values obtained by the subtraction method contain significant systematic errors due in large part to the angular response (or cosine) errors of the unshaded pyranometer and do not have the

FIGURE 6.4 (See color insert.) An example of a shaded-disk measurement of DHI. The arm holds disks that shade each of three black-and-white pyranometers.

accuracy to establish a reliable diffuse value. Most—maybe all—of the correction methods that have been developed used the subtraction method for diffuse values as the standard for comparison. It is clear that more research is needed to establish a more reliable shadowband adjustment factor; however, fixed shadowband DHI values will always have high uncertainties associated with mathematical shadowband corrections. Our recommendation, as will become clear, is to measure diffuse with a tracking disk such as the one shown in Figure 6.4.

6.2.2 CALCULATED DIFFUSE IRRADIANCE VERSUS SHADE DISK DIFFUSE

DHI can be calculated from GHI and DNI data. The difficulty with this approach stems from using the difference of the global horizontal and direct beam as the best estimate of diffuse horizontal irradiance, that is,

$$diffuse = global - direct \cdot \cos(sza) \tag{6.1}$$

where sza is the solar zenith angle. A very good measurement of the DNI has a 95% uncertainty of around +/–2%, and a very good global measurement has a 95% uncertainty of +/–3% or higher because of additional uncertainties associated with thermal offsets and imperfect cosine responses (Stoffel et al., 2010). The uncertainty of calculating diffuse irradiance from GHI and DNI is significant. To illustrate the uncertainty in calculating diffuse in this way, assume a value for DNI and GHI of 1000 Wm^{-2} each at a solar zenith angle of 30°. The calculated diffuse would be

$$diffuse = 1000 - 1000 \times \cos(30°) = 134[Wm^{-2}] \tag{6.2}$$

with an *expanded* uncertainty of

$$U_{95} = \sqrt{(30)^2 + (20 \times 0.866)^2} = 35[Wm^{-2}] \tag{6.3}$$

or a 26% uncertainty in the calculated diffuse using two state-of-the-art measurements.

A vast improvement in uncertainty is possible if the diffuse is, instead, measured with a tracking disk, or ball, shading a pyranometer. The 3% uncertainty of the pyranometer measurement, using this example, would amount to only 4 Wm^{-2}, an improvement of a nearly a factor of 10 in uncertainty. The preferred method to evaluate fixed shadowband corrections, discussed in Section 6.2.1, is to use tracking disk diffuse measurements for the reference irradiance. However, if diffuse measurements are important, efforts should be made to purchase an automatic tracker because diffuse values obtained from fixed shadow bands or by the subtraction method have high uncertainties.

First-class pyranometers often have thermal offsets that are large compared with diffuse irradiance levels. These thermal offsets can be 20% or more of the DHI value. In fact, pyranometers that are uncorrected for thermal offsets have reported diffuse irradiances lower than pure Rayleigh scattering atmospheres (Cess, Qian, and Sun, 2000). Therefore, it is actually preferable to use second-class black-and-white pyranometers with very small thermal offsets, as discussed in Chapter 5, for DHI measurements. High-cost, secondary-standard pyranometers also have minimal thermal offsets and better cosine responses, so the best DHI measurements may be made using secondary-standard pyranometers. A comparison of DHI measurements made using most of the commercially available pyranometers can be found in Michalsky et al. (2003).

6.2.3 ROTATING SHADOWBAND DIFFUSE MEASUREMENTS

The rotating shadowband pyranometer offers a third option for the measurement of diffuse irradiance with at least four suppliers of these types of instruments (see Appendix F). Irradiance, Inc. (http://www.irradiance.com) offers the rotating shadowband radiometer (RSR), which uses a rapidly moving band to alternately expose and shade the diffuser of a silicon photodiode pyranometer with 10 μsec time response. Hundreds of measurements are made during the band sweep with many near, but not quite shading, the diffuser. The measurements made in the solar aureole, the bright disk surrounding the sun, are used to derive an approximation for the excess skylight that is blocked by the band during the shaded measurement. This is akin to a real-time fixed shadowband correction, which was discussed earlier, but uses actual measurements for the correction. The most significant issue with the RSR is that the silicon photodiode sensor has a nonuniform spectral response that significantly underestimates clear-sky diffuse if the instrument is calibrated for global or direct irradiance (see Section 5.4.2). Software to correct this shortcoming is included with the instrument. The effectiveness of this or other correction schemes is under investigation.

Solar Millennium AG (http://www.solarmillennium.de/) offers a very similar device to the RSR just discussed. It offers to acquire and correct the data from the site of the installation and deliver a finished data product to the user, as does Irradiance, Inc.

Yankee Environmental Systems (YES), Inc. (http://www.yesinc.com) makes a seven-channel multifilter rotating shadowband radiometer (MFR-7) and a single detector radiometer (SDR) that makes a spectral and broadband measurement of diffuse irradiance. Until recently, they, too, used a silicon photodiode detector and had to deal with most of the same measurement issues previously discussed for the RSR. YES now offers a single-channel thermopile shadowband radiometer (TSR), and the MFR-7 uses a fast-response thermopile in place of the unfiltered silicon cell (with six additional channels of narrowband wavelength measurements). The performance of this thermopile device is still under study since this thermopile-based radiometer is a relatively new development.

Prede, Inc. (http://www.prede.com) makes the rotating shadow blade (RSB-100), which works with commercial pyranometers (thermopile or silicon cell versions) to make diffuse and global irradiance measurements. There is little information on the design and measurement performance of this new device.

6.3 CALIBRATION OF DIFFUSE PYRANOMETERS

Michalsky et al. (2007) provide a thorough discussion of the calibration and characterization of pyranometers with the goal of defining a standard for DHI measurements. The preferred, but labor-intensive, method for calibrating any pyranometer is to alternately shade and unshade the pyranometer while measuring the reference DNI with an absolute cavity radiometer. The pyranometer response, which is the inverse of the calibration factor (a multiplier), is then calculated using

$$R = \frac{(V_{unshaded} - V_{shaded})}{DNI \cdot \cos(sza)} \tag{6.4}$$

where V_x are the voltage readings of the pyranometer unshaded and shaded, and sza is the solar zenith angle at the time of measurement. These calibration measurements should be made when the sun is at a solar zenith angle of $45° \pm 3°$. The details of performing this calibration procedure are in Michalsky et al. (2007) and are discussed in detail in Chapter 5.

Why calibrate at 45°? If skylight were isotropic (i.e., uniform in all directions), then the diffuse irradiance falling on a horizontal surface could be calculated as

$$DHI = R_{sky} \int_0^{\frac{\pi}{2}} \int_0^{2\pi} \cos(\theta)\sin(\theta)\,d\psi\,d\theta \tag{6.5}$$

where R_{sky} is the constant radiance of the isotropic sky. The term $\sin(\theta)d\theta d\phi$ comes from spherical geometry (see Figure 6.5), and the $\cos(\theta)$ term comes from the Lambert cosine law, which states that the irradiance flux on a surface is proportional to the cosine of the angle of incidence. Since $\sin(\theta)\cos(\theta) = \sin(2\theta)/2$, it

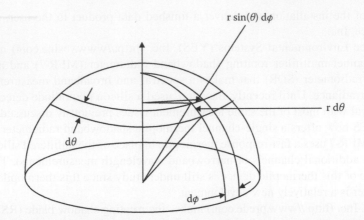

FIGURE 6.5 The geometry used to integrate the radiance distribution of skylight to calculate diffuse horizontal irradiance.

is clear that the largest contribution to the diffuse irradiance is from the annulus of skylight near 45° and falls monotonically and symmetrically to zero at 0° and 90°. This means that the effective angle of incidence for diffuse radiation is 45°, which does not change throughout the day for isotropic skies. Skylight is not isotropic, but contributions from all parts of the sky are such that the effective incident angle for diffuse is never far from the nominal 45° and changes little throughout the day (Michalsky et al., 2007).

Calibrating pyranometers using the shade–unshade technique eliminates problems with thermal offsets since offsets for the global and diffuse measurements are approximately equal and are eliminated by subtraction of the two components, as can be deduced from Equation 6.4. However, when making diffuse measurements with a single black detector thermopile pyranometer, even if calibrated with the shade–unshade method, a correction must be made to subtract the thermal offset.

Having calibrated a reference pyranometer for diffuse irradiance measurements through the rather onerous procedure previously described, it is possible to calibrate similar pyranometers for diffuse irradiance by simply comparing to this standard under totally overcast, low-cloud conditions. While not strictly isotropic, the sky is uniform enough to yield a reliable calibration. Further, the offset for these cloud-cover conditions will be near zero since the effective sky temperature will be close to the instrument temperature resulting in low values of net infrared (IR) and, consequently, low thermal offsets.

6.4 VALUE OF ACCURATE DIFFUSE MEASUREMENTS

Because there are systematic errors with all pyranometers currently available, the most accurate GHI values can best be obtained from DNI and DHI measurements. As with the example for calculating diffuse from GHI and DNI, the uncertainty of GHI can also be estimated when it comes from DNI and DHI. Under clear-sky conditions with a properly shaded pyranometer used for the diffuse measurement with a +/–3% uncertainty, the calculated global would be

$$GHI = 134 + 1000 \times \cos(30°) = 1000[Wm^{-2}] \tag{6.6}$$

with an uncertainty of

$$U_{95} = \sqrt{(4)^2 + (20 \times 0.866)^2} = 18[Wm^{-2}] \tag{6.7}$$

The uncertainty in the GHI is then around +/–2%, which is better than the +/–3% obtainable from most well-maintained GHI measurements. The reason for the improved uncertainty is that most of the irradiance comes from the DNI contribution. Under cloudy conditions, the DNI is zero, and the uncertainty in the GHI is +/–3%, the same as the uncertainty in the DHI.

Because of the relationship between DHI, DNI, and GHI (Equation 4.1), a three-parameter comparison is the best way to validate the accuracy of solar irradiance measurements. If rain or soiling affects the measurements, then the three-component check can check the reliability of the measurements.

Accurate DHI measurements are also essential for the calibration of pyranometers using the summation technique. This method allows the simultaneous calibrations of multiple pyranometers as described in Chapter 5. Diffuse irradiance contributes to all flat-plate solar collectors and is used in all performance calculations. Therefore, any accurate performance analysis requires good DHI measurements. Because DHI is much smaller (typically less than 20%) than the DNI contribution on sunny days, the uncertainty in the DHI contribution is often neglected. However, as the importance of precision measurements increases for analysis of solar system performance, accurate DHI data are becoming indispensable.

QUESTIONS

1. On a clear day, what are two of the several indirect sources (the sun is not a correct answer) of diffuse light on a horizontal surface?
2. What is the most accurate way to measure DHI?
3. Why is it possible to measure smaller DHI values with a shaded pyranometer based on a single-black detector than a theoretical pure Rayleigh sky (no clouds or aerosols) would produce?
4. What are the two sources of measurement error when using a fixed band to measure DHI?
5. What is the major advantage of the black-and-white pyranometer in measurements of DHI?
6. On a clear day with some aerosols present, which is likely to be brighter. (1) a point in the sky 90° from the sun in the same vertical plane as the sun and the zenith, or (2) a point near the horizon?
7. What is the most significant problem with using a photodiode pyranometer for DHI measurements?
8. Under what circumstances is skylight isotropically distributed during the daylight hours?

REFERENCES

Battles, F. J., L. Alados-Arbodelas, and F. J. Olmo. 1995. On shadow band correction methods for diffuse irradiance measurements. *Solar Energy* 54:105–114.

Cess, R. D., T. Qian, and M. Sun. 2000. Consistency tests applied to the measurement of total, direct, and diffuse shortwave radiation at the surface. *Journal of Geophysical Research*, 105:24,881–24,887. doi:10.1029/2000JD900402

Drummond, A. J. 1956. On the measurement of sky radiation. *Archives for Meteorology, Geophysics, and Bioclimatology–Series B: Climatology, Environmental Meteorology, Radiation Research.* 7:413–436.

Gueymard, C. A. and D. R. Myers. 2009. Evaluation of conventional and high-performance routine solar radiation measurements for improved solar resource, climatological trends, and radiative forcing. *Solar Energy* 83: 171–185.

Kudish, A. I. and E. G. Evseev. 2008. The assessment of four different correction models applied to the diffuse radiation measured with a shadow ring using global and normal beam radiation measurements for Beer Sheva, Israel. *Solar Energy* 82: 144–156.

LeBaron, B. A., J. J. Michalsky, and R. Perez. 1990. A new simplified procedure for correcting shadow band data for all sky conditions. *Solar Energy* 44:249–256.

Michalsky, J. J., R. Dolce, E. G. Dutton, M. Haeffelin, G. Major, J. A. Schlemmer, D. W. Slater, J. R. Hickey, W. Q. Jeffries, A. Los, D. Mathias, L. J. B. McArthur, R. Philipona, I. Reda, and T. Stoffel. 2003. Results from the first ARM diffuse horizontal short-wave irradiance comparison. *Journal of Geophysical Research*, 112:D16111, 10 pp. doi:10.1029/2007JD008651

Michalsky, J. J., C. Gueymard, P. Kiedron, L. J. B. McArthur, R. Philipona, and T. Stoffel. 2007. A proposed working standard for the measurement of diffuse horizontal short-wave irradiance. *J. Geophys. Res.* 112:D16112. doi:10.1029/2007JD008651

Moon, P. and D. Spencer. 1942. Illumination from a non-uniform sky. *Illuminating Engineering* 37:707–726.

Muneer, T. and X. Zhang. 2002. A new method for correcting shadow band diffuse irradiance data. *Journal of Solar Energy Engineering* 124:34–43.

Myers, D. R., K. A. Emery, and T. L. Stoffel. 1989. Uncertainty estimates for global solar irradiance measurements used to evaluate PV device performance. *Solar Cells* 27:455–464.

Stoffel, T., D. Renné, D. Myers, S. Wilcox, M. Sengupta, R. George, and C. Turchi. 2010. *Concentrating solar power: Best practices handbook for the collection and use of solar resource data.* NREL/TP-550-47465. Golden, CO: National Renewable Energy Laboratory. http://www.nrel.gov/docs/fy10osti/47465.pdf

7 Rotating Shadowband Radiometers

The total assembly is called the "dial" radiometer both because the rotating shade resembles the face of a dial with a revolving arm and because the device is used to measure "D-and-I-AL" quantities (Diffuse, Incident direct-beam, and globAL irradiance).

Marvin L. Wesely
(1944–2003)

7.1 INTRODUCTION

A simple rotating shadowband radiometer was introduced by Wesely (1982), as discussed in Chapter 3; it used a LI-COR LI-200 photodiode pyranometer because a rapid response time (10 μsec for the LI-200) was needed as the shading band passed over the detector to block direct sunlight. The band rotated at constant speed and completely shaded the diffuser every 4 or 5 minutes depending on the motor selected. The output from the pyranometer was a current; therefore, a low-temperature coefficient resistor was used in the circuit to produce a voltage that was output to a chart recorder. An upper envelope of the global horizontal irradiance (GHI) bound the output on a clear day, and a lower envelope represented the diffuse horizontal irradiance (DHI) (Figure. 7.1). Direct normal irradiance (DNI) could be calculated from the difference in the global and diffuse horizontal irradiance after division by the cosine of the solar zenith angle. The labor-intensive process of analyzing the chart manually limited the reduction of these paper charts to useful data.

Michalsky, Berndt, and Schuster (1986) built a microprocessor-controlled version of the rotating shadowband radiometer that recorded four measurements from the photodiode pyranometer for each rotation of the stepper motor-driven band. The measurements are made with the band in the nadir position to measure unobstructed global horizontal irradiance, blocking the sun to measure the diffuse horizontal irradiance, and with the band offset symmetrically slightly before and slightly after the sun-blocking position to estimate the excess diffuse sky radiation blocked by the band during the blocked-sun measurement. The blocked-sun position of the band prevents the direct sun from reaching the receiver but also blocks circumsolar diffuse skylight. An estimate of this correction is calculated using the two side-band measurements.

Yankee Environmental Systems, Inc. (http://www.yesinc.com) manufactures a seven-channel rotating shadowband radiometer (MFR-7) (Figure 7.2). Six of the channels are 10 nm–wide spectral channels in the visible and near infrared (NIR); the seventh channel is a broadband solar sensor. It uses the same band rotation and sampling algorithm as the Michalsky et al. (1986) design.

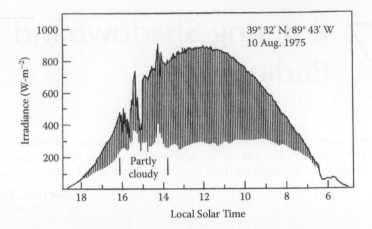

FIGURE 7.1 Chart recorder output from a Wesely-type constant-speed rotating shadow-band radiometer. The bottom of the envelope represents the diffuse horizontal irradiance, and the upper envelope the global horizontal irradiance. The direct can be calculated from these two measurements and is proportional to the length of the vertical lines. (Courtesy of the American Meteorological Society.)

Edward Kern developed a version of the rotating shadowband radiometer that is controlled by a commercial data logger (Figure 7.3). This instrument rotates a band at high speed with simultaneous and rapid sampling of the photodiode pyranometer. The minimum signal is used as the first-order diffuse signal, and measurements just outside this minimum value are used to estimate a shadowband correction for excess circumsolar diffuse sunlight blockage. Its operation is described in the *RSR2 Installation and Operation Manual* (available at http://www.irradiance.com/rsr.html).

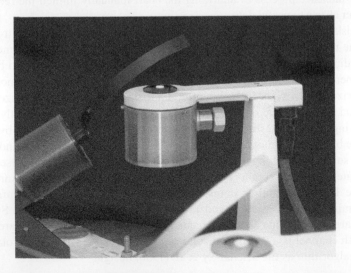

FIGURE 7.2 **(See color insert.)** The Yankee Environmental Systems, Inc. seven-channel multifilter rotating shadowband radiometer model MFR-7.

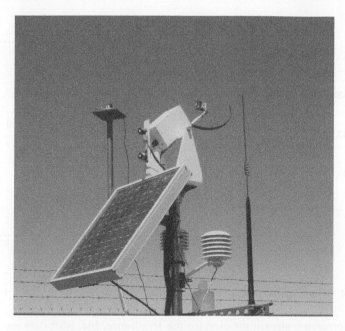

FIGURE 7.3 The Irradiance, Inc. rotating shadowband radiometer version 2 (RSR2). (Courtesy of Ed Kern.)

Recently, Solar Millennium AG has developed a version of the rotating shadowband radiometer (http://www.solarmillennium.de/technology/meteostations/mode-of-operation/index.html) that appears very similar to Irradiance, Inc.'s instrument. Both companies offer complete installation and data reduction services.

Prede Co. Ltd. of Japan (http://www.prede.com) makes the rotating shadow blade (RSB-100), which is advertised to work with either thermopile or photodiode sensors of any manufacturer. The authors have no experience with this device.

Rotating shadowband radiometers are popular because they provide all three solar components for a much lower cost than do traditional thermopile radiometers that require an expensive tracker to perform the direct normal and diffuse horizontal measurements. To date, their major drawback is that they require several corrections that limit their accuracy compared with the accuracy of thermopile radiometers. When a few percent accuracy matter, as in multimillion-dollar power plants, then a thermopile measurement system is the preferred choice. The next two sections describe the two fundamentally different implementations of the rotating shadowband radiometer.

7.2 ROTATING SHADOWBAND RADIOMETER

This section describes instruments similar to that developed by Kern that uses a commercial photodiode pyranometer, a commercial data-logger system, and a band driven to shade the pyranometer on command. The hardware for the Solar Millennium instrument appears very similar, but nothing is known regarding the corrections applied to it since the correction algorithms are proprietary; however,

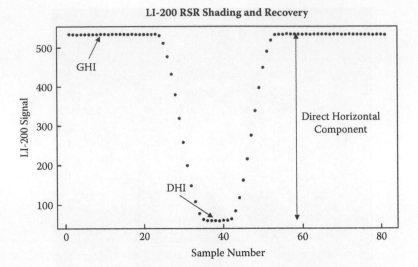

FIGURE 7.4   The response of the LI-COR, Inc. LI-200 to shading by the RSR2 band.

the basic correction algorithms used for the Irradiance, Inc. RSR data are available in Augustyn et al. (2004).

The need to automatically measure the direct and diffuse irradiance inexpensively led to the development of the RSR, which was originally called the rotating shadowband pyranometer (RSP). As the shadowband arm sweeps past the pyranometer, the data logger goes into a fast data-gathering mode and measures the minimum irradiance during the sweep of the arm (Figure 7.4). This minimum irradiance is the initial estimate of diffuse irradiance. Measurements just before and after shading are used to estimate the diffuse irradiance blocked by the band during the measurements when the solar disk is totally eclipsed. This provides a positive correction that is added to the minimum irradiance for a corrected diffuse horizontal irradiance (DHI) estimate. The average reading taken when the band is far from shading the diffuser is the global irradiance. The difference between the global horizontal irradiance (GHI) and DHI is the direct horizontal irradiance. The direct normal irradiance (DNI) is calculated by dividing the direct horizontal irradiance by the cosine of the incident angle, which is the solar zenith angle for a horizontal pyranometer. (Note that the standard abbreviation for diffuse horizontal irradiance, DHI, should not be confused with the *direct* horizontal irradiance.)

Direct normal irradiance (DNI) is calculated using

$$DNI = direct\_horizontal\_irradiance/\cos(sza) \tag{7.1}$$

where *sza* is the solar zenith angle.

In principle, the installation and operation of the RSR is simple and straightforward. The RSR can run unattended in the field for long periods, and the operation of the instrument does not require a complex and, hence, expensive tracker to point the instrument at the sun. The instrument is easy to align and is easily integrated into an automated monitoring station (see Figure 7.3).

Although only one pyranometer is required to obtain the GHI, DHI, and DNI measurements, an auxiliary pyranometer is an option for automated data quality control, and some models integrate an unshaded pyranometer into the design. The National Renewable Energy Laboratory (NREL) has developed a quality control check for an RSR with a primary pyranometer for the GHI, DHI, and DNI measurements and a secondary pyranometer for GHI measurements. If the performance of either pyranometer degrades because of soiling, precipitation, or other problems, then the output ratio between the two pyranometers changes. The auxiliary pyranometer can also be tilted at other angles such as by the station latitude. The tilted pyranometer tends to be less affected by dirt, moisture, frost, or snow, especially if it is tilted at 90°. However, the quality control check becomes less dependable because ground-reflected light has a significant wavelength dependence, as does the photodiode-based pyranometer used in the Irradiance, Inc. RSR. For tilted pyranometers, quality control checks are not reliable for hours during the months when the sun rises and sets behind the tilted pyranometer.

In practice, the accuracy of the RSR is only as good as the pyranometer used, and the uncertainties in the measured DHI and computed DNI are greater than the uncertainties associated with the GHI measurements. There are large statistical errors associated with infrequent sampling and large systematic errors associated with using a photodiode pyranometer, especially for the DHI measurements. To record the DHI when the band sweeps in front of the pyranometer, the pyranometer needs a fast response time. This limits the choice of pyranometers. Thermopile pyranometers are generally better instruments for broadband shortwave measurement because they have a uniform response over most solar wavelengths; however, their response is too slow for shadowband operation. Consequently, photodiode pyranometers are used in most RSRs.

A random error is introduced by the lower sampling rates of GHI and DHI using the RSR compared with frequent sampling possible with thermopile instruments. Variable cloudiness will increase the scatter of the RSR values for DNI and DHI when compared with pyrheliometers and pyranometers that sample continuously over the measurement period. Newer models of the Irradiance, Inc. RSR take several readings during a minute and initiate more frequent band rotation and sampling during periods of rapidly changing GHI, whereas older models took a single reading, typically every minute. This scatter in the measured data is random, and when averaged over longer periods the two measurements approach the same mean value. Figure 7.5 clearly indicates the reduction in scatter when 1-minute data are compared with 15-minute data for the same data set.

A systematic error in some of the original RSRs was the calculation of the cosine of the solar zenith angle. Because of limited memory in the data acquisition and control system, an approximate formula for the solar position was used. This led to uncertainties in the cosine of the solar zenith angle, and these errors are imbedded in the calculated values of DNI (Vignola, 2006).

Systematic errors in the DHI occur because the shadowband obscures more than just the sun when it passes in front of the pyranometer. Samples near the shading minimum are used to correct for the excess sky blockage by the shadowband (Figure 7.4).

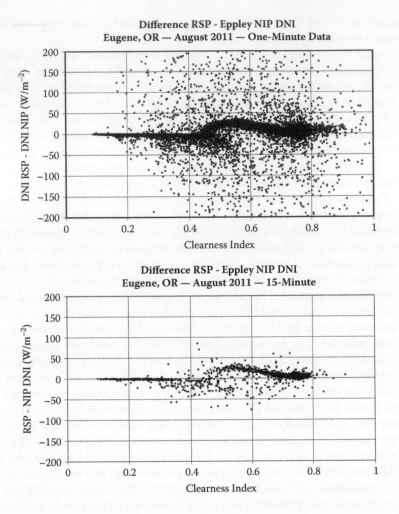

FIGURE 7.5 While there is much scatter caused by the less frequent sampling of the photodiode pyranometer, the mean of the photodiode DNI measurements approaches the thermopile results; note the reduction in scatter with time averaging. Clearly, there are further improvements possible for clearness indices around 0.55.

The photodiode pyranometer has a spectral dependence, and when it is used to measure DHI the responsivity determined for GHI measurements produces DHI values that match the GHI or are lower by as much as 30% depending on whether the sky is completely cloudy or completely clear. Said another way, the DHI responsivity of the photodiode pyranometer is almost 30% less for a clear blue sky than for an overcast sky. The systematic reduction of the DHI for a clear blue sky results in an overestimate of the DNI by about 5% during clear episodes. The deviation from true cosine response and the temperature dependence of the photodiode pyranometer also cause systematic errors. King, Boyson, Hansen, and Bower (1998) characterized many of the systematic errors of the LI-COR Model LI-200 pyranometer that is used in the Irradiance, Inc. RSR instruments. Typical deviations from true cosine response were shown in Figure 5.19. In the RSR, the angular response and temperature corrections to the LI-200 measurements of GHI are based on King et al.'s analysis of these systematic errors. When plotted against time of day, the LI-200's responsivity takes on a shape similar to a cat's front profile with peaks occurring near a solar zenith angle of 81° (Augustyn et al., 2002). This *cat ears* effect (Figure 5.25) is similar for all LI-COR pyranometers and is related to the shape of the diffuser and artificial horizon on the pyranometer. This diffuser produces a good cosine response over most of the range of solar zenith angles but has difficulties producing a precise cosine response at solar zenith angles greater than 75° (as do all pyranometers whether they are photodiode or thermopile based). The GHI corrections in the Irradiance, Inc. RSR include the corrections suggested by King et al. and the *cat ears* corrections in Augustyn et al. (2004).

Vignola (1999) published a study on the diffuse spectral responsivity of the LI-COR pyranometer. This study showed the magnitude of the deviation and also provided a model of the diffuse responsivity as a function of the clearness index. In 2005, NREL funded a study to see if the results from an RSR with a LI-COR pyranometer could be improved by removing the systematic errors (Vignola, 2006). The RSR was similar to the previous RSR, but a correction algorithm was added to the data-logging program in an attempt to remove the systematic errors that were produced. First the GHI and the DHI values were corrected to eliminate the effects of temperature, and the *cat ears* corrections helped improve the cosine response. Next, the diffuse spectral responsivity was corrected using an algorithm based on Vignola's (1999) study but using the corrected GHI values. Finally, a more accurate algorithm was used to calculate the cosine of the solar zenith angle, which was then used to calculate the DNI from the direct horizontal values.

Other improvements in design and operation were introduced at the same time. The sampling frequency has been increased from 1 per minute to as many as 20 per minute. The higher sampling rates should produce more accurate results during rapidly varying atmospheric conditions. Temperature measurements on the pyranometer were also introduced to better determine the temperature correction to the responsivity. Tests show that the corrections to the RSR data improve the accuracy of the measurements significantly. Comparisons between Schenk (black-and-white pyranometer) DHI and DHI measured with RSRs agree to within a few percent over the full range of clearness index (Figure 7.6). Figure 7.6 also indicates the improvement in scatter with averaging time, as was shown for the DNI in Figure 7.5. It is especially evident in Figure 7.5 that there is still room for improvement in the DNI corrections.

Two common measurement artifacts can appear in the RSR data. The first relates to the LI-COR pyranometer. During some mornings, a bead of moisture collects on the diffuser of the pyranometer. This bead directs sunlight striking at low angles onto the photodiode. This slightly increases the GHI but does not affect the DHI as it comes from all parts of the sky. When the DHI is subtracted from the GHI, the slight increase in GHI increases the calculated DNI significantly because early in the morning the cosine of the large solar zenith angle is small, and when dividing this into the direct horizontal value the error is amplified.

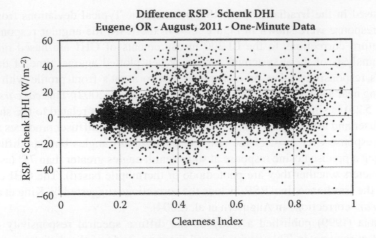

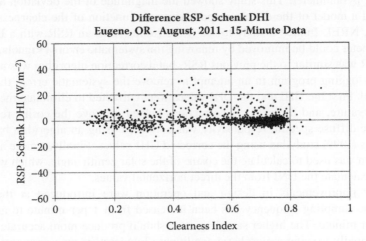

FIGURE 7.6 Comparison of corrected RSR DHI and Schenk black-and-white pyranometer DHI measurements; note the reduction in scatter with time averaging.

Occasionally the band does not rotate when it should, and the DHI value exhibits a sudden spike while the DNI has a sudden dip. This problem is most evident on clear days. In the original RSPs this problem appears to increase as the instrument ages. A software modification has removed this problem from the latest version of the RSR.

There are ongoing efforts to improve the measurements of RSRs. The correction method described in this chapter works for RSR instruments using LI-COR pyranometers. Other instrument manufacturers use their own proprietary algorithms to improve their instrument performance. Of course, corrections removing systematic errors can be applied only to well-maintained instruments with known calibration histories. If corrections are applied to instruments with unknown or poor calibration records or to instruments that are soiled, corrections may actually make the values deviate further from the truth. To address the soiling problem, the development of an automatic cleaning for RSR instruments is under way.

7.3 MULTIFILTER ROTATING SHADOWBAND RADIOMETER

The multifilter version of the rotating shadowband radiometer (MFRSR) was developed in the early 1990s (Harrison, Michalsky, and Berndt, 1994). The instrument retained a broadband silicon channel and added measurements at six additional wavelengths with full widths at half maximum transmission of around 10 nm. Five of the filters are between strong absorption bands and are used to calculate column aerosol optical depths near 415, 500, 615, 673, and 870 nm. One is in the water vapor band centered near 940 nm that can be used to calculate water vapor column. The MFRSR differs from the Irradiance, Inc. RSR in that the rotating shadowband makes three stops on its rotation: one stop just short of shading the sensor, one stop that shades the sensor, and a final stop just after shading the sensor. The extra stops that do not quite block the direct sun on the diffuser allow the diffuse irradiance that is shaded by the large band on the sun-shaded stop to be estimated.

The focus of the next part of this section is on improving the retrievals of global and diffuse horizontal and direct normal irradiance made with the MFRSR's photodiode detector. Corrections have to be made for the pyranometer's angular response departures from a perfect cosine response (Lambertian response), for small temperature departures from the nominal set temperature in the MFRSR, and, most significantly, for the spectral dependence of a photodiode's response.

Before deployment in the field, each MFRSR head is mounted in a laboratory angular response bench so that the center of rotation of the head is on the axis and perpendicular to a nearly collimated beam of a xenon arc lamp located about 5 meters from the MFRSR (Michalsky, Harrison, and Berkheiser, 1995). Measurements are made every 1° from the horizon to the zenith in the north, south, east, and west directions. These measurements are normalized to the zenith measurement producing a cosine response for each of the seven MFRSR channels.

The "raw" measurements from the MFRSR operating outdoors are the uncorrected global horizontal irradiance and the blocked diffuse horizontal irradiance, which is corrected by using the average of the two side-band measurements subtracted from the uncorrected global horizontal to correct for the excess skylight that is blocked during the sun-blocked measurement. This modified diffuse is subtracted from the uncorrected global horizontal irradiance to calculate the direct beam on a horizontal surface. In equation form

$$modified\_diffuse = global - \frac{(sideband1 + sideband2)}{2} + blocked \qquad (7.2)$$

$$direct\_horizontal = global\_horizontal - modified\_diffuse \qquad (7.3)$$

A cosine correction is applied to the *direct_horizontal* using interpolated measurements of the cosine response for the sun's position in the sky. A cosine correction is applied to the diffuse assuming an isotropic distribution of skylight. While this may not be a realistic distribution for most skies, calculations show that realistic distributions functions for clear and cloudy skies give the same correction as isotropic skies to within 1%. The corrected global horizontal irradiance is then calculated by adding the cosine-corrected *direct_horizontal* and cosine-corrected diffuse horizontal irradiances. The DNI is obtained by dividing the cosine-corrected *direct_horizontal* by the cosine of the solar zenith angle.

Augustine et al. (in preparation) showed that although the original design of the MFRSR allowed controlling the filter and detector chamber to a set point usually near 40°C, in field use, the head fails to hold this temperature to better than about ±4°C over the course of a year. This temperature departure from the set point results in corrections as large as ±4% for the narrowband filter measurements but should not be an issue for the silicon photodiode detector.

The spectral correction is the most significant of all adjustments because of the significant change in the spectral response of photodiode pyranometers with wavelength and the major change in skylight's spectral irradiance between cloudy- and clear-sky conditions. If, as in current practice, the global horizontal irradiance on a clear day is used to calibrate a photodiode pyranometer, the diffuse measurement of clear skylight will be seriously underestimated. If this underestimated diffuse is subtracted from the global to get direct irradiance as in a rotating shadowband radiometer, the direct will be overestimated.

Finding spectral corrections for the unfiltered silicon sensor in the MFRSR is subtly different from a standard photodiode pyranometer because the corrections for cosine response, temperature response, and spectral response are intertwined for the latter instrument. For the MFRSR the cosine and temperature corrections are performed independently. Michalsky (2001) developed a method for correcting the spectral response issues associated with using the unfiltered silicon channel of the MFRSR for broadband solar measurements. The results indicated almost zero bias for all components in the best methods and large root mean square (RMS) differences associated with differences in sampling frequency (1 Hz versus 1/20 Hz) under partly cloudy skies. Later, Michalsky, Augustine, and Kiedron (2009) used the filtered measurements and the unfiltered photodiode channel to find channel combinations that explained most of the variance in a multivariate approach that consisted of regressing the photodiode's measured solar components against the same reference thermopile solar components. They found that it was *unnecessary* to use the unfiltered photodiode channel to produce equivalent broadband estimates; therefore, it was possible to eliminate the unfiltered photodiode channel for another useful filter if needed. Surprisingly, using only three carefully chosen filtered channels produced equivalent broadband estimates to those produced by using all six filtered channels. This can be plausibly explained because two spectrally separated filters outside strong absorption bands of the shortwave spectrum capture the broad spectral extinction caused by Rayleigh and aerosol extinction. Water vapor absorption causes the largest molecular extinction in the shortwave spectrum; therefore, including 940 nm filter measurements captures this variability. It is important to note that since the filter measurements are over a very narrow range of the spectrum, they do not require spectral corrections.

These calculations of GHI, DNI, and DHI depend on the accuracy of the narrowband filter measurements. The transmissions of some of the narrowband filters degrade over time, and this affects the measured irradiance values. These changes can be automatically monitored by use of clear-day Langley plots. According to Beer's law,

$$I_\lambda = I_{\lambda o} \cdot \exp^{-m\tau_\lambda} \qquad (7.4)$$

where I_λ is the beam irradiance at a wavelength λ, $I_{\lambda o}$ is the extraterrestrial beam irradiance at wavelength λ, m is air mass, and τ_λ is optical depth at wavelength λ. The air mass m can be calculated accurately, and I_λ is measured with the MFRSR. By taking the logarithm of each side of Equation 7.4 a linear equation in optical depth is obtained:

$$\ln(I_\lambda) = \ln(I_{\lambda o}) - m\tau_\lambda \qquad (7.5)$$

On clear days a Langley plot of $ln(I_\lambda)$ versus m can be made. If τ_λ is constant, a linear plot results, and extrapolating to zero air mass yields $ln(I_{\lambda o})$. Changes in $ln(I_{\lambda o})$ can be used to determine the change in filter transmission. In practice there is considerable scatter in $ln(I_{\lambda o})$ (see Section 12.5.1), but a robust determination can be made with a sufficient number of Langley plots to follow the changes in filter transmission and make adjustments as needed. Periodic lamp calibrations have confirmed that this method tracks filter degradation well.

Recently, the manufacturer of the MFRSR, Yankee Environmental Systems, Inc., developed a thermopile detector that replaces the photodiode detector. The small thermopile has millisecond response times, making it suitable for use in a rotating shadowband radiometer. For the thermopile detector to function properly, controlling the temperature of the detector housing was paramount. Measurements during March temperature swings in Boulder, Colorado, found the housing temperature was stable to ±0.02°C 95% of the time. This impressive temperature stability has the added benefit that it eliminates the need for any temperature correction to the measurements, as needed for the earlier Yankee MFRSR versions (Augustine et al., in preparation).

Despite the flat broadband response of the thermopile sensor, a spectral dependence of the system's response remains. This is associated with the spectral dependence of the transmission of the Spectralon diffuser to the thermopile detector. Scattering from Spectralon is greater in the shorter wavelengths, giving rise to underestimates of diffuse and overestimates of direct as discussed already but with different wavelength dependencies from those of the photodiode pyranometers. Spectralon's wavelength-dependent transmission can be found at http://host.web-print-design.com/labsphere/products/images/graphs/SpectralonDiffuser.gif.

With the six spectral channels of the MFRSR, it becomes possible to retrieve aerosol optical depth as a function of wavelength and to retrieve water vapor, the most significant molecular absorber of shortwave radiation. With a second down-viewing MFRSR of matching filters, spectral surface reflectivity can be measured. With these quantities it is possible to use radiative transfer codes to calculate the spectral irradiance of direct normal and diffuse horizontal radiation. This broader capability of the MFRSR has not been exploited; however, spectrally sensitive technologies like photovoltaics (PV) are becoming increasingly important, and it will be necessary to measure the solar resource spectrally to evaluate and optimize PV efficiencies.

QUESTIONS

1. What was the major disadvantage of Wesely's (1982) simple rotating shadowband radiometer compared with modern instruments?
2. Explain the difference in the measurement algorithms used by Yankee Environmental Systems, Inc. and Irradiance, Inc. for their distinct versions of the rotating shadowband radiometer.
3. What is the major reason for the popularity of rotating shadowband radiometers? What is their major shortcoming?
4. Why are traditional thermopile pyranometers not used in rotating shadowband radiometers?
5. What are two of the four corrections that have to be performed for rotating shadowband instruments?

REFERENCES

Augustyn, J., T. Geer, T. Stoffel, R. Kessler, E. Kern, R. Little, and F. Vignola. 2002. Improving the accuracy of low cost measurement of direct normal irradiance. Proceedings of the American Solar Energy Society, R. Campbell-Howe and B. Wilkins-Crowder (eds.), American Solar Energy Society, Boulder, Colorado, USA.

Augustyn, J., T. Geer, T. Stoffel, R. Kessler, E. Kern, R. Little, F. Vignola, and B. Boyson. 2004. Update of algorithm to correct direct normal Irradiance measurements made with a rotating shadow band pyranometer. Proceedings of the Amercian Solar Energy Society, R. Campbell-Howe and B. Wilkins-Crowder (eds.), Amercian Solar Energy Society, Boulder, Colorado, USA.

Harrison, L., J. Michalsky, and J. Berndt. 1994. Automated multifilter rotating shadow-band radiometer: An instrument for optical depth and radiation measurements. *Applied Optics* 33:5118–5125.

King, M. L., W. E. Boyson, B. R. Hansen, and W. I. Bower. 1998. *Improved accuracy for low-cost solar irradiance sensors.* Paper presented at the 2nd world conference and exhibition of photovoltaic solar energy conversion, July 6–10, Vienna, Austria.

Michalsky, J. J. 2001. *Accuracy of broadband shortwave irradiance measurements using the open silicon channel of the MFRSR.* Paper presented at the 11th ARM science team meeting, Atlanta, GA. (Available at http://www.arm.gov/publications/proceedings/conf11.)

Michalsky, J. J., J. A. Augustine, and P. W. Kiedron. 2009. Improved broadband solar irradiance from the multi-filter rotating shadowband radiometer. *Solar Energy* 83:2144–2156.

Michalsky, J. J., J. L. Berndt, and G. J. Schuster. 1986. A microprocessor-controlled rotating shadowband radiometer, *Solar Energy* 36:465–470.

Michalsky, J. J., L. C. Harrison, and W. E. Berkheiser III. 1995. Cosine response characteristics of some radiometric and photometric sensors. *Solar Energy* 54: 397–402.

Vignola, F. 1999. Solar cell based pyranometers: Evaluation of diffuse responsivity. In *Proceedings of the American Solar Energy Society,* R. Campbell-Howe, ed., American Solar Energy Society, Boulder, CO.

Vignola, F. 2006. Removing systematic errors from rotating shadowband pyranometer data. In *Proceedings of the American Solar Energy Society,* R. Campbell-Howe, ed., American Solar Energy Society, Boulder, CO.

Wesely, M. L. 1982. Simplified techniques to study components of solar radiation under haze and clouds. *Journal of Applied Meteorology* 21:373–383.

8 Measuring Solar Radiation on a Tilted Surface

The real question now at issue is simply the variability of daily or quite frequent observed values of solar intensity...

Charles F. Marvin
(1858–1943)

8.1 INTRODUCTION

When evaluating the performance of a solar energy conversion system, it is necessary to know the irradiance incident on the collector. The incident solar irradiance either can be estimated from the direct normal irradiance (DNI) and diffuse horizontal irradiance (DHI) values, or the incident irradiance can be measured. For concentrating solar power systems (CSP) the DNI is the only irradiance measurement or model estimate needed to represent the energy input to the collector. The global tilted irradiance (GTI) is needed for nonconcentrating (flat plate) solar collectors. GTI is the incident irradiance on a surface with a given tilt and azimuth orientation. Surfaces of interest can vary from building façades facing north to flat rooftops to any collector tilt and orientation in between.

When specifying collector tilts and orientations, the notation used in this book will be GTI (*Tilt, Orientation*), where *Tilt* is given in degrees from the horizontal, and *Orientation* is in degrees measured eastward from true north (0°). For example, total irradiance on a surface tilted up 30° from a horizontal surface and facing true south (180°) is designated GTI(30°,180°).

It is impractical to assess the solar resource by measuring solar irradiance on a surface at every possible tilt and orientation. Therefore, models have been developed that use measured or estimated DNI and DHI components to calculate an approximation for tilted surface GTI. A tilted surface will receive irradiance reflected from the ground and will receive reflected irradiance from nearby objects that are within its field of view. The ground-reflected irradiance is often difficult to calculate because the reflectance is not spatially uniform and the absorption and reflectance properties of the surface change the spectral composition of the reflected irradiance from that of the incident solar spectral distribution. For a specularly reflecting surface, the angle of reflectance is equal to the angle of incidence. However, natural surfaces are generally granular in character, so an approximation is used to estimate the ground-reflected irradiance. A generic formula is often used to calculate GTI is

$$GTI = DNI \cdot \cos(sza_T) + DHI \cdot (1 + \cos(T))/2 + GHI \cdot \rho \cdot (1 - \cos(T))/2 \quad (8.1)$$

where sza_T is the angle of incidence of the DNI with respect to the tilted surface, T is the tilt angle of the surface from the horizontal, and ρ is the average albedo or reflectivity of the ground near the tilted surface. While the DNI can be measured with considerable accuracy and sza_T can be calculated precisely (Equation 2.7), the other two terms in the equation are approximations (see Appendix A for further discussions on modeling GTI). Therefore, to know the solar radiation incident on a tilted solar collector with any precision, the GTI must be measured with a pyranometer.

Pyranometers were originally designed for the measurement of global horizontal irradiance (GHI) and installed on horizontal surfaces. It is only with the need to measure the performance of solar energy systems that measurements on other than horizontal surfaces were considered and pyranometers were increasingly used on tilted surfaces. When tilted, the measurement performance of the pyranometer changes. The nature and magnitude of the change depend on the design of the pyranometer. The change in behavior of three types of pyranometers will be discussed:

1. Secondary reference and other single black detector thermopile pyranometers
2. Black-and-white pyranometers
3. Photodiode pyranometers

8.2 EFFECT OF TILT ON SINGLE BLACK DETECTOR PYRANOMETERS

In single black detector thermopile pyranometers the conduction of heat from the black detector through the thermopile to the body of the pyranometer is measured to determine the radiant energy (solar irradiance) incident on the instrument. Extraneous heat flow by conduction, convection, and radiative transfer systematically reduces heat flow through the thermopile and, hence, affects the measurements. Each pyranometer model has its own design features to minimize other conductive losses as well as convective and radiative losses within the pyranometer. Thermal isolation and insulation help reduce the conductive losses, and the use of double domes over the black detector help to reduce the convective thermal losses. Still, there are some losses due to conduction, radiation, and convection in these pyranometers.

Tilting a pyranometer from the horizontal affects the convective losses and radiative losses (thermal offsets) of thermopile pyranometers. For single black thermopile pyranometers the conductive losses remain about the same as when they are used horizontally since orientation typically does not affect conduction. The convective losses change because the heat flow patterns within the domes change. Thermal offsets actually decrease as the pyranometer is tilted at a steeper angle because the pyranometer sees more of the warmer ground and less of the colder sky. Figure 8.1 shows the thermal offsets from Eppley precision spectral pyranometers (PSPs) at different tilts (Vignola, Long, and Reda, 2007). As the pyranometer tilt increases, the observed thermal offsets decrease. The pyranometers facing the ground (i.e., inverted) show almost no thermal offset. Of course, the amount of radiative loss depends on the difference between the target temperature (e.g., the sky

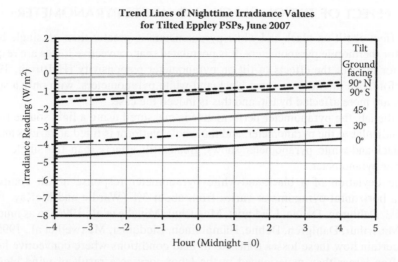

FIGURE 8.1 Trend lines of nighttime values for tilted PSPs for June 2007. Note that as the pyranometer's tilt increases, the nighttime values get smaller. This means that the infrared loss becomes less as the pyranometer "sees" more of the ground.

dome temperature) and ambient temperature (instrument temperature) as discussed in Chapter 5. Many factors influence sky temperature—primarily cloud cover and precipitable water vapor. Sky temperature can vary over time scales of seconds and minutes with underlying seasonal trends. Therefore, pyranometers with minimal thermal offset losses are better suited for tilted measurements.

Tests of the effect of tilt on pyranometer performance are typically performed indoors under controlled conditions and, consequently, do not show thermal offset effects. For single black detector thermopile instruments, the change in responsivity with tilt is minimal (on the order of ±1% to ±2%). It is presumed that these changes are related to changes in convective losses, but the exact nature of the change and the magnitude vary with pyranometer model. Some pyranometers show a slight responsivity increase with tilt, while others may have a slight decrease.

While there are accurate measurements of tilt effects indoors, outdoor evaluations of tilt effects are very difficult because of the variation in ground reflection on the instrument and the absence of an absolute pyranometer to provide a reference measurement. In addition, the reduction of thermal offsets as the pyranometer is tilted and the imperfect Lambertian response of the pyranometer further complicate the analysis. Typical uncertainties quoted for the tilt effect vary from 0.5% for secondary-standard instruments and up to 2% for some first-class instruments (Hulstrom, 1989). While these values were determined indoors and provide an estimate of the uncertainties, they may not be as accurate under actual field operating conditions. Note that the uncertainty caused by tilting a pyranometer must be added in quadrature to the uncertainty determined for the pyranometer when it is installed horizontally. Therefore, tilted measurements from a given pyranometer will always have a larger uncertainty than GHI measurements from the same instrument.

8.3   EFFECT OF TILT ON BLACK-AND-WHITE PYRANOMETERS

The effect of tilt on black-and-white pyranometers is greater than on single black detector thermopile pyranometers. It is postulated that convective losses are responsible for most of the effects of tilt on pyranometer responsivity (Hulstrom, 1989). Therefore, one might expect that black-and-white pyranometers, with only a single dome, are more affected by tilt, and this is indeed the case.

In the lab, the pyranometer performance is measured using a light source that is perpendicular to the pyranometer, and the whole platform is tilted at various angles; the black-and-white pyranometer shows greater tilt effects than the single black detector pyranometer.

The deviation of a black-and-white pyranometer response at tilt compared with a horizontal pyranometer varies from about 3% (Wardle and McKay, 1984; Wardle, Dahlgren, Dehne, Liedquist, McArthur, Miyake et al., 1996) to as much as 8% (McArthur, Dahlgren, Dehne, Hämäläinen, Liedquist, Maxwell et al., 1995). It is uncertain how these losses translate to field conditions where convective losses are often larger than experienced in the laboratory as a result of wind blowing across the pyranometer dome and the naturally changing temperatures and relative humidity of the air. The lack of an absolute pyranometer makes evaluation of tilt effects in the field difficult to measure and characterize. Furthermore, it is difficult to isolate tilt effects from other issues in the field. Even the best pyranometers have at least 2% to 3% absolute uncertainty and a nonuniform Lambertian response. In addition, the responsivity of black-and-white pyranometers has an azimuthal dependence as well as an imperfect Lambertian response. Isolating a 2% to 8% tilt effect from the other effects is fraught with uncertainty. The change in responsivity as a function of tilt is shown in Figure 8.2 for a Schenk Star pyranometer.

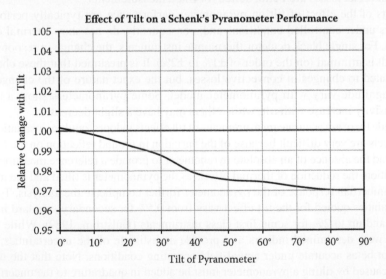

FIGURE 8.2   Relative effect of tilt on a Schenk pyranometer. Data taken indoors. (From Andersson, H.E.B., L. Liedquist, J. Lindblad, and L. A. Norsten, Report SP-PAPP 1981: 7, Statens Provningsanstalt [National Testing Institute], Borås, Sweden, 1981).

The uncertainty resulting from the tilt effect has to be added in quadrature to the typical black-and-white instrument uncertainty of ±5%. Since this can lead to estimated measurement uncertainties of over 9%, it is not considered good practice to use black-and-white pyranometers on tilted surfaces.

8.4 EFFECT OF TILT ON PHOTODIODE PYRANOMETERS

One would not expect photodiode pyranometers to be affected by tilt because they are not thermal instruments. Photodiode pyranometers, in fact, are often used to test the effect of tilt on thermopile pyranometers under controlled laboratory conditions where they can serve as the control. Photodiode pyranometers have a limited spectral response. In the field, the change in spectral composition of reflected light affects the measurement of tilted irradiance by photodiode pyranometers. In general, ground-reflected irradiance, especially from vegetation, is shifted more to the infrared portion of the spectrum. On a 90° tilted surface, the ground-reflected irradiance is roughly equal to one-half the GHI times the reflectivity of the surface (see Equation 8.1). Typical ground reflectivity is approximately 0.2 for non-snow-covered surfaces. Hence, the contribution of the ground-reflected irradiance to a 90° tilted pyranometer is about 10% of the GHI, and for a 45° tilted pyranometer the reflected contribution is only about 3% of the GHI. Therefore, even a 30% change in responsivity resulting from a spectral shift in the irradiance would lead to an increase in uncertainty in the GTI measurement of only about 1% for the 45° tilt case. Of course, the initial absolute uncertainty in the photodiode reading is about 5%. A clear-day comparison of thermopile and photodiode pyranometers is shown in Figure 8.3 for a southern orientation with tilts of 0°, 40°, and 90° and in Figure 8.4 for 90° tilted pyranometers facing north, east, south, and west. Data are taken from the NREL's Solar Radiation Research Laboratory's Baseline Measurement System using PSP and LI-200 pyranometers (www.nrel.gov/midc/srrl_bms). The difference between the photodiode and thermopile pyranometers is divided by the thermopile reading to obtain the percent difference. Average nighttime readings were subtracted from the daytime values to reduce the effect of the thermal offset on the values. Differences vary between −5% and +5% for the south-facing pyranometers and 0% to 15% for the 90° tilted pyranometers, depending on the time of day and the tilt and orientation of the instruments. The pattern of difference changes for every orientation. It is difficult to separate cosine response differences from tilt responses from shifts caused by the spectral reflectivity changes throughout the day for these outdoor measurements. Again, the absence of an absolute pyranometer makes it difficult to precisely determine the error caused by the tilt alone.

8.5 RECOMMENDATIONS FOR TILTED IRRADIANCE MEASUREMENTS

The accuracy of tilted irradiance measurements will always be less than the accuracy of horizontal measurements because the changes in pyranometer performance caused by tilting must be added to the uncertainty of the performance when the

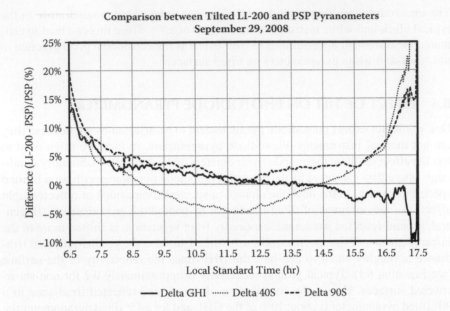

FIGURE 8.3 Clear-day comparison of thermopile Eppley PSP with a LI-COR LI-200 photodiode pyranometer on south-facing surfaces tilted 0°, 40°, and 90°. Solar Radiation Research Laboratory at NREL on September 29, 2008. The thermopile instrument agrees rather well over the day with the photodiode response increasing in the morning and afternoon. At tilts of 40° and 90° the behavior is different with deviations due to cosine response, spectral reflectivity response and tilt response. It is difficult to isolate and quantify the various sources of error.

pyranometer is horizontal. Most broadband thermopile pyranometers are subject to almost negligible spectral shifts of ground-reflected irradiance since glass domes used on thermopile instruments block none or only a small portion of the near-infrared irradiance. For the very best measurements on tilted surfaces, single black detector thermopile pyranometers that use double domes made of material that uniformly pass a broad range of wavelengths yield the most accurate broadband measurements of GTI. Single black detector thermopile pyranometers with double domes that pass a slightly narrower band of wavelengths are the second choice. Photodiode pyranometers have no tilt effect, and, if used to evaluate a photovoltaic (PV) panel with a similar spectral response, it might be the best choice for the tilted measurement if the ground-reflected contribution is small (i.e., the tilt is less than 45°).

The specifications and measurement characteristics of the pyranometer on the horizontal surface helps determine the quality of the tilted measurements that the pyranometer could make. Pyranometers that make good GHI measurements have a likelihood of making good GTI measurements. Of course, the shallower the tilt, the less likely there will be a significant tilt effect for any pyranometer. Black-and-white pyranometers have significant tilt effects and are not recommended for use on tilted surfaces.

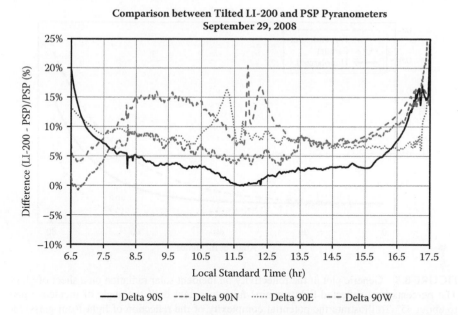

FIGURE 8.4   Clear-day comparison of thermopile Eppley PSP pyranometer with a LI-COR LI-200 photodiode pyranometer on north, east, south, and west surfaces tilted 90°. Data taken at the Solar Radiation Research Laboratory at NREL on September 29, 2008. Differences range from 0 to 15% over the day. The peak differences of the east- and west-facing pyranometers around solar noon is likely the *cat ears* effect of the LI-200 pyranometer discussed in Chapter 5.

8.6   NOTES ON MODELING PV SYSTEM PERFORMANCE WITH DATA FROM PHOTODIODE PYRANOMETERS

Photodiode pyranometers have spectral responses similar to silicon-based PV modules, and, consequently, these pyranometers may correlate better with the output of a PV system than a broadband thermopile pyranometer. To model the performance of a photovoltaic system, not only is it necessary to know the total incident solar irradiance on the module, but also the incident solar radiation must be separated into its diffuse and beam components to properly calculate the transmission of the solar radiation through the glazing. The separation of diffuse and beam components presents a special problem for tilted measurements made with photodiode pyranometers because the spectral composition of the beam and diffuse components are significantly different on clear days and the responsivity of the photodiode is different for the beam and diffuse components.

The glazing on the front of a solar module reflects an increasing percentage of the irradiance as the angle of incidence increases (Fresnel reflection) (see Figure 8.5). Notice that the glazing does not exhibit a Lambertian response, which would be uniform with angle of incidence. The difference from a Lambertian response is especially divergent for angles of incidence larger than 50°. DNI values and the glazing's refractive index (often modified by an antireflective coating) are required for accurate

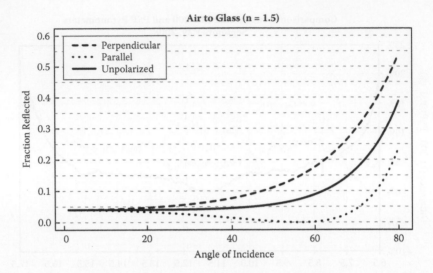

FIGURE 8.5 Generic plot of the reflectivity on incident solar radiation on a sheet of glass. The percentage of reflected light remains fairly constant until the angle of incidence gets to about 45°. To illustrate the potential complexity of the reflection of light from glass, the reflection of polarized light is also shown. Antireflective coatings are used with photovoltaic modules, and these coatings can reduce the reflected light by a factor of two or more. (From Andersson, H.E.B., L. Liedquist, J. Lindblad, and L.A. Norsten, Calibration of Pyranometers, Report SP-PAPP, 1981. With permission.)

estimates of the transmission through the glazing to the solar cell. PV system performance models are developed and tested with DNI measurements obtained from pyrheliometers and GTI measurements obtained using high-quality pyranometers. The PV system performance models incorporate the effects of changes in the spectral composition incident on the solar cell. This is the reason that there is a better spectral match of the photodiode pyranometer response; however, using photodiode pyranometer data causes the performance model to overcorrect for the changes in spectral composition. This is analogous to counting the effect of changes in the spectral composition twice.

The need to separate the DNI contribution from the GTI measurement to properly account for the reflectivity of the glazing further complicates the performance estimates if a photodiode pyranometer is used for GTI measurements. DNI values can be obtained from measurements made with pyrheliometers, estimated from rotating shadowband radiometer (RSR) measurements, or modeled from measured GHI values. As discussed in Chapter 5, photodiode pyranometers systematically underestimate the DHI value on clear days by 20 to 30%, and if a pyrheliometer is used to obtain the DNI value, the diffuse and ground-reflected irradiance reading of the photodiode pyranometer will underestimate contributions of the diffuse component to the total energy incident on the solar cells. If not taken into account, this underestimation of the incident energy can overestimate the efficiency of the solar system.

If the DNI is obtained with an RSR, it is important to thoroughly understand how the DHI and DNI values are obtained and whether responsivities have been corrected for differences in the spectral composition of DNI and DHI irradiance. If the

DNI is obtained from a RSR that has been corrected for the underestimation of the diffuse component, the DNI can be treated as if it came from a pyrheliometer, except the uncertainty in the DNI value is increased. If the DHI values have not been spectrally corrected, then an algorithm must be applied to correct for the effects caused by the underreporting of the DHI on clear days.

Correlations between DNI and tilted measurements are crude at best (Vignola, 2003). Therefore, if DNI measurements are not available, GHI measurements are needed to estimate the DNI value required for modeling photovoltaic module performance. Models correlating DNI with GHI are usually derived from GHI measurements made with high-quality pyranometers using single black thermopile detectors. Because each type of pyranometer has its characteristic systematic errors, using correlations derived from one type of pyranometer with measurements made with a different model or type of pyranometer increases the uncertainty in the results and may introduce systematic errors. Therefore, if a photodiode-based pyranometer is used for the GHI measurement, then a DNI–GHI correlation derived using GHI measurements with a photodiode pyranometer should be employed. The same applies to GHI measurements obtained from thermopile-based pyranometers.

Spectral effects from ground-reflected irradiance are similar for a tilted photodiode pyranometer and solar PV cells. So while good correlations with solar module performance can be obtained, the absolute uncertainty in module efficiency is larger than the uncertainty obtained when irradiance measurements are made with thermopile-based pyrheliometers and pyranometers.

QUESTIONS

1. What is the best pyranometer to use for tilted surface measurements?
2. What type of pyranometer should not be used for tilted measurements?
3. Why do the design qualities of a photodiode pyranometer both make it good and bad for tilted measurements?

REFERENCES

Andersson, H. E. B., L. Liedquist, J. Lindblad, and L. A. Norsten. 1981. *Calibration of pyranometers*. Report SP-PAPP 1981: 7. Statens Provningsanstalt (National Testing Institute), Borås, Sweden.

ASTM Standard E1036-08. 2008. *Standard test methods for electrical performance of non-concentrator terrestrial photovoltaic modules and arrays using reference cells*. West Conshohocken, PA: ASTM International.

ASTM Standard E948-09. 2009. *Standard test method for electrical performance of photovoltaic cells using reference cells under simulated sunlight*. West Conshohocken, PA: ASTM International.

ASTM Standard E1362-10. 2010. *Standard test method for calibration of non-concentrator photovoltaic secondary reference cells*. West Conshohocken, PA: ASTM International.

Flowers, E. C. 1984. Solar radiation measurements. In *Proceedings of the recent advances in pyranometry, International Energy Agency IEA Task IX solar radiation and pyranometer studies*. Downsview, Ontario, Canada: Atmospheric Environment Service.

Hulstrom, R. L. (ed.). 1989. *Solar resources*. Cambridge, MA: MIT Press.

International Electrotechnical Commission (IEC). 2007. IEC 60904-2: *Photovoltaic devices – Part 2: Requirements for reference solar cells*. London: British Standards Institution.

McArthur, L. J. B., L. Dahlgren, K. Dehne, M. Hämäläinen, L. Liedquist, G. Maxwell, C. V. Wells, D. I. Wardle. 1995. *Using pyranometers in tests of solar energy converters*. International Energy Agency, Solar Heating and Cooling Programme, Task 9, IEA-SNCP-9F1. Downsview, Ontario, Canada: Experimental Studies Division, Atmospheric Environmental Service.

Vignola, F. 2003. *Beam-tilted correlations*. Paper presented at the solar 2003 American Solar Energy Society Conference, Austin, TX.

Vignola, F., C. N. Long, and I. Reda. 2007. *Evaluation of methods to correct for IR loss in Eppley PSP diffuse measurements*. Paper presented at the SPIE conference, San Diego, CA.

Wardle, D. I., L. Dahlgren, K. Dehne, L. Liedquist, L. J. B. McArthur, Y. Miyake, O. Motschka, C. A. Velds, and C. V. Wells. 1996. *Improved measurements of solar irradiance by means of detailed pyranometer characterisation*. International Energy Agency, Solar Heating and Cooling Programme, Task9, IEA-SHCP-9C-2. Downsview, Ontario, Canada: National Atmospheric Radiation Centre, Atmospheric Environmental Service.

Wardle, D. L. and D. C. McKay (eds.). 1984. *Recent advances in pyranometery*. Paper presented at the symposium proceedings of the Swedish Meteorological and Hydrological Institute, Norrkoping, Sweden.

9 Albedo

When you're thinking of putting on a new roof, make it white.

Steven Chu
Secretary of the U.S. Department of Energy (2009)

9.1 INTRODUCTION

A surface reflects and scatters a fraction of the incident solar irradiance that falls on it, including the direct normal irradiance (DNI) and the scattered sunlight that arrives from all parts of the sky, that is, the diffuse horizontal irradiance (DHI). However, a surface reflects the DNI and DHI components very differently, as will be discussed in this chapter on albedo. In general, albedo is defined as the total reflected irradiance divided by the total incident irradiance. Snowfields and deep oceans represent extremes in albedo. For newly fallen snow, 90% or more of the incident radiation may be reflected at visible wavelengths, but as snow ages, melts, and refreezes the reflectivity decreases. For the deep ocean the albedo is around 4% for calm seas at all solar wavelengths. However, if the seas have breaking waves, the albedo increases. If one looks toward the sun at the angle of incidence, a glint is detectable from the ocean surface. This glint obeys Fresnel's equations. Albedos of surfaces can be very different spectrally with some surfaces showing higher reflectivity in the visible than in the near-infrared while other surfaces can show the opposite behavior. The albedos of vegetated areas have considerable spectral structure and change appreciably as the plants go through stages of growth, flowering, and senescence. Measurements by Bowker, Davis, Myrick, Stacy, and Jones (1985) resulted in cataloged reflectances for many common surface types, and this remains a primary reference for surface albedo for real surfaces.

The next section discusses the most frequent albedo measurements of broadband solar and photosynthetically active radiation (PAR), followed by a section that explains the usefulness of important but infrequently made spectral albedo measurements. The third section of this chapter is devoted to explaining the bidirectional reflectance function and why it is important to understanding the nature of solar radiation. The chapter concludes with a section on how these measurements should be made, including calibration tips.

9.2 BROADBAND ALBEDO

The most frequent measurement of albedo uses broadband solar radiation detectors, that is, a pair of horizontally oriented pyranometers: one mounted facing the ground for measuring the reflected (upwelling) irradiance; and another mounted facing skyward to measure the downwelling irradiance, as shown in Figure 9.1. The most frequent measurement of albedo is the total solar spectrum albedo, and PAR is probably the second most frequent measurement of broadband albedo.

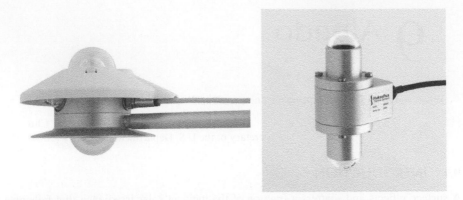

FIGURE 9.1   Two examples of commercial albedometers for measuring broadband albedos: (left) Kipp & Zonen's CMA albedometer; (right) the Hukseflux SRA01.

Broadband albedos are ever changing. They depend on whether the clouds hide the sun, and if they do not, then it depends on the sun angle with respect to the zenith; early morning and late afternoon direct solar radiation (DNI) produces a higher albedo than solar noon direct radiation. If the day is overcast, the albedo is nearly constant. If surfaces are wet, they appear darker than when they are dry; this is very obvious for PAR, which covers the wavelengths 400–700 nm, where the eye can clearly detect the effect.

Understanding surface albedo characteristics is important in renewable energy applications, atmospheric physics, agricultural research, oceanography, and climate studies. Yang et al. (2008) found that snow-free surface albedos used in the National Centers for Environmental Prediction (NCEP) Global Forecast Systems (GFS) model had too little dependence on solar zenith angle and were too low at all solar zenith angles. Further, they found that MODIS (NASA satellite) parameterizations of albedo gave overestimates for high sun angles and underestimates for low sun angles. These comparisons were based on long-term surface-based observations of broadband albedos at three U.S. Department of Energy Atmospheric Radiation Measurement (ARM) Program sites (Ackerman and Stokes, 2003; Stokes and Schwartz, 1994) and at six of the National Oceanic and Atmospheric Administration Surface Radiation (SURFRAD) Network sites (Augustine et al., 2000).

PAR albedos have been found to be robust predictors of relative water content in maize leaves (Zygielbaum, Gitelson, Arkebauer, and Rundquist, 2009). It is reasonable to assume that this would apply to other plants and serve as an early predictor of plant stress. As the plant becomes water stressed, the PAR albedo increases in a very predictable way. The authors suggest that this effect of increasing vegetation albedos has not been incorporated in climate modeling. While broadband albedos are useful and the measurements required are relatively straightforward, much more can be learned from albedo measurements with even moderate spectral resolution.

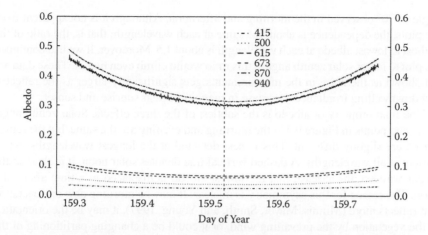

FIGURE 9.2 Illustration of how the albedo changes on a clear day over a vegetated surface as a function of time and as a function of wavelength. The wavelengths in the legend are in nanometers. Since the data are for June 8 in the northern hemisphere, the sun is high in the sky at solar noon where the albedo is the lowest. Since it is a clear day, the solar zenith angle dependence is associated with the direct sun's incidence angle on the surface.

9.3 SPECTRAL ALBEDO

Surface radiation is not reflected equally across the solar spectrum. For many surfaces, the wavelength dependence is dramatic. Figure 9.2 is an example of measured albedos (fraction of reflected radiation to incident radiation) as a function of local time in fractional day at six visible and near-infrared wavelengths. The clear-day albedos are for June 8, 2006, at the ARM Southern Great Plains (SGP) site between Lamont and Billings, Oklahoma (36.604°N, 97.485°W). The albedo, in this case, was calculated as an equally weighted average of two measurements over a pasture and over a fallowed field; the two measurements of albedo were used because these were the dominant surface types for the area that represents the central ARM SGP site. Figure 9.2 illustrates at least three properties of spectral albedo: the wavelength dependence, the solar zenith angle dependence, and a slight asymmetry in solar zenith angle dependence.

The reflectance is lowest at the four shortest wavelengths (those in the PAR region of the spectrum). Reflectivity in the *visible* window of the solar spectrum (400–700 nm) typically peaks around 550 nm and decreases on either side of this peak. The reflectivity at 615 nm is slightly higher than at 673 nm. The reflectivity at 500 nm and 415 nm are very low. Relative to the visible, there is a threefold increase in reflectivity at the 870 and 940 nm wavelengths that are beyond the infrared jump that is seen in all green vegetation. Vegetation has evolved to reflect this infrared energy that cannot be use for photosynthesis and would heat the plant with deleterious effects if this longwave energy were absorbed in any significant amount.

Another property of albedo that is immediately obvious in Figure 9.2 is the large and smooth dependence on the solar zenith angle. For the SGP station on June 8, the solar zenith angle at solar noon is 14°, and the end points in this plot are at a solar zenith angle of 70° observed in the morning and afternoon. Although it is not apparent from the plots, the dependence is about the same at each wavelength; that is, the ratio of the highest to lowest albedo at each wavelength is about 1.5. Moreover, if we had continued the plot to larger solar zenith angles, this ratio would climb even higher. These data are not shown as the noise in the measurements gets significantly larger as the reflected and downwelling irradiances decrease in magnitude near sunrise and sunset.

The final property of albedo is the subtlest of the three effects. Solar zenith angle at the end points in Figure 9.2 in the morning and evening are the same, but the reflectances are slightly different. This is best detected at the longest wavelengths, but is present at all wavelengths. A dashed vertical line denotes solar noon. It is clear, at the longer wavelengths at least, that the lowest albedo is displaced to a time after solar noon. The reason for this asymmetry is unclear: It may be morning dew on vegetation that reflects more (Minnis, Mayor, Smith, and Young, 1997), it may be the orientation of the vegetation by the prevailing wind, or it could be a changing partitioning of the diffuse and direct sunlight throughout the day, which will be discussed later.

While useful spectral information is gleaned from Figure 9.2, more spectral coverage with higher spectral resolution allows an examination of the adequacy of discrete spectral sampling of surface albedo. The data to calculate albedo in Figure 9.3 were taken with a portable spectrometer, the LI-COR LI-1800. This instrument, whose manufacture was discontinued in 2003, takes 30 seconds to acquire continuous spectral data with 6 nm resolution between 330 and 1100 nm. The data in Figure 9.3 were

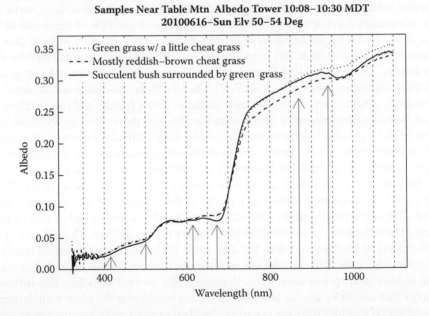

FIGURE 9.3 Three measurements of spectral albedo made using a spectrometer under a 10 m albedo tower. The measurements were chosen to examine the visual extremes under the tower. Clearly, there are only subtle differences. The notable features are the peak at 550 nm, the sharp rise near 700 nm, and the liquid absorption in plants near 975 nm.

acquired in three locations from a height of about 1.5 m above ground level under a 10 m albedo tower on Table Mountain, 12 km north of Boulder, Colorado, which is used for continuous measurement of spectral albedo at the same wavelengths as in Figure 9.2. The three sample areas were chosen to represent the extremes in visual diversity of vegetation under the tower. The measurement sequence was to move quickly to acquire downwelling measurements before and after an upwelling measurement. The two downwelling measurements were averaged and divided into the upwelling measurement to calculate albedo. Irradiance conditions were stable with the sun high in the sky and only a few clouds on the western horizon. The 10 m tower site was surrounded by some grasses that were green, although past their peak, cheat grass (*Bromus tectorum*) that was well on its way to senescence, some taller succulent bushes, and a few exposed rocks, but no bare soil. From Figure 9.3, it is clear that albedo is a smooth function over most of these wavelengths. The noisy albedo values in the ultraviolet are caused by very low upwelling irradiance and losses in the fiberoptic cable that is used to transfer sunlight to the LI-1800 spectrometer. Some notable features are the peak near 550 nm, the sharp rise beyond 700 nm, and the absorption band near 1000 nm. The peak spectral albedo over the PAR region occurs because the predominant absorption features are in the blue and red regions of the spectrum, giving a green reflective peak as a result. The rise in spectral albedo beyond 700 nm was discussed earlier, but the sharpness of the increase is clearly illustrated in Figure 9.3. Liquid water in the plants causes the absorption band around 975 nm (Pu, Ge, Kelly, and Gong, 2003). The gray arrows indicate the central wavelengths of the spectral measurements made in Figure 9.2.

Although Figures 9.2 and 9.3 represent albedos from different locations on different days, they are typical of growing vegetation. Their spectral dependencies, in fact, closely match if we consider the high sun (low solar zenith angle) albedos in Figure 9.2. The only exception is that the 940 nm albedo in Figure 9.3 slightly exceeds the 870 nm albedo, while in Figure 9.2 the opposite is true. This suggests that even with sparse spectral sampling, using the six wavelengths of Figure 9.2, and knowledge that the surface was vegetative, the continuous spectral dependence over the 330–1100 nm region in Figure 9.3 could be approximated with some confidence that one had captured the general spectral behavior. Obviously, the details would not be uncovered without the higher spectral resolution data. McFarlane, Gaustad, Long, Mlawer, and Delamere (2011) developed a technique for approximating the continuous albedo using discrete spectral measurements and information on the surface type and state.

Figure 9.4 illustrates the change in albedo as vegetation transforms from vigorous growth to complete senescence. The discrete measurements were taken at the SGP site in Oklahoma that was described in the beginning of this section, and continuous albedo spectra were estimated using McFarlane et al.'s (2011) method, for four different times of the year as listed in the legend of Figure 9.4. The first measurements on May 14 were over green pasture and a green wheat field. The next measurements on June 8 showed the grasses and wheat still green, but they were beginning to dry and the wheat ripening, respectively. By July 14 the wheat was harvested and the grasses showed almost no hint of green. Also, a portion of the surface measured under one of the two towers had been tilled resulting in a surface of exposed soil. On October 11 the surfaces were dead grasses and weeds. Figure 9.4 illustrates the changes in albedo with the clearest green peak and highest near-infrared reflectance on May 14. The green

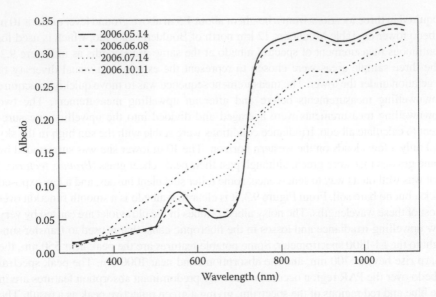

FIGURE 9.4 Illustration of how the spectral albedo of vegetation changes from the vigorous growth stage through senescence and death. The peak at 550 nm and the sharp rise near 700 nm disappear.

peak is less distinct and the reflectivity in the near-infrared lower by June 8. After harvest on July 14, there is only a slight red gradient after 700 nm. By October 11 the only spectral signature is the monotonic increase with wavelength.

In Figure 9.5 the albedos of four distinct soil samples are examined. Dirt is not dirt. Soil does not look the same to our eye, and, indeed, it does not look the same spectrally. Bowker et al. (1985) pointed out two of the spectral features of soils in the 300–1100 nm spectral region that depend on their organic content and on their iron content. Even when organic and iron content is low, there are some general differences that depend on water content and the specific mixture of sand, clay, and silt. In Figure 9.5 the line labeled "dry soil" has both the organic feature that shows higher reflectivity starting at about 0.55 μm. Iron absorption is the broad feature in the 0.9–1.0 μm area. The other soil albedos are monotonically increasing with wavelength with no distinct features. The wet sample has lower overall reflectivity, but is higher than black loam until around 0.7 μm. The samples labeled "soil" and "black loam" are similar, but they show different slopes with wavelength. Note that these three soil samples resemble the wavelength dependence of the dead vegetation albedo on October 11, 2006, in Figure 9.4.

The final class of examined surfaces is the snow-covered surface. Determining the reflectivity of snow is challenging because albedo changes with the condition of the snow. Newly fallen fine snow has an extremely high albedo as shown in Figure 9.6. Granular or corn snow gives the next highest reflectivity, and melting snow shows high reflectivity but less than the other two conditions. Clearly, snow has little spectral dependence in the ultraviolet (UV) and visible, but begins to indicate absorption in the near-infrared wavelengths.

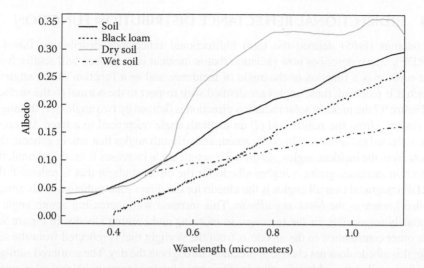

FIGURE 9.5 Albedos for four different soils from Bowker et al. (1985). Note the sharp rise near 500 nm due to organics and the depression center near 950 nm due to iron in the dry soil albedo spectrum. The other samples are monotonically increasing at different rates. (From Bowker, D. E., R. E. Davis, M. L. Myrick, K. Stacy, and W. T. Jones, *Spectral Reflectances of Natural Targets for Use in Remote Sensing Studies,* NASA Ref. Publ. 1139, 1985.)

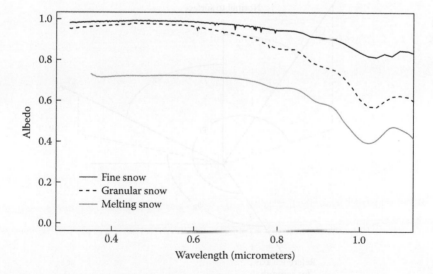

FIGURE 9.6 The albedo of snow depends on how recently it was deposited on the surface. The older the snow, the less reflective it is. As it starts to melt there is a noticeable drop in reflectivity. The visible spectral dependence is almost uniform, but the solid ice absorption starts to be a factor beyond 800 nm.

9.4 BIDIRECTIONAL REFLECTANCE DISTRIBUTION FUNCTION

Nicodemus (1965) defined the term bidirectional reflectance distribution function (BRDF), which describes how radiation that is incident on a surface will scatter from that surface as a function of the angle of incidence and as a function of the angle to which it is reflected; these angles are defined with respect to the normal to the surface. In Figure 9.7 the incident solar radiation direction is defined by two angles: (1) the angular distance from the zenith; and (2) an azimuth angle referenced to a fixed direction. The reflected radiation is defined by zenith and azimuth angles that are, in general, different from the incident angles. As the solar zenith angle increases from the normal, the reflection increases giving a higher albedo for the direct sunlight that is reflected; the BRDF integrated over all angles is the albedo for the direct solar radiation. This albedo is also known as the *black-sky albedo*. This increase with increasing zenith angle is primarily responsible for the increases in morning and evening albedos in Figure 9.2. The other contribution to the albedo is from the skylight that is reflected from the surface; this albedo does not change appreciably throughout the day. The scattered sunlight albedo is called the *white-sky albedo*. The total albedo changes with the solar zenith angle and, therefore, depends on the partitioning of the DNI and DHI.

Why does the white-sky albedo remain approximately constant throughout the day? This was discussed in Chapter 6 on diffuse measurements but is repeated

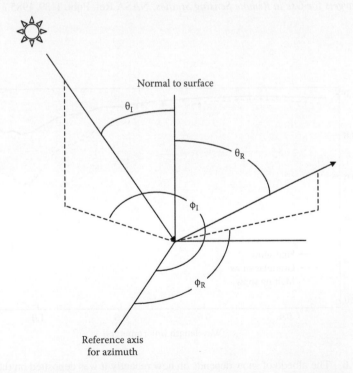

FIGURE 9.7 Bidirectional reflectance from a surface depends on the incident angle and azimuth of the incoming light and on the reflectance angle and azimuth from which this reflected light is viewed.

verbatim here. If skylight radiances were isotropic, that is, uniform in all directions, then the diffuse irradiance falling on a horizontal surface could be calculated as

$$DHI = R \int_0^{\frac{\pi}{2}} \int_0^{2\pi} \cos(\theta)\sin(\theta)\,d\phi\,d\theta, \tag{9.1}$$

where R is the constant radiance of the isotropic sky. The term $\sin(\theta)d\theta d\phi$ comes from spherical geometry that defines the radiance from every direction over the sky dome, and the $\cos(\theta)$ term comes from the assumption that the surface receives radiation with an exact cosine response. Since $2\sin(\theta)\cos(\theta)=\sin(2\theta)$, it is clear that the largest contribution to the diffuse irradiance is from the annulus near 45° and falls monotonically and symmetrically to zero at 0° and 90°. This means that the effective angle of incidence for diffuse radiation (DHI) is 45°, which does not change throughout the day for isotropic skies. However, skylight is not strictly isotropic but has contributions from all parts of the sky, especially horizon-brightening and circumsolar radiation; however, the effective incident angle for diffuse is never far from the nominal 45° and changes little throughout the day.

Measuring the white- and black-sky albedos separately under clear-sky conditions is difficult. However, a reasonable estimate of the black-sky albedo can be obtained from the measurement of albedo and irradiance on cloudy-sky and clear-sky days that occur just a few days apart. The measurements need to be separated by no more than a few days so that the surface does not change appreciably between the measurements. The estimate of the black-sky albedo is based on the reasonable assumption that the cloudy-sky albedo is about the same as the clear-sky albedo. With measurements of direct sunlight and diffuse skylight on the clear day and diffuse skylight on the cloudy day, the white-sky albedo on the cloudy day can be calculated and can be used to subtract the diffuse skylight albedo on the clear day to derive the black-sky albedo on the clear day. In equation form

$$albedo_{white\_sky} = \frac{DHI \uparrow}{DHI \downarrow}(on\_cloudy\_day) \tag{9.2}$$

$$albedo_{black\_sky} = \frac{GHI \uparrow - albedo_{white\_sky} \cdot DHI \downarrow}{DNI \downarrow \cdot \cos(sza)}(on\_clear\_day) \tag{9.3}$$

Where the up arrow indicates reflected irradiance and the down arrow indicates downward irradiance.

9.5   ALBEDO MEASUREMENTS

9.5.1   BROADBAND ALBEDO

Broadband albedo measurements are best made with two broadband pyranometers of the same make and model mounted with one looking upward and one looking downward (see Figure 9.1). It is important to use a pyranometer with a good cosine response for the downwelling measurement. It is less critical for the upwelling (reflected)

FIGURE 9.8   (See color insert.) Photograph of the SURFRAD albedo tower near Sioux Falls, South Dakota. Note the shields that ensure the direct sunlight will not reach the upwelling detector even for minor horizontal leveling errors.

measurement. For both instruments the horizontal alignment is critical, especially for the low sun angle albedos where direct beam radiation could be detected in the upwelling measurement if alignment is not adequate. A shade ring around the upwelling pyranometer should be used to ensure that direct sun is never able to irradiate this pyranometer. Figure 9.8 illustrates the solar shading used in SURFRAD for all of their installations; this is at the Sioux Falls, South Dakota, site with the solar measurement on the right, which is the south side of the tower. The instrument on the left, which is also shielded, is for the upwelling infrared. For the albedo measurement the typical shield used for downwelling measurements is not used, and the inner shield is painted with flat black paint to reduce scattered solar light from the inside of the shield. The height chosen for the albedo measurements for the SURFRAD sites is 9 m. The height of the albedo measurement should be chosen so that the surface area seen at 45 degrees to the nadir direction represents the surrounding surface area. This is important since the highest contribution to the upwelling radiation is from this direction. However, that said, it is best if the area under the measurement is similar out to about four times the measurement height.

An ideal albedo measurement would be made using one instrument rotated to periodically face upward and downward to guarantee the same broadband spectral response. Of course, the downwelling solar radiation would need to be stable during this measurement sequence and the time-response of the pyranometer would need to be consistent with the data acquisition intervals.

If these instruments were to be used only for albedo measurements, the only calibration needed would be to force their agreement when both are pointed upward since albedo is a relative measurement. This ideal measurement ignores one important point: the thermal offset issue. For the pyranometer used to measure upwelling irradiance, the thermal offset is almost absent. When it looks at the clear sky, the thermal offset is negative and must be corrected as discussed in Chapter 4 and Chapter 5.

9.5.2 SPECTRAL ALBEDO

Spectral albedo is becoming more important for climate science and satellite remote-sensing research; however, there are few options for all-weather measurements. As seen in Figure 9.3 through Figure 9.6, the spectral behavior of albedo varies greatly, but high spectral resolution is not necessary to capture the key features of the wavelength dependence. Because the wavelength dependence is not highly structured like the solar spectrum, it is possible to judiciously sample the albedo spectrum with discrete filters and interpolate, for example, using the technique described in McFarlane et al. (2011). The Yankee Environmental Systems, Inc. seven-channel rotating shadowband radiometer (MFR-7) is an all-weather instrument that can be used for the discrete filter measurement in the visible and near-infrared. For UV filter measurements there are several choices including the NILU-UV (NILU Products AS), the GUV-511 and GUV-541 (Biospherical Instruments, Inc.), and the UVMFR-7 (Yankee Environmental Systems, Inc.). The EKO Instruments Co., Inc. MS-700 is an all-weather grating spectrometer that measures between 350 and 1050 nm with continuous resolution of 3.3 nm. None of these instruments measure beyond 1100 nm, and commercial all-weather instruments that measure beyond 1100 nm do not currently exist. In purchasing these instruments, response stability (2% per year), temperature control (±2°C) or correction, and cosine response (better than 10% out to incidence angles of 80°) are the most important qualities to consider.

QUESTIONS

1. How is the broadband albedo of a surface measured?
2. Are albedos wavelength independent; that is, are they the same at all wavelengths? Name a surface whose albedo is almost wavelength independent in the visible between 400 and 700 nm.
3. Does the white-sky albedo change with sun angle? Does the black-sky albedo?
4. Which surface type shows the greatest spectral diversity? A field of alfalfa? A field of wheat stubble?
5. What would cause the ocean surface albedo to become higher?
6. Can you think of a way to measure albedo quickly with a single sensor?

REFERENCES

Ackerman, T. and G. M. Stokes. 2003. The atmospheric radiation measurement program. *Physics Today* 56:38–45.

Augustine, J. A., J. J. DeLuisi, and C. N. Long. 2000. SURFRAD—A national surface radiation budget network for atmospheric research. *Bulletin of the American Meteorological Society* 81:2341–2357.

Bowker, D. E., R. E. Davis, D. L. Myrick, K. Stacy, and W. T. Jones. 1985, June. *Spectral reflectances of natural targets for use in remote sensing studies.* NASA Reference Publication RP-1139. NASA Scientific and Technical Information Branch.

McFarlane, S., K. Gaustad, C. Long, E. Mlawer, and J. Delamere. 2011. Development of a high resolution spectral surface albedo product for the ARM Southern Great Plains Central Facility. *Atmospheric Measurement Technology*, 1713–1733.

Minnis, P., S. Mayor, W. L. Smith, Jr., and D. F. Young. 1997. Asymmetry in the diurnal variation of surface albedo. *IEEE Transactions on Geoscience and Remote Sensing.* 35:879–891.

Nicodemus, F. 1965. Directional reflectance and emissivity of an opaque surface. *Applied Optics* 4:767–775. doi:10.1364/AO.4.000767

Pu, R., S. Ge, N. M. Kelly, and P. Gong. 2003. Spectral absorption features as indicators of water status in coast live oak (Quercus agrifolia) leaves. *International Journal of Remote Sensing* 24:1799–1810.

Stokes, G. M. and S. E. Schwartz. 1994. The atmospheric radiation measurement (ARM) program: Programmatic background and design of the cloud and radiation test bed. *Bulletin of the American Meteorological Society* 75:1201–1221.

Yang, F., K. Mitchell, Y.-T. Hou, Y. Dai, X. Zeng, Z. Wang, and X.-Z. Liang. 2008. Dependence of land surface albedo on solar zenith angle: Observations and model parameterization. *Journal of Applied Meteorology and Climatology* 47:2963–2982. doi:10.1175/2008JAMC1843.1

Zygielbaum, A. I., A. A. Gitelson, T. J. Arkebauer, and D. C. Rundquist. 2009. Non-destructive detection of water stress and estimation of relative water content in maize. *Geophysical Research Letters* 36:L12403, 4 pp. doi:10.1029/2009GL038906

10 Infrared Measurements

The world Infrared Standard Group has been operating continuously since 2003, showing an excellent relative stability between individual instruments of ± 1 W m^{-2}.

Julian Gröbner
(2009)

10.1 INTRODUCTION

A body with a temperature above 0 K emits radiation. For the earth's surface and atmospheric temperatures, this radiation is in the infrared (IR). The Stefan–Boltzmann law states that the power per unit area L emitted by a body of absolute temperature T is

$$L = \varepsilon \sigma T^4 \tag{10.1}$$

In this equation σ is the Stefan–Boltzmann constant, equal to 5.670×10^{-8} W m^{-2} K^{-4}, and ε is the emissivity of the body. If the emitting body is a *gray body*, then the emissivity is < 1; for a *black body*, ε is equal to 1. For example, if a pyrgeometer, which measures infrared radiation (see Section 10.2) is used to look at the surface of a parking lot, the temperature of the tarmac can be calculated very accurately, assuming an emissivity of 1. When looking at a clear sky, the temperature of the sky is more difficult to measure because there is no longer a near-perfect black body or gray body with a known emissivity. Figure 10.1 illustrates black-body spectral irradiance in power per unit area per unit wavelength at five temperatures: 250, 257, 275, 299 and 300 K. Looking at the two clear-sky curves in this figure, the spectral distribution of infrared radiation does not follow any of these curves at all wavelengths but does follow the 257 and 299 K black-body distributions at some of the wavelengths. So, what is going on in the sky? Parts of the infrared spectrum are completely absorbed by the molecules in the earth's atmosphere; water vapor and carbon dioxide are especially significant absorbers. If there is little molecular absorption or there is only weak broadband extinction caused by aerosols, then the infrared spectrum will have "windows" that allow outgoing infrared radiation to escape. The "299 K Sky" plot in Figure 10.1 is for a modeled tropical clear sky. The "257 K Sky" plot in the same figure is for a modeled subarctic clear sky. Both tropical and subarctic clear skies exhibit significantly reduced irradiance at around 10 μm. This hole, or window if one is thinking of IR radiation from the earth passing through the atmosphere, is caused by the reduction in the number of molecules that emit, or absorb, radiation at these wavelengths, mainly, water vapor. Note that the subarctic clear sky is very transparent around 10 μm and the 20 μm region is semitransparent, whereas the latter region is black-body-like for the tropical sky. If the sky is filled with thick clouds, the spectrum fills in (the windows are no longer there) and the irradiance from the sky takes on the distribution of a black body with a temperature that is the effective temperature of the bottom of the clouds as viewed from the surface of the earth.

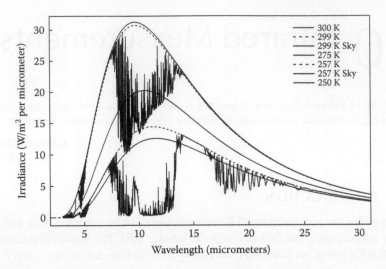

FIGURE 10.1 Black-body irradiances for true black bodies at the temperatures in the legend plus clear-sky irradiances for a tropical sky and a subarctic sky. Note the 10 mm window for both skies and the semitransparent window for the subarctic sky in the 20 mm region.

Ground-based, spectrally resolved infrared measurements are not discussed in this book. The best measurements of spectral infrared are made using Fourier transform spectrometers. These measurements are radiance measurements as opposed to the irradiance measurements that are the focus of this book. The University of Wisconsin's Space Science and Engineering Center has led the way in developing ground-based spectral infrared measurements (Knuteson et al., 2004a, 2004b). Feltz, Smith, Howell, Knuteson, Woolf, and Revercomb (2003) described continuous retrievals of temperature and moisture that are possible using these measurements; however, a wealth of information on other atmospheric trace species such as carbon dioxide and ozone can be found in the literature.

In this chapter we first discuss broadband instruments for measuring infrared radiation, including how they function and three similar but different equations that have been developed to produce measured infrared irradiance from the same inputs. Next, laboratory calibrations using a black-body calibrator are discussed along with two different methods for using the same black-body to obtain calibrations. In Section 10.4, recent developments in calibrating pyrgeometers are discussed. These methods use outdoor measurements against a standard group of well-calibrated pyrgeometers for the derivation of the governing equations. Section 10.5 outlines information about other pyrgeometer manufacturers than the ones presented in this chapter. The final section contains some operational considerations for using a pyrgeometer and estimates uncertainties that one may obtain in routine operations.

10.2 PYRGEOMETERS

Downwelling and upwelling infrared radiation measurements in the broadband are made with instruments called pyrgeometers (see, e.g., Figure 3.17). Broadband instruments use a thermopile detector covered by a silicon dome whose underside

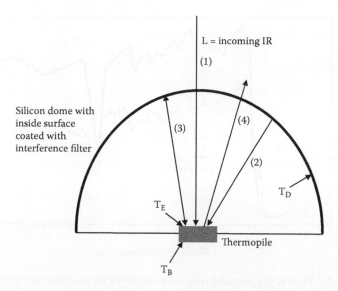

FIGURE 10.2 Schematic of a pyrgeometer thermopile detector under a hemispherical interference filter designed to block solar (short wave) radiation. Typically, the dome and pyrgeometer body temperatures are measured (TD and TB). The thermopile signal depends on the difference in temperatures between the exposed surface T_E and the pyrgeometer body temperature T_B. Generally, the body temperature is warmer than the exposed surface because the exposed surface cools with exposure to the dome cooled by exposure to the cooler sky. The pyrgeometer equation is derived by equating the three incoming contributions to the exposed thermopile surface (1–3) to the one outgoing term from this surface (4). (See Equation 10.2.)

is coated with an interference filter (Figure 10.2). The shortwave cutoff of the transmission of the filter is generally in the 3–5 μm range with a longwave cutoff at 30–50 μm. Sample transmission curves for the Eppley Laboratory, Inc. model PIR and the Kipp & Zonen model CG 4 are given in Figure 10.3. Obviously they are not spectrally flat, and their short wavelength cutoffs are similar but not the same. What this implies with regard to infrared measurement under changing infrared spectral conditions will be discussed in Section 10.4.

In this discussion of pyrgeometer measurements, it is assumed that the dome temperature and the thermopile junction body temperature of the Eppley PIR are measured and that the less accurate practice of using a battery-powered circuit to compensate for the radiation loss from the exposed junction of the thermopile is not used. The equation for determining the incident infrared irradiance on a pyrgeometer is derived by adding the three most significant contributing terms to the received flux on the exposed junction of the thermopile (see Figure 10.2) and subtracting the term that represents the outgoing flux from the exposed junction of the thermopile; this summation of four terms is equal to the net flux at the exposed junction of the thermopile. The three incoming terms are (1) the incoming infrared that is transmitted through the dome and is absorbed at the exposed junction of the thermopile, (2) the infrared that is emitted from the dome and is absorbed by the exposed junction of the thermopile if they are not at the same temperature, and (3) the infrared emitted by the exposed junction of the

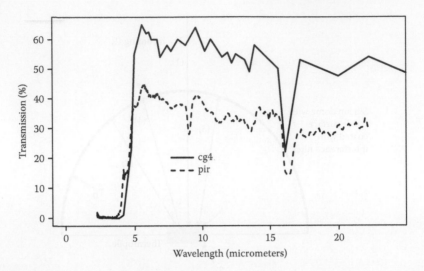

FIGURE 10.3 Typical pyrgeometer transmission curves for two different manufacturers' domes. Note that the cut-on wavelengths vary somewhat and depend on temperature to some extent.

thermopile and reflected back to the exposed junction of the thermopile after reflection from the dome. The outgoing flux (4) is the IR radiation emitted from the exposed junction of the thermopile that passes through the dome. Note that the exposed junction of the thermopile is the emitting surface; however, there is no temperature measurement at the exposed junction, but there is a temperature measurement at the unexposed junction of the thermopile, referred to here as the body (B) temperature. The temperature of the exposed junction is calculated using $T_E = T_B + \alpha V$, where α is a product of the Seebeck coefficient for the thermopile type, the number of junctions, and the thermopile efficiency factor. V is the voltage produced by the temperature difference between the unexposed and exposed junctions of the thermopile.

Derivations of the pyrgeometer equation to calculate the incident infrared flux are usually expressed in terms of the measurable dome and body temperatures; therefore, only the highest-order term that includes the product of αV and T_B^3 is kept since αV is very small. For comparison, a large temperature difference between the exposed and hidden junctions of a thermopile would be on the order of 0.5 K, while the dome and case temperatures typically are in the range 260 to 310 K for midlatitudes.

Albrecht, Poellot, and Cox (1974) derived the basic equation for how an Eppley PIR is used to measure incoming infrared radiation. Their Equation (1) as given in Albrecht and Cox (1977) using modified symbols is

$$L = U_{emf}\left(c_1 + c_2 T_B^3\right) + \varepsilon_o \sigma T_B^4 - k\sigma\left(T_D^4 - T_B^4\right) \tag{10.2}$$

This has an analytic form similar to Philipona, Fröhlich, and Betz's (1995) Equation (11):

$$L = \frac{U_{emf}}{C}\left(1 + k_1 \sigma T_B^3\right) + k_2 \sigma T_B^4 - k_3 \sigma\left(T_D^4 - T_B^4\right) \tag{10.3}$$

In these equations L is the measured longwave irradiance, T_D is the dome temperature, T_B is the unexposed thermopile junction (body) temperature, and U_{emf} is the measured voltage across the thermopile. Albrecht and Cox (1977) considered $c_1 \gg c_2 T_B^3$ in Equation 10.2 and ignored the T_B^3 dependence. Therefore, the most common form of Albrecht and Cox's formula is written

$$L = \frac{U_{emf}}{c} + \sigma T_B^4 - k\sigma\left(T_D^4 - T_B^4\right) \tag{10.4}$$

where ε_o in Equation 10.2 is assumed to be unity, and c is the responsivity factor, a constant. Reda, Hickey, Stoffel, and Myers (2002) derived a form of the pyrgeometer equation that used the estimated thermopile exposed junction temperature T_E as given above rather than T_B

$$L = k_0 + k_1 U_{emf} + k_2 \sigma T_E^4 - k_3 \sigma\left(T_D^4 - T_E^4\right) \tag{10.5}$$

The addition of an offset k_0 is unique among pyrgeometer equations and was added by Reda et al. (2002) to account for the offset of the pyrgeometer.

10.3 CALIBRATION

The Eppley PIR is calibrated at the Eppley Laboratory, Inc., using a circulating water bath black body that can be held at a constant, uniform temperature. The PIR is raised into the Parsons black-painted recessed hemisphere of this black body. After about 1 minute the instrument reaches equilibrium at 5°C and later at 15°C. From Equation 10.4, c can be obtained from

$$L = \sigma T_{BB}^4 = \frac{U_{emf}}{c} + \sigma T_B^4 \Rightarrow c = \frac{U_{emf}}{\sigma\left(T_{BB}^4 - T_B^4\right)} \tag{10.6}$$

where the thermopile and dome temperatures are assumed to be equal or negligibly different once equilibrium is obtained, and T_{BB} is the temperature of the laboratory black-body. The values of c at 5 and 15°C are averaged for the final determination of the calibration. The value of k in Equation 10.4 is not provided by Eppley but is assumed to be 4 for KRS-5 domes; however, a typical value of k equal to 3.5 is suggested for the silicon domes now used for the Eppley PIR (Thomas Kirk, personal communication).

Laboratory calibration black bodies, not unlike the Eppley black body, have been used in a procedure that determines both the thermopile responsivity c and the dome correction factor k. One technique that yields both c and k in Equation 10.4 is described in Albrecht and Cox (1977). Their black-body calibrator is made from a copper cylindrical block that has a black recessed cone. The calibrator is chilled to a low temperature overnight, that is, lower than the environment where it will be used, and then allowed to warm to room temperature throughout the several-hour calibration process. Before placing the pyrgeometer upside down in the cone, the dome of each pyrgeometer to be calibrated is first heated so that it is a few degrees above the pyrgeometer body temperature. Measurements of the black-body calibrator temperature are made in two places to establish the black body's temperature and uniformity; typical agreement is within

0.2°C. Thermopile body and dome temperatures and the thermopile voltage output are measured simultaneously with the calibrator temperature. The dome will cool to have a lower temperature than the pyrgeometer body passing through a point where the dome and body temperatures are the same. This is repeated several times during calibration as the black body warms to room temperature over a few hours. The thermopile sensitivity c can be determined from Equation 10.6 at the body and dome temperature-crossing points by averaging these, or it can be calculated from the slope of a plot of the thermopile voltage U_{emf} versus $\sigma(T_{BB}^4 - T_B^4)$ at the crossing points. With c determined, k can be calculated from an average of multiple determinations of k using

$$k = \left[\sigma\left(T_{BB}^4 - T_B^4\right) - \frac{U_{emf}}{c} \right] \bigg/ \sigma\left(T_D^4 - T_B^4\right) \qquad (10.7)$$

or from the slope of $\sigma(T_{BB}^4 - T_B^4) - \frac{U_{emf}}{c}$ versus $\sigma(T_D^4 - T_B^4)$. Either method permits the calculation of the uncertainty of c and k from the standard deviation of the multiple determinations of k or the standard deviation in the slope of the fit.

At the National Oceanic and Atmospheric Administration (NOAA) this same black body is operated in much the same way to obtain the basic data; however, c and k are estimated by using all of the simultaneous measurement points. Equation 10.4 is used in a multivariate linear regression to solve for the two parameters. Dutton (1993) demonstrated that the thermopile coefficient obtained in this way is within 2% of the Eppley thermopile coefficient provided by the manufacturer. Further, in a round-robin comparison of five Eppley PIRs sent to 11 laboratories, Philipona et al. (1998) identified six laboratory calibrations of the thermopile coefficient, including those of the NOAA and the Eppley laboratories, that agreed to within 2% of the median value; the other five laboratory calibrations showing inconsistent behavior with mean differences all greater than 5% and many larger than 10% different from the median.

Thus far the discussion has been about the Eppley PIR. There is a different pyrgeometer designed by Kipp & Zonen, the CG 4, for which the calibration strategy is different. A photo of the Kipp & Zonen CG 4 is shown in Figure 10.4. The CG 4 could be calibrated using a black body, but the manufacturer asserts that its outside calibration using sky irradiance is superior. The CG 4 to be calibrated is compared to a reference CG 4 that is calibrated in Davos, Switzerland, at the World Radiation Center against the interim World Infrared Standard Group (WISG), which is discussed in the following section. The manufacturer contends that the thermal contact between the dome and the body of the CG 4 is such that there is never a significant difference in temperatures between dome and the body of the CG 4. This means that the thermopile sensitivity can be easily calculated using Equation 10.4 with the last term set to zero and L set to the irradiance measured by the reference CG 4.

10.4   IMPROVED CALIBRATIONS

Rolf Philipona, formerly of the World Radiation Center, and his colleagues have been responsible for most of the improvement in infrared radiometry in the last several years. Foremost among his contributions was the development of an infrared sky-scanning radiometer (Philipona, 2001). The sky scanner is calibrated by pointing a

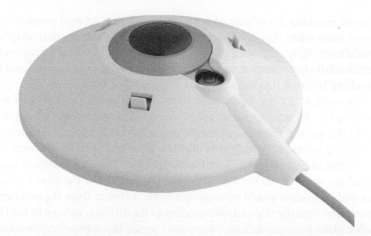

FIGURE 10.4   The Kipp & Zonen CG 4 is constructed to minimize the difference in temperature between the dome and pyrgeometer body. This has the effect of removing the last term of Equations 10.3 to 10.5 and making the measurement problem simpler.

6° field of view radiometer into a well-characterized black body, then scanning the sky at 32 points (four elevation and eight azimuth positions) with a dwell time at each position of 30 seconds, and then pointing into the black body at the end of the sky scan. The 32-point scan takes about 24 minutes to complete; therefore, a stable sky is required. The instrument is unfiltered (i.e., designed as an absolute instrument with no optical surface affecting incoming radiation) so it can be used only at night. The integral of the radiance measurements over the sky provides an absolute irradiance calibration for pyrgeometers. Two International Pyrgeometer and Absolute Sky-scanning Radiometer Comparisons (IPASRCs) have been held. IPASRC-I (Philipona et al., 2001) was at midlatitudes in the early autumn at the U.S. Department of Energy Atmospheric Radiation Measurement (ARM) Program's Radiometer Calibration Facility at the Southern Great Plans site in northern Oklahoma. Measurements were obtained over a range of 260 to 400 Wm^{-2} downwelling infrared for clear to cloudy skies and irradiances typical of late summer temperatures. IPASRC-II was conducted at the ARM North Slope of Alaska site near Barrow, Alaska, during late winter (Marty et al., 2003). The longwave irradiances varied between 120 and 240 Wm^{-2} with frequent clear skies. The two comparisons included 15 and 14 pyrgeometers, in IPASRC-I and -II with many, but not all pyrgeometers, the same.

   In both comparisons better agreement among the pyrgeometers for outdoor measurements was realized using the calibrations based on the sky-scanning radiometer than using calibrations performed on all of the instruments using a good black-body calibration system. In turn, the uniform black-body calibration on the instruments provided better agreement among the pyrgeometers than using calibrations that came with the instruments either from the manufacturer or their own laboratories. In these studies the measurements were also compared to infrared models of the atmosphere including MODTRAN (Berk et al., 2000) and the line-by-line radiative transfer model (LBLRTM) (Clough, Iacono, and Moncet, 1992). Both models were slightly low with respect to pyrgeometer and scanning radiometer measurements; however, neither model included the effects of aerosols because no aerosol measurements in the infrared were made that could be used as inputs to these models. Any aerosol effect would have likely improved the agreement, but it is not clear by how much. In summary, black-body-calibrated, sky-scanner-calibrated measurements, and models of downwelling infrared irradiance agreed to within 1–2 Wm^{-2} at night.

The improvement in the agreement when the pyrgeometer calibrations were obtained using the sky scanner versus the black body can be plausibly explained when we consider the spectral response of the pyrgeometer and the spectral distribution of the downwelling infrared. From Figure 10.1 it is clear that spectral infrared sky irradiance is not that of a black body. From Figure 10.3 it is clear that the filters that transmit infrared are not uniform with wavelength. The product of the filter transmission and black-body emission would be expected to be different from the product of the filter transmission and the sky emission leading to the differences seen in the IPASRC results when black-body calibrations were used versus sky-scanner calibrations.

The previous discussion of the absolute sky scanner and improvement in infrared measurements made with pyrgeometers using sky-scanner calibrations suggested a methodology to improve the standard for infrared measurements. The interim World Infrared Standard Group (WISG) for the pyrgeometer infrared standard (http://www. pmodwrc.ch/pmod.php?topic=irc) consists of four pyrgeometers: an Eppley PIR and a Kipp & Zonen CG 4 that participated in both IPASRCs plus another of each type of these pyrgeometers that were calibrated using these two IPASRC participant pyrgeometers.

The current best practice for calibrating pyrgeometers is by outdoor comparison to this standard group (WISG) or, alternatively, by outdoor comparison to pyrgeometers that have been calibrated at the World Radiation Center in Davos referenced to the WISG. Regressions may be performed using any of Equations 10.3 to 10.5. Alternatively, Reda et al. (2006) demonstrated a pyrgeometer calibration method where the nighttime atmosphere is used as the infrared reference.

Julian Gröbner, who is now responsible for the WISG at the World Radiation Center, is developing an absolute pyrgeometer that has been compared to WISG (Groebner, 2012). The comparisons are quite close and indicate that there may be room for slight improvements to the infrared standard, as it exists. Ibrahim Reda, at the National Renewable Energy Laboratory, in collaboration with colleagues at the National Institute of Standards and Technology and Labsphere, has developed a new Absolute Cavity Pyrgeometer (ACP) to measure IR irradiance with traceability to the international system of units (Reda, Zeng, Scheuch, Hanssen, Wilthan, Meyers et al., 2012). The radiation measurement community recognizes the importance of combining the design experiences of multiple sources to ultimately improve the international measurement references.

10.5   OTHER PYRGEOMETER MANUFACTURERS

There are other pyrgeometer manufacturers, but the previous discussion covers the two basic designs for a commercially available pyrgeometer. Other commercial pyrgeometers known to the authors are the Kipp & Zonen CG 3, which differs from the CG 4 in that the field of view of the CG 3 is 150° as opposed to 180°. Hukseflux makes the IR02 pyrgeometer that is based on a PT100 temperature sensor with a 150° field of view (http://www.hukseflux.corn/products/radiationMeasurement/ir02_pyrgeometer.html). EKO Instruments Co., Ltd. makes the MS-202 pyrgeometer, which appears to function much like the Eppley PIR including circuitry to compensate for radiation loss from the exposed junction of the thermopile (the σT_B^4 in Equation 10.4). However, thermistors or, alternatively, platinum resistance thermometers, to measure the body and dome temperatures are available as options.

10.6 OPERATIONAL CONSIDERATIONS

Figure 10.5 is a photograph of three Eppley PIRs in ventilated housings. The three pyrgeometers are mounted on an automatic solar tracker to keep the instruments shaded from the direct solar beam at all times of the day. This is the preferred way to operate a pyrgeometer, ventilated and shaded (McArthur, 2004). The ventilation, with optional heating, helps prevent dew and frost formation and keeps the dome cleaner than it would be without ventilation. Shading with a tracking ball or disk minimizes the effects of shortwave leaks associated with the interference filter cut-on wavelength and with possible pinholes in the interference filter, plus the heating of the dome by sunlight is kept to a minimum. Operating without a shaded pyrgeometer will, as expected, result in more uncertainty for the daytime measurements.

Two (for CG 4-type instruments) or three (for PIR-type instruments) voltages need to be measured from a pyrgeometer, and one (CG 4) or two (PIR) voltages need to be supplied to excite the thermistor circuits used for temperature measurements.

The thermopile voltage is produced by a temperature difference between the exposed and unexposed junctions of the thermopile. This cumbersome language

FIGURE 10.5 **(See color insert.)** Three Eppley PIR pyrgeometers operated in the optimum manner; the instruments are continuously shaded from direct sunlight using an automatic tracker and are placed in heated ventilators to minimize dew formation and dust deposition on the instrument domes.

(exposed and unexposed) is substituted for the usual terms *hot* and *cold* junctions because of the reversal of the roles these junctions play in the pyrgeometer relative to their usual use in temperature measurements. The unexposed junction is at the body temperature of the pyrgeometer, and the exposed junction is generally at a lower temperature because it is cooled by exposure to the dome that radiates to the colder sky. This results in a negative voltage output from the thermopile most of the time. In fact, a good test to verify that the thermopile is connected correctly is to briefly cover the dome with your warm hand and watch the signal become positive.

The thermistor circuit typically used (rarely are platinum resistance thermometers used for these measurements) to measure the temperature of the body and dome may differ in the thermistors used, but most use a configuration similar to that in Figure 10.6. The thermistor's resistance has a reproducible behavior with temperature. Slight imperfections in thermistor resistance versus temperature are included in the calibration coefficients (k's) in Equations 10.3 to 10.5. The equations and algebra in

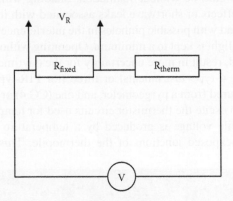

Given that

$$V = i \times R_{fixed} + i \times R_{therm}$$

where i is the current through the low temperature coefficient fixed resistor R_{fixed} and the temperature dependent variable resistor R_{therm}, it can be shown with a little algebra that:

$$R_{therm} = R_{fixed} \left(\frac{V - V_R}{V_R} \right)$$

Therefore, the resistance of the thermistor placed in contact with the body of the pyrgeometer or the dome can be calculated by measuring the voltage drop across the fixed resistor. Consequently, this resistance can be related to the temperature using the Steinhart-Hart equation:

$$1/T = C1 + C2 \times \ln (R_{therm}) + C3 \times \ln (R_{therm})^3$$

where C_1, C_2, and C_3 are the Steinhart-Hart coefficients for the thermistor type used in the pyrgeometer.

FIGURE 10.6 One method to derive the dome or pyrgeometer body temperatures using a thermistor circuit.

Figure 10.6 permit the temperature of the dome and body to be determined. For the CG 4 there is only one thermistor for the body temperature, since the manufacturer asserts that the dome and case are nearly the same temperature and does not normally supply a dome thermistor. In support of this claim, Marty et al. (2003) suggested that the dome and body temperature difference term is 20 times smaller for the CG 4 than it is for an Eppley PIR. In an unpublished study, Ellsworth Dutton (personal communication) finds that more than 95% of these dome–body corrections for the PIR are smaller than 30 Wm^{-2}; this suggests that the CG 4 may have a bias, but it is less than 1.5 Wm^{-2}.

Philipona et al. (2001) and Reda et al. (2012) provide guidance on the uncertainty of pyrgeometer measurements. The uncertainty depends on the provider of the calibration, the formula used to reduce the data to irradiances, and whether a calibration is from a black body or from a comparison to the WISG infrared standard. Furthermore, daytime uncertainties are generally twice the nighttime uncertainties. Philipona et al. (2001) show that the nighttime 95% level of confidence uncertainties range from 0.8 to 4.6 Wm^{-2} and daytime uncertainties range from 1.6 to 8.2 Wm^{-2}. The uncertainty of the WISG needs to be included in the total estimated uncertainty, which is estimated to be approximately ±1.7 Wm^{-2} (Julian Gröbner, head of the infrared radiometry section of the World Radiation Center, personal communication). Adding these uncertainties as the square root of the sums of the square implies nighttime uncertainties between 1.9 and 4.9 Wm^{-2} and daytime uncertainties between 2.3 and 8.4 Wm^{-2} depending on the calibration procedure followed.

QUESTIONS

1. What distinguishes a black body from a gray body?
2. At about what central wavelength is the atmospheric window that keeps a clear atmosphere from resembling a perfect black body?
3. What is the shortest wavelength that typical pyrgeometer domes transmit?
4. What is the current working calibration standard for broadband infrared radiation?
5. Why is the thermopile signal from a pyrgeometer only rarely positive?
6. Why are two temperature measurements required for the Eppley PIR and only one for the Kipp & Zonen CG 4?
7. A human body should emit like a black body at what temperature?
8. Why do the best daytime measurements of infrared use a sun-shaded instrument?

REFERENCES

Albrecht, B. and S. K. Cox. 1977. Procedures for improving pyrgeometer performance. *Journal of Applied Meteorology* 16:188–197.

Albrecht, B., M. Poellot, and S. K. Cox. 1974. Pyrgeometer measurements from aircraft. *Review of Scientific Instruments* 45:33–38.

Berk, A., G. P. Anderson, P. K. Acharya, J. H. Chetwynd, L. S. Bernstein, E. P. Shettle, M. W. Matthew, and S. M. Adler-Golden. 2000. *Modtran4 user's manual*. Air Force Res. Lab., Space Vehicle Dir., Air Force Mater. Command, Hanscom Air Force Base, MA.

Clough, S. A., M. J. Iacono, and J.-L. Moncet. 1992. Line-by-line calculations of atmospheric fluxes and cooling rates: Application to water vapor. *Journal of Geophysical Research* 97:15761–15785.

Dutton, E. G. 1993. An extended comparison between LOWTRAN7-computed and observed broadband thermal irradiances: Global extreme and intermediate surface conditions. *Journal of Atmospheric and Oceanic Technology* 10:326–336.

Eppley Laboratory, Inc. 1995. *Blackbody calibration of the precision infrared radiometer, model PIR. Tech. Procedure 05.* Available from Eppley Laboratories, P.O. Box 419, Newport, RI 02840.

Fairall, C. W., P. O. G. Persson, E. F. Bradley, R. E. Payne, and S. P. Anderson. 1998. A new look at calibration and use of Eppley precision infrared radiometers. Part I: Theory and application. *Journal of Atmospheric and Oceanic Technology* 15:1229–1242.

Feltz, W. F., W. L. Smith, H. B. Howell, R. O. Knuteson, H. Woolf, and H. E. Revercomb. 2003. Near-continuous profiling of temperature, moisture, and atmospheric stability using the atmospheric emitted radiance interferometer (AERI). *Journal of Applied Meteorology* 42:584–597. doi: 10.1175/1520-0450

Groebner, J. 2012. A transfer standard radiometer for atmospheric longwave irradiance measurements. *Meteorologia* 49: S105–S111. doi:10.1088/0026-1394/49/2/S105.

Knuteson, R. O., H. E. Revercomb, F. A. Best, N. C. Ciganovich, R. G. Dedecker, T. P. Dirkx, S. C. Ellington, W. F. Feltz, R. K. Garcia, H. B. Howell, W. L. Smith, J. F. Short, and D. C. Tobin. 2004a. Atmospheric emitted radiance interferometer. Part I: Instrument design. *Journal of Atmospheric and Oceanic Technology* 21:1763–1776.

Knuteson, R. O., H. E. Revercomb, F. A. Best, N. C. Ciganovich, R. G. Dedecker, T. P. Dirkx, S. C. Ellington, W. F. Feltz, R. K. Garcia, H. B. Howell, W. L. Smith, J. F. Short, and D. C. Tobin. 2004b. Atmospheric emitted radiance interferometer. Part II: Instrument performance. *Journal of Atmospheric and Oceanic Technology* 21:1777–1789.

Marty, C., R. Philipona, J. Delamere, E. G. Dutton, J. Michalsky, K. Stamnes, R. Storvold, T. Stoffel, S. A. Clough, and E. J. Mlawer. 2003. Downward longwave irradiance uncertainty under arctic atmospheres: Measurements and modeling. *Journal of Geophysical Research* 108:4358–43669. doi:10.1029/2002JD002937

McArthur, L. J. B. 2004. Baseline Surface Radiation Network (BSRN). Operations Manual. WMO/TD-No. 879, World Climate Research Programme/World Meteorological Organization.

Payne, R. E. and S. P. Anderson. 1999. A new look at calibration and use of Eppley precision infrared radiometers. Part II: Calibration and use of the Woods Hole Oceanographic Institution improved meteorology precision infrared radiometer. *Journal of Atmospheric and Oceanic Technology* 16:739–751.

Philipona, R. 2001. Sky-scanning radiometer for absolute measurements of atmospheric longwave radiation. *Applied Optics* 40: 2376-2383.

Philipona, R., E. G. Dutton, T. Stoffel, J. Michalsky, I. Reda, A. Stifter, P. Wendling, N. Wood, S. A. Clough, E. J. Mlawer, G. Anderson, H. E. Revercomb, and T. R. Shippert. 2001. Atmospheric longwave irradiance uncertainty: Pyrgeometers compared to an absolute sky-scanning radiometer, atmospheric emitted radiance interferometer, and radiative transfer model calculations. *Journal of Geophysical Research* 106:28129–28141.

Philipona, R., C. Fröhlich, and C. Betz. 1995. Characterization of pyrgeometers and the accuracy of atmospheric long-wave radiation measurements. *Applied Optics* 34:1598–1605.

Philipona, R., C. Fröhlich, K. Dehne, J. Deluisi, J. A. Augustine, E. G. Dutton, D. W. Nelson, B. Forgan, P. Novotny, J. Hickey, S. P. Love, S. Bender, B. McArthur, A. Ohmura, J. H. Seymour, J. S. Foot, M. Shiobara, F. P. J. Valero, and A. W. Strawa. 1998. The baseline surface radiation network pyrgeometer round-robin calibration experiment. *Journal of Atmospheric and Oceanic Technology* 15:687–696.

Reda, I., J. R. Hickey, J. Grobner, A. Andreas, and T. Stoffel. 2006. Calibrating pyrgeometers outdoors independent from the reference value of the atmospheric longwave irradiance. *Journal of Atmospheric and Solar-Terrestial Physics* 68:1416–1424.

Reda, I., J. R. Hickey, T. Stoffel, and D. Myers. 2002. Pyrgeometer calibration at the National Renewable Energy Laboratory (NREL). *Journal of Atmospheric and Solar-Terrestrial Physics* 64:1623–1629.

11 Net Radiation Measurements

Changes in the atmospheric abundance of greenhouse gases and aerosols, in solar radiation and in land surface properties alter the energy balance of the climate system. These changes are expressed in terms of radiative forcing, which is used to compare how a range of human and natural forces drive warming or cooling influences on global climate.

Intergovernmental Panel on Climate Change (2007)

11.1 INTRODUCTION

Net radiation (R_n) is the algebraic sum of the downwelling broadband solar, the same as GHI, also called shortwave (S↓), the downwelling broadband infrared, also called longwave (L↓), the upwelling broadband shortwave (S↑), and the upwelling broadband longwave radiation (L↑), that is,

$$R_n = S\downarrow - S\uparrow + L\downarrow - L\uparrow \tag{11.1}$$

Net radiation is the energy that evaporates water or sublimates ice, produces sensible heat that raises the temperature of the atmosphere, and heats surface soil or water. A small amount of net energy is used to photosynthesize carbon dioxide and water into organic compounds, especially sugars, and oxygen in plants. Measuring the net energy with the highest accuracy is important because this is the energy that drives weather and climate.

Figure 11.1 plots all of the variables in Equation 11.1 for two of the National Oceanic and Atmospheric Administration Surface Radiation (SURFRAD) Network sites on the same day as a function of the local standard time of the site (http://www.srrb.noaa.gov/surfrad/index.html). The downwelling components are in solid lines, and the upwelling components are dashed lines. Obviously, it was clear at both sites with global horizontal irradiance (GHI) reaching nearly the same maximum values. Since the desert has a higher albedo, the reflected GHI is higher than the vegetative surroundings at Goodwin Creek. Upwelling infrared is higher at both sites because the surface temperature and emissivity are greater than the atmosphere's temperature and emissivity. The difference in upwelling and downwelling infrared at Goodwin Creek is smaller than the difference at Desert Rock because of the higher atmospheric water vapor content at Goodwin Creek (midday relative humidities were 10% and 40%, respectively). At both sites, the net energy is negative at night because the ground is warmer than the atmosphere and there is no solar radiation contribution to the energy budget. The daily integrated net energy is larger at Goodwin Creek than at Desert Rock with higher values during the day and night.

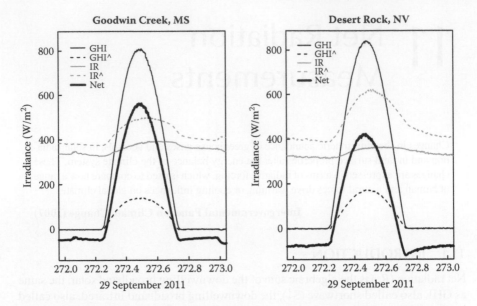

FIGURE 11.1 Plots of all of the components of net radiation at the surface and the calculated net radiation throughout the same clear day at two SURFRAD sites.

Four different types of commercial instruments are used to measure net radiation.
The most basic instrument measures net radiation by determining the thermopile
voltage created by connecting the hot and cold junctions of a thermopile to the top
and bottom absorbing surfaces of the radiometer, respectively. The thermopile is
covered with a dome that transmits most wavelengths of the shortwave and longwave
spectrum, often polyethylene. A second type of net radiometer consists of two sensors that measure downwelling and upwelling radiation separately. The third type
of net radiometer measures net shortwave and net longwave separately. The fourth
method to measure net radiation is to construct net radiometers using four sensors
that separately measure the four components of net radiation. These instruments and
their accuracies will be discussed in the next three sections.

11.2 SINGLE-SENSOR (ALL-WAVE) NET RADIOMETERS

The simplicity of concept and construction of a single-unit-construction net radiometer
would appear to be a good design. The hot junction of the thermopile is connected to
the top of the detector, which receives downwelling shortwave and longwave radiation,
and the cold junction is connected to the bottom of the detector, which receives upwelling shortwave and longwave radiation. The difference in temperature of these matched
receivers is proportional to the net radiation. To reduce the effects of wind, dust, and precipitation a material that transmits both solar and infrared radiation protects the surfaces,
usually polyethylene. Since thermopile responses are sensitive to temperature, it is best
to include a measurement of the thermopile temperature needed to make this correction.
A bubble level is used to ensure the proper horizontal orientation of the instrument, and

Photo by: Tilden Meyers

FIGURE 11.2   This is an example of an all-wave net radiometer. It is a Radiation and Energy Balance Systems (REBS) model Q-7.1. Note the polyethylene domes (note the bottom dome is barely visible), the built-in level, and the air tube that keeps the domes inflated.

a desiccant holder located within the housing is often included to eliminate internal condensation. The cost of this sensor tends to be much lower than the multisensor net radiometers. Examples of commercially available instruments of this type of net radiometer are the Radiation and Energy Balance Systems (REBS) Q-7.1, the Middleton CN1-R, the EKO Instruments MF-11, and the Kipp & Zonen NR Lite2. Figure 11.2 is an example of a single-sensor (all-wave) instrument in its most common form.

The polyethylene domes transmit most of the radiation in the short and long wavelengths. The bubble levels allows for a proper (horizontal) orientation. Desiccated air flows through the support arm. Figure 11.3 is a novel design of the all-wave net

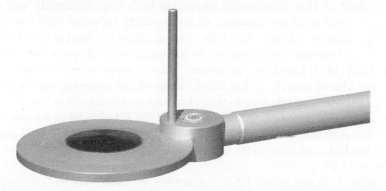

FIGURE 11.3   This net radiometer is an alternate design for an all-wave net radiometer. It is a Kipp & Zonen NR Lite2 that uses black Teflon-coated conical absorbers over the thermopile. (Courtesy of Kipp & Zonen.)

radiometer using a rugged shield that does not require constant airflow to keep the protective shield inflated. The vertical shaft is to discourage birds from perching on this instrument.

There are design and operational issues when using these types of net radiometers. Many sensors of this type use a flexible polyethylene dome to protect the detector from dust, wind, and precipitation yet transmit most of the solar and infrared wavelengths. These domes use a continuous desiccated air or nitrogen stream to keep the domes inflated and free of condensation on the interior surfaces (Figure 11.2). Some manufacturers now use thicker polyethylene domes so that continuous inflation is no longer required, but the air inside continues to need desiccation. Manufacturers suggest that the domes be changed every 3 to 6 months, mainly because they can be easily scratched, and they become brittle and crack under freezing conditions. The domeless NR Lite2 (Figure 11.3), which has a black Teflon conical absorber covering the sensors, is more rugged than the polyethylene domes, but this instrument is the most sensitive all-wave net radiometer to wind speed. All of these sensors show some dependence on wind speed (Brotzge and Duchon, 2000; Smith et al., 1997), and manufacturers provide corrections for wind-speed dependency. Further, Brotzge and Duchon (2000) found that the effects of precipitation are pronounced in the NR Lite. Imperfect Lambertian responses (responses that are proportional to the cosine of the angle of incidence) are not thoroughly discussed in the literature, although Brotzge and Duchon pointed out some problems with the NR Lite, the predecessor to NR Lite2. Calibration is problematic because there is no accepted standard for this measurement. Calibrations are discussed at the end of the chapter.

11.3 TWO-SENSOR NET RADIOMETERS

As mentioned in the introduction, two types of net radiometers employ two separate sensors; however, their mode of operation is fundamentally different. The Schulze-Däke net radiometer is composed of two separate pyrradiometers (2π steradian field of view instruments measuring both long-wavelength and short-wavelength irradiance) that measure all downwelling radiation with one and all upwelling radiation with the other. Polyethylene domes are used to shield the sensors' surfaces from the weather but are thick enough to be self-supporting. Halldin and Lindroth (1992) found this instrument "showed superior performance, and its output was almost entirely within the accuracy of the reference measurements" (p. 762). This was a comparison to the net radiation instruments of the day, many of which still exist in some form. It appears that the Schulze-Däke net radiometer is no longer available. The Schenk 8111 pyrradiometer is similarly configured as the Schulze-Däke net radiometer and is available from Ph. Schenk GmbH Wien (see Figure 11.4).

The Kipp & Zonen model CNR 2 measures net shortwave and net longwave using separate sensors (Figure 11.5). The CNR 2 is a relatively new instrument. It uses glass domes over the shortwave detector and silicon windows over the longwave detector. The infrared sensor field of view is 150°. The Hukseflux RA01 is

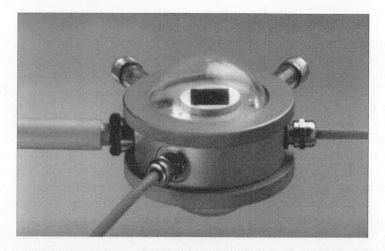

FIGURE 11.4 This is the Schenk 8111. This instrument separately measures the downwelling and upwelling longwave plus shortwave. It is similar to the Schulze-Däke net radiometer that is no longer available. It uses thick polyethylene domes. (Courtesy of Schenk GmbH Wien & Co.)

similar in its configuration with the infrared measurements also limited to a 150° field of view. Blonquist, Tanner, and Bugbee (2009) recently compared the CNR 2 with other types of net radiometers. They concluded that the CNR 2 data were less accurate than four-component net radiometers but were better than the single-sensor net radiometers.

FIGURE 11.5 This instrument is the Kipp & Zonen CNR 2. It has two separate signals like the Schenk 8111, but it measures net shortwave and net longwave. The domes are durable glass and silicon windows. (Courtesy of Kipp & Zonen.)

11.4   FOUR-SENSOR NET RADIOMETERS

The Kipp & Zonen CNR 1 is the oldest of this genre, but the manufacturer no longer offers it. Its replacement from Kipp & Zonen is the CNR 4. Hukseflux's NR01 is the other four-component sensor available for net radiation measurements (see Figure 11.6). The NR01 infrared measurements have a field of view that is 150°, as was the case for the old Kipp & Zonen CNR 1. The 150° field of view instruments use a flat surface for better deposition of the interference filter for the infrared compared with the convex surfaces of the 180° instruments. These systems have the advantage that each sensor can be calibrated separately, and the downwelling short-wave sensor's cosine response can be measured. The sensors provide details of the incident radiation for each of the components of net radiation. These four-component instruments do not, however, use the manufacturers' top-of-the-line pyranometers or pyrgeometers because of the high cost of such a system.

In a very careful study using the current and best pyranometer and pyrgeometer calibration techniques, Michel, Philipona, Ruckstuhl, Vogt, and Vuilleumier (2008) found that a heated and ventilated CNR 1 could be field calibrated to yield net radiation uncertainties smaller than 10% as claimed by the manufacturer. However, using the manufacturer's indoor calibrations led to significantly higher uncertainties. They found that even with field calibrations that an unheated and unventilated CNR 1 could not measure net radiation with uncertainties under 10%.

FIGURE 11.6   This instrument is an example of a four-component net radiometer, the NR1 by Hukseflux. Each of the four components of net radiation is available separately. (Courtesy of Hukseflux USA, Inc.)

11.5 ACCURACY OF NET RADIOMETERS

Blonquist et al. (2009) addressed the question of the accuracy for many of the currently manufactured net radiometers. They found that, in general, the accuracy increases with the cost of the net radiometer. The four-component systems were the best, followed by the two-component systems and then the all-wave net radiometers. However, one problem with this comparison is that there was no true reference. As a proxy for the reference, this study used the average of each separate sensor on the two Kipp & Zonen CNR 1's and the three Hukseflux NR01's. The most significant problem is the lack of a longwave, or broadband infrared, standard. The lack of a broadband infrared standard has been of concern for a long time and has been mentioned early and often in the literature (Blonquist et al., 2009; Brotzge and Duchon, 2000; Halldin and Lindroth, 1992; Ohmura et al., 1998). Recently, an interim standard for the measurement of infrared irradiance was developed at the World Radiation Center in Davos, Switzerland. The interim World Infrared Standard Group (WISG) (http://www.pmodwrc.ch/pmod.php?topic=irc) consists of four pyrgeometers that have been calibrated using an absolute sky-scanning radiometer and black-body characterizations (Marty et al., 2003; Philipona et al., 2001). This infrared standard, along with the World Radiometric Reference (WRR) for solar radiation (Finsterle, 2006), presents an excellent opportunity to develop a true net radiation standard.

11.6 A BETTER NET RADIATION STANDARD

With improvements in the measurement of the diffuse and direct components of solar radiation (Michalsky et al., 2007, 2011) and the establishment of the interim World Infrared Standard Group (WISG) at the World Radiation Center in Davos, Switzerland, for the measurement of infrared (http://www.pmodwrc.ch/pmod.php?topic=irc), there is the potential to develop an accurate working standard for net radiation.

The very best measurement of net radiation requires four first-class horizontal radiometers (two pyranometers for shortwave and two pyrgeometers for infrared) plus a first-class pyrheliometer and a solar tracker with shading for the two downwelling, horizontally mounted radiometers. These measurements are not widely made, but several Baseline Surface Radiation Network (BSRN) sites, which field first-class radiation sensors (Ohmura et al., 1998), are making these measurements. Due to the complexities associated with a 2π steradian field of view and other pyranometer design considerations, the most difficult measurement of shortwave irradiance is the downwelling component. Adding the independently measured direct normal component and the diffuse horizontal irradiance to compute the total horizontal shortwave irradiance minimizes the cosine response error common to current pyranometers. The thermal offset correction for properly calibrated pyranometers must also be made (Michalsky et al., 2007). Further, by shading the downwelling infrared from direct solar radiation, the solar correction to the infrared irradiance measurement is minimized.

A worthwhile study that should be undertaken in light of the recent improvements to solar and infrared radiometry is the characterization of commercial net radiometers using any one of the fully equipped BSRN sites.

11.7 NET RADIOMETER SOURCES

This list identifies known manufacturers that continue to produce net radiometers:

- EKO Instruments Co., Ltd.   (http://www.eko.co.jp)
- Hukseflux Thermal Sensors B.V.   (http://www.hukseflux.com/)
- Kipp & Zonen BV   (http://www.kippzonen.com/)
- Middleton Solar   (http://www.middletonsolar.com/)
- Ph. Schenk GmbH Wien   (http://www.schenk.co.at/schenk/)
- Radiation and Energy Balance Systems (no website; phone: (206) 624-7221)

QUESTIONS

1. What is considered the most accurate method for measuring net radiation at the surface?
2. What are the four different types of net radiometers?
3. Why is polyethylene a popular dome material for net radiometers?
4. Why is the fractional accuracy of net radiation lower than the individual radiation measurements?
5. Why is the measurement of net radiation important?

REFERENCES

Blonquist, Jr., J. M., B. D. Tanner, and B. Bugbee. 2009. Evaluation of measurement accuracy and comparison of two new and three traditional net radiometers. *Agricultural and Forest Meteorology* 149:1709–1721.

Brotzge, J. A. and C. E. Duchon. 2000. A field comparison among a domeless net radiometer, two four-component net radiometers, and a domed net radiometer. *Journal of Oceanic and Atmospheric Technology* 17:1569–1582.

Finsterle, W. 2006. *WMO International Pyrheliometer Comparison IPC-X*. September 26–October 14, Davos, Switzerland, Final Report 91, WMO/TD No. 1320, PMOD/WRC Internal Report, Davos.

Halldin, S. and A. Lindroth. 1992. Errors in net radiometry: Comparison and evaluation of six radiometer designs. *Journal of Oceanic and Atmospheric Technology* 9:762–783.

Kustas, W. P., J. H. Prueger, L. E. Hipps, J. L. Hatfield, and D. Meek. 1998. Inconsistencies in net radiation estimates from use of several models of instruments in a desert environment. *Agricultural and Forest Meteorology* 90:257–263.

Marty, C., R. Philipona, J. Delamere, E. G. Dutton, J. Michalsky, K. Stamnes, R. Storvold, T. Stoffel, S. A. Clough, and E. J. Mlawer. 2003. Downward longwave irradiance uncertainty under arctic atmospheres: Measurements and modeling. *Journal of Geophysical Research* 108:4358–4369. doi:10.1029/2002JD002937

Michalsky, J., E. G. Dutton, D. Nelson, J. Wendell, S. Wilcox, A. Andreas, P. Gotseff, D. Myers, I. Reda, T. Stoffel, K. Behrens, T. Carlund, W. Finsterle, and D. Halliwell. 2011. An extensive comparison of commercial pyrheliometers under a wide range of routine observing conditions. *Journal of Atmospheric and Oceanic Technology* 28:752–766.

Michalsky, J. J., C. Gueymard, P. Kiedron, L. J. B. McArthur, R. Philipona, and T. Stoffel. 2007. A proposed working standard for the measurement of diffuse horizontal shortwave irradiance. *Journal of Geophysical Research* 112:D16112. 10 PP. doi:10.1029/2007JD008651

Michel, D., R. Philipona, C. Ruckstuhl, R. Vogt, and L. Vuilleumier. 2008. Performance and uncertainty of CNR1 net radiometers during a one-year field comparison. *Journal of Oceanic and Atmospheric Technology* 25:442–451.

Ohmura, A., H. Gilgen, H. Hegner, G. Müller, M. Wild, E. G. Dutton, B. Forgan, C. Fröhlich, R. Philipona, A. Heimo, G. König-Langlo, B. McArthur, R. Pinker, C. H. Whitlock, and K. Dehne. 1998. Baseline Surface Radiation Network (BSRN/WCRP): New precision radiometry for climate research. *Bulletin of the American Meteorological Society* 79:2115–2136.

Philipona, R., E. G. Dutton, T. Stoffel, J. Michalsky, I. Reda, A. Stifter, P. Wendling, N. Wood, S. A. Clough, E. J. Mlawer, G. Anderson, H. E. Revercomb, and T. R. Shippert, 2001. Atmospheric longwave irradiance uncertainty: Pyrgeometers compared to an absolute sky-scanning radiometer, atmospheric emitted radiance interferometer, and radiative transfer model calculations. *Journal of Geophysical Research* 106:28129–28141.

Smith, E. A., G. B. Hodges, M. Bacrania, H. J. Cooper, M. A. Owens, R. Chappell, and W. Kincannon. 1997. *BOREAS net radiometer engineering study.* NASA Contractor Report (NASA Grant NAG5-2447), NASA Goddard Space Flight Center, Greenbelt, MD, pp. 51+.

12 Solar Spectral Measurements

I shall without further ceremony acquaint you that in the year 1666, I procured me a triangular glass prism, to try therewith the celebrated phenomena of colors.

Sir Isaac Newton
(1642–1727)

12.1 INTRODUCTION

Solar radiation is not uniformly distributed in wavelength. The spectral distribution of solar radiation as it arrives at the top of the atmosphere is equivalent to a black-body spectrum with an effective temperature of 5777 K that is modified by selective absorption and emissions in the sun's own atmosphere. This radiation is further and significantly modified by scattering and absorption processes as it passes through the earth's atmosphere before it reaches the surface.

This chapter begins with a discussion of the extraterrestrial solar spectrum and its modest variability with the 11-year solar cycle. Its variability with sun–earth distance is covered in Chapter 2. Clear-sky interactions of solar radiation with the atmosphere are also discussed in this chapter including molecular, or Rayleigh, scattering, aerosol scattering and absorption, and absorption by atmospheric gases. Scattering and absorption by clouds is a very complex topic and is not addressed in this handbook. The authors suggest the textbook by Liou (1992) for an introduction to this subject.

Some useful *broadband spectral* measurements including photometric, photosynthetically active radiation (PAR), and ultraviolet photometry are covered in this chapter. A discussion of *narrowband photometry* follows with a focus on its use in retrieving aerosol optical depth and water vapor (the most important clear-sky constituents affecting the spectral distribution of solar radiation at the earth's surface). Instruments for making these measurements are then described with some advice on the specifications to consider if purchasing these instruments. The chapter concludes with brief sections on moderate spectral resolution spectrometry and moderate resolution spectral irradiance models.

12.2 THE EXTRATERRESTRIAL SOLAR SPECTRUM

The solar spectrum covers a wide wavelength range; however, for practical purposes virtually all of the sun's energy that reaches the earth's surface lies between the wavelengths of 280 and 4000 nm. Figure 12.1 is a plot of the spectral irradiance in W/(m^2-nm) versus wavelength in nm (Gueymard, 2004). The integral of this spectral distribution over all wavelengths, if measured just above the earth's atmosphere for

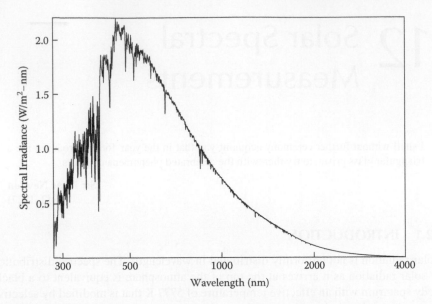

FIGURE 12.1 Extraterrestrial solar spectral irradiance distribution at mean sun–earth distance.

one 11-year solar cycle, yields 1366 W/m^2 at the mean distance between the sun and the earth (149,597,887 km or 92,955,818 miles). This mean distance is defined as one *astronomical unit* (1 AU). Fröhlich (2006) summarized how extraterrestrial irradiance has varied over the last two 11-year solar cycles. The maximum to minimum solar cycle variation averages about 0.1% of the total energy with short-term excursions about three times this amount. The total irradiance monitor (TIM) on the Solar Radiation and Climate Experiment (SORCE) satellite (Rottman, 2005), however, measured a smaller mean value for the integrated spectral irradiance at 1 AU of 1361 W/m^2. To date these differences, which are statistically significant, have not been completely resolved (Kopp and Lean, 2011; Kopp, Fehlman, Finsterle, Harber, Heureman, and Wilson, 2012).

Harder, Fontenla, Pilewskie, Richard, and Woods (2009) used the solar irradiance monitor (SIM) on the SORCE satellite to investigate the variations in portions of the solar spectrum. Separately assessing the variability in the spectral bands from 201–300, 300–400, 400–691, 691–972, 972–1630, and 1630–2423 nm, they found maximum changes of about 0.6 Wm^{-2} with the largest changes in the shortest wavelength bands. The measurements spanned a period of only 4 years or about one-half of the maximum change in total irradiance expected over a solar cycle; however, even twice this change would amount to a small fraction of 1 Wm^{-2} by the time the radiation reaches the surface of the earth. Consequently, it is reasonable to assume that the sun's inherent variability is lower than can reasonably be detected at the surface given the scattering and absorption that occurs within the earth's atmosphere. These interactions with the earth's atmosphere are discussed in the next section.

12.3   ATMOSPHERIC INTERACTIONS

When solar radiation passes through the atmosphere, complex interactions occur depending on the wavelength of the radiation and the composition of the atmosphere at the time. More specifically, solar radiation either passes unscathed to the surface, or it is scattered or absorbed by molecules, aerosols, cloud water droplets, and ice crystals.

12.3.1   RAYLEIGH SCATTERING

Molecular scattering is the elastic scattering of solar radiation by the molecules in the atmosphere first explained by Lord Rayleigh (1871). Many papers have been written describing the scattering of solar radiation by molecules, but it is not a dead subject (Eberhard, 2010). Bodhaine, Wood, Dutton, and Slusser (1999) thoroughly examined the problem of calculating the molecular scattering optical depth as a function of wavelength. *Optical depth*, in general, is a measure of the wavelength dependent extinction (by scattering or absorption) that occurs as a beam of radiation propagates through a medium. It can be defined using

$$I(\lambda)/I_0(\lambda) = e^{-\tau(\lambda)m} \tag{12.1}$$

where I_0 is the strength of the source (e.g., the spectral solar irradiance) before it enters the atmosphere, I is the strength after it has passed through the atmosphere to the surface, τ is the optical depth in a vertical path, m is the amount of atmosphere traversed relative to the vertical path, and λ is wavelength. A good approximation to optical depth by Rayleigh scattering is given by Hansen and Travis (1974):

$$\tau_R = \frac{P}{P_0} 0.008569\lambda^{-4}(1 + 0.0113\lambda^{-2} + 0.00013\lambda^{-4}) \tag{12.2}$$

where wavelength λ is in μm, P is the pressure at the measurement site within the earth's atmosphere in kilopascals, and P_0 is the standard pressure at sea level equal to 101.325 kilopascals (100 kilopascals = 1000 millibars). It is evident from examining Equation 12.2 that Rayleigh scattering is most effective in the shortest wavelengths of the solar spectrum and decreases dramatically at longer wavelengths (approximately as λ^{-4}). It is also evident that at higher elevations there is less Rayleigh scattering because of lower atmospheric pressure.

12.3.2   AEROSOL SCATTERING AND ABSORPTION

Aerosols are always present in the atmosphere. They manifest themselves as the haze often noticed when looking toward the horizon. Generally, aerosols are concentrated near the surface but may appear in layers especially when transported into a region from long distances. When volcanic eruptions occur that are strong enough to inject dust and gases into the stratosphere, such as Mount Pinatubo did in 1991, a stratospheric aerosol layer may persist for years from the sulfur gases that are chemically transformed into aerosols (see Figure 4.9). The dust particles from an eruption, however, are large and therefore are removed from the atmosphere quickly

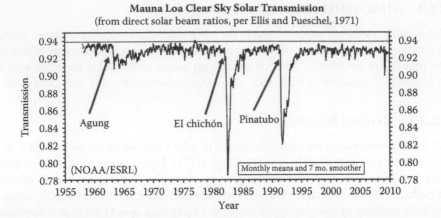

FIGURE 12.2 Transmission record from Mauna Loa Observatory. The large dips in transmission are associated with stratospheric aerosols from three volcanoes. Note that the transmission is suppressed for around three years after the initial eruptions.

because of gravitational settling. Figure 12.2 is the long-term record of transmission from Mauna Loa Observatory. The largest excursions are caused by the three named volcanic eruptions. Lower transmission persists for years after the events.

Aerosols are capable of both scattering and absorbing solar radiation depending on their chemical makeup and on the wavelength of the incident solar radiation. The probability of a photon scattering from an aerosol is called the *single scattering albedo*, ϖ_0. Often ϖ_0 exceeds 0.9 at visible wavelengths for typical continental aerosols. For sea salt aerosols, ϖ_0 is close to 1.0 at visible wavelengths. The term *co-albedo* is used for the quantity $(1 - \varpi_0)$ and is the probability of the photon being absorbed upon encountering an aerosol. Therefore, sea salt aerosols would have a co-albedo of zero in the visible spectral region. Another quantity that describes the scattering of aerosols is the *asymmetry parameter*, often denoted g. The probability of scattering in any given direction is given, approximately, by the Henyey–Greenstein function (Henyey and Greenstein, 1941):

$$p(\cos\theta) = \frac{1}{2}\frac{1-g^2}{(1+g^2-2g\cos\theta)^{3/2}} \tag{12.3}$$

where θ is $0°$ in the direction of photon propagation and $180°$ in the opposite direction. This probability function exhibits these properties:

$$\int_{-1}^{1} p(\cos\theta)d(\cos\theta) = 1 \tag{12.4}$$

$$\int_{-1}^{1} p(\cos\theta)\cos\theta\, d(\cos\theta) = g \tag{12.5}$$

The larger g is, the more the scattering into the forward hemisphere. A g value of zero would imply *isotropic scatter* or equal probability of scattering in all directions. Typical values of g are between 0.5 and 0.7 for continental aerosols. Sea salt aerosols tend to be large and have large values of asymmetry parameter, while anthropogenic pollution aerosols are small with lower asymmetry parameters. The use of these two parameters ϖ_0 and g as inputs to spectral models of solar radiation is discussed in Section 12.6.2.

12.3.3 Gas Absorption

The dominant gas absorber affecting the spectral distribution of the incident solar irradiance in the solar spectrum is water vapor with several major absorption bands in the near-infrared. For example, Figure 2.15 indicates the strong water vapor absorption band in the global horizontal irradiance (GHI) spectrum centered near 940 nm. This is of particular importance to many solar photovoltaic (PV) collectors since the peak response of crystalline silicon PV is near the strong water vapor band around 940 nm. Ozone is another important absorber in the atmosphere. However, its strongest absorption is below 380 nm in the Hartley–Huggins bands. The Chappuis ozone band is broad but has modest absorption centered near 610 nm. The Wulf bands in the near-infrared are even less absorbing. Other gases such as O_2 and CO_2 are ever-present but less significant in the sense that they do not remove a large fraction of the total solar radiation. Nitrogen dioxide (NO_2) is an important gas when air pollution concentrations are high. This gas produces a reddish-brown tinge to the skylight by preferentially absorbing blue solar radiation. The major production of NO_2 is by motor vehicles, which is also the main source of ozone production in cities that exceed the U.S. Environmental Protection Agency (EPA) standards for this molecule.

Optical depths associated with ozone absorption τ_{ozone} may be calculated using

$$\tau_{ozone} = \eta_{ozone} \cdot \alpha_{ozone} \qquad (12.6)$$

where η_{ozone} is the *ozone abundance*, and α_{ozone} is the *ozone absorption coefficient*. Typically, for midlatitudes, the ozone column (the amount of ozone present from the surface to the top of the atmosphere) is in the neighborhood of 300 Dobson units but varies typically between 200 and 400 Dobson units. A Dobson unit (DU) of ozone is equal to 2.687×10^{16} molecules per cm^2. This is 1/1000 of Loschmidt's number, or the number of molecules in a cubic centimeter of a gas under standard temperature and pressure. A typical value of 300 DUs of ozone at the standard temperature of 273.15 K and standard pressure of 101.325 kilopascals would form a column of ozone 0.3 cm high. Current ozone data from NASA's Ozone Monitoring Instrument (OMI) satellite are posted on the NASA website (http://toms.gsfc.nasa.gov/teacher/ozone_ overhead_v8.html) 1 or 2 days after they are acquired. Ozone absorption coefficients α_{ozone} in units of molecules^{-1} at 1 nm resolution between 407 and 1086 nm for the visible and near-infrared (Chappuis and Wulf bands, respectively) based on Shettle and Anderson (1995) are given in Table 12.1. As an example, using Equation 12.6 we find that 300 DU will yield a vertical optical depth at 602 nm of

$$\tau_{ozone(602nm)} = 300 \times 2.687 \cdot 10^{16} \times 5.190 \cdot 10^{-21} = 0.04184 \qquad (12.7)$$

TABLE 12.1
Ozone Absorption Coefficients (molecules^{-1}) @ the Given Wavelengths (nm)

Wavelength	Absorption coefficient	Wavelength	Absorption coefficient	Wavelength	Absorption coefficient	Wavelength	Absorption coefficient
407	2.001e-24	577	4.706e-21	747	4.262e-22	917	1.862e-23
408	4.739e-24	578	4.665e-21	748	4.216e-22	918	1.795e-23
409	6.727e-24	579	4.623e-21	749	4.118e-22	919	1.737e-23
410	8.178e-24	580	4.576e-21	750	3.947e-22	920	1.682e-23
411	8.984e-24	581	4.532e-21	751	3.740e-22	921	1.657e-23
412	8.255e-24	582	4.490e-21	752	3.544e-22	922	1.650e-23
413	7.163e-24	583	4.452e-21	753	3.337e-22	923	1.633e-23
414	7.526e-24	584	4.423e-21	754	3.139e-22	924	1.626e-23
415	9.642e-24	585	4.401e-21	755	2.977e-22	925	1.637e-23
416	1.233e-23	586	4.385e-21	756	2.827e-22	926	1.646e-23
417	1.442e-23	587	4.382e-21	757	2.709e-22	927	1.642e-23
418	1.553e-23	588	4.382e-21	758	2.612e-22	928	1.628e-23
419	1.569e-23	589	4.395e-21	759	2.536e-22	929	1.645e-23
420	1.561e-23	590	4.414e-21	760	2.473e-22	930	1.651e-23
421	1.655e-23	591	4.456e-21	761	2.421e-22	931	1.669e-23
422	1.954e-23	592	4.507e-21	762	2.379e-22	932	1.662e-23
423	2.466e-23	593	4.575e-21	763	2.607e-22	933	1.622e-23
424	3.148e-23	594	4.659e-21	764	2.586e-22	934	1.631e-23
425	3.828e-23	595	4.742e-21	765	2.552e-22	935	1.703e-23
426	4.229e-23	596	4.838e-21	766	2.522e-22	936	1.834e-23
427	4.203e-23	597	4.931e-21	767	2.492e-22	937	2.025e-23
428	3.928e-23	598	5.013e-21	768	2.503e-22	938	2.284e-23
429	3.722e-23	599	5.084e-21	769	2.513e-22	939	2.564e-23
430	3.707e-23	600	5.139e-21	770	2.515e-22	940	2.900e-23
431	4.002e-23	601	5.173e-21	771	2.570e-22	941	3.277e-23
432	4.690e-23	602	5.190e-21	772	2.626e-22	942	3.623e-23
433	5.316e-23	603	5.184e-21	773	2.696e-22	943	3.913e-23
434	5.506e-23	604	5.156e-21	774	2.795e-22	944	4.049e-23
435	5.465e-23	605	5.115e-21	775	2.924e-22	945	4.019e-23
436	5.501e-23	606	5.058e-21	776	3.027e-22	946	3.848e-23
437	5.812e-23	607	4.983e-21	777	3.092e-22	947	3.493e-23
438	6.625e-23	608	4.906e-21	778	3.141e-22	948	3.135e-23
439	7.903e-23	609	4.821e-21	779	3.155e-22	949	2.740e-23
440	9.554e-23	610	4.730e-21	780	3.119e-22	950	2.435e-23
441	1.110e-22	611	4.641e-21	781	3.035e-22	951	2.129e-23
442	1.217e-22	612	4.554e-21	782	2.935e-22	952	1.938e-23
443	1.299e-22	613	4.475e-21	783	2.791e-22	953	1.683e-23
444	1.299e-22	614	4.395e-21	784	2.642e-22	954	1.512e-23
445	1.237e-22	615	4.321e-21	785	2.478e-22	955	1.512e-23
446	1.158e-22	616	4.251e-21	786	2.359e-22	956	1.436e-23
447	1.120e-22	617	4.181e-21	787	2.266e-22	957	1.368e-23
448	1.150e-22	618	4.122e-21	788	2.169e-22	958	1.271e-23

TABLE 12.1 (CONTINUED)
Ozone Absorption Coefficients (molecules⁻¹) @ the Given Wavelengths (nm)

Wavelength	Absorption coefficient	Wavelength	Absorption coefficient	Wavelength	Absorption coefficient	Wavelength	Absorption coefficient
449	1.260e-22	619	4.062e-21	789	2.068e-22	959	1.161e-23
450	1.391e-22	620	4.011e-21	790	1.999e-22	960	1.066e-23
451	1.478e-22	621	3.963e-21	791	1.919e-22	961	9.900e-24
452	1.514e-22	622	3.911e-21	792	1.833e-22	962	9.293e-24
453	1.501e-22	623	3.866e-21	793	1.764e-22	963	8.866e-24
454	1.504e-22	624	3.821e-21	794	1.696e-22	964	8.599e-24
455	1.572e-22	625	3.774e-21	795	1.632e-22	965	8.295e-24
456	1.735e-22	626	3.721e-21	796	1.597e-22	966	7.920e-24
457	2.001e-22	627	3.673e-21	797	1.571e-22	967	7.137e-24
458	2.326e-22	628	3.620e-21	798	1.553e-22	968	6.496e-24
459	2.658e-22	629	3.563e-21	799	1.522e-22	969	6.058e-24
460	2.986e-22	630	3.510e-21	800	1.489e-22	970	5.446e-24
461	3.224e-22	631	3.457e-21	801	1.477e-22	971	5.003e-24
462	3.326e-22	632	3.403e-21	802	1.472e-22	972	4.651e-24
463	3.312e-22	633	3.347e-21	803	1.469e-22	973	4.341e-24
464	3.135e-22	634	3.290e-21	804	1.473e-22	974	4.207e-24
465	2.934e-22	635	3.231e-21	805	1.498e-22	975	3.977e-24
466	2.834e-22	636	3.172e-21	806	1.525e-22	976	3.942e-24
467	2.935e-22	637	3.111e-21	807	1.569e-22	977	3.827e-24
468	3.128e-22	638	3.048e-21	808	1.638e-22	978	3.777e-24
469	3.300e-22	639	2.989e-21	809	1.713e-22	979	3.701e-24
470	3.475e-22	640	2.926e-21	810	1.795e-22	980	3.676e-24
471	3.558c-22	641	2.867e-21	811	1.866e-22	981	3.841e-24
472	3.601e-22	642	2.814e-21	812	1.937e-22	982	4.032e-24
473	3.594e-22	643	2.761e-21	813	2.009e-22	983	4.367e-24
474	3.707e-22	644	2.710e-21	814	2.072e-22	984	4.989e-24
475	3.990e-22	645	2.662e-21	815	2.114e-22	985	5.915e-24
476	4.457e-22	646	2.616e-21	816	2.135e-22	986	7.191e-24
477	4.997e-22	647	2.569e-21	817	2.127e-22	987	8.732e-24
478	5.624e-22	648	2.527e-21	818	2.089e-22	988	1.041e-23
479	6.256e-22	649	2.487e-21	819	2.028e-22	989	1.246e-23
480	6.932e-22	650	2.445e-21	820	1.946e-22	990	1.802e-23
481	7.408e-22	651	2.405e-21	821	1.839e-22	991	3.049e-23
482	7.730e-22	652	2.366e-21	822	1.717e-22	992	2.713e-23
483	7.779e-22	653	2.324e-21	823	1.591e-22	993	2.195e-23
484	7.552e-22	654	2.283e-21	824	1.471e-22	994	2.195e-23
485	7.243e-22	655	2.242e-21	825	1.359e-22	995	2.102e-23
486	7.026e-22	656	2.201e-21	826	1.252e-22	996	1.724e-23
487	6.901e-22	657	2.161e-21	827	1.154e-22	997	1.345e-23
488	6.963e-22	658	2.119e-21	828	1.071e-22	998	1.134e-23

(continued)

TABLE 12.1 (CONTINUED)
Ozone Absorption Coefficients (molecules⁻¹) @ the Given Wavelengths (nm)

Wave-length	Absorption coefficient	Wave-length	Absorption coefficient	Wave-length	Absorption coefficient	Wave-length	Absorption coefficient
489	7.225e-22	659	2.083e-21	829	1.001e-22	999	9.501e-24
490	7.590e-22	660	2.046e-21	830	9.396e-23	1000	7.456e-24
491	7.815e-22	661	2.009e-21	831	8.855e-23	1001	7.291e-24
492	7.963e-22	662	1.973e-21	832	8.463e-23	1002	5.993e-24
493	8.033e-22	663	1.935e-21	833	8.143e-23	1003	5.603e-24
494	8.198e-22	664	1.899e-21	834	7.848e-23	1004	5.164e-24
495	8.383e-22	665	1.858e-21	835	7.679e-23	1005	4.140e-24
496	8.752e-22	666	1.817e-21	836	7.605e-23	1006	4.452e-24
497	9.275e-22	667	1.774e-21	837	7.514e-23	1007	3.733e-24
498	9.929e-22	668	1.732e-21	838	7.446e-23	1008	3.303e-24
499	1.074e-21	669	1.692e-21	839	7.426e-23	1009	3.501e-24
500	1.170e-21	670	1.653e-21	840	7.458e-23	1010	2.965e-24
501	1.269e-21	671	1.614e-21	841	7.562e-23	1011	2.673e-24
502	1.367e-21	672	1.575e-21	842	7.712e-23	1012	2.804e-24
503	1.454e-21	673	1.537e-21	843	7.874e-23	1013	2.559e-24
504	1.517e-21	674	1.501e-21	844	8.120e-23	1014	2.322e-24
505	1.561e-21	675	1.470e-21	845	8.523e-23	1015	2.200e-24
506	1.570e-21	676	1.437e-21	846	9.008e-23	1016	2.250e-24
507	1.557e-21	677	1.404e-21	847	9.562e-23	1017	2.078e-24
508	1.533e-21	678	1.373e-21	848	1.024e-22	1018	1.892e-24
509	1.500e-21	679	1.347e-21	849	1.105e-22	1019	1.889e-24
510	1.480e-21	680	1.322e-21	850	1.196e-22	1020	1.808e-24
511	1.470e-21	681	1.300e-21	851	1.290e-22	1021	1.631e-24
512	1.491e-21	682	1.278e-21	852	1.368e-22	1022	1.532e-24
513	1.522e-21	683	1.260e-21	853	1.412e-22	1023	1.452e-24
514	1.559e-21	684	1.239e-21	854	1.415e-22	1024	1.463e-24
515	1.594e-21	685	1.217e-21	855	1.388e-22	1025	1.406e-24
516	1.628e-21	686	1.191e-21	856	1.347e-22	1026	1.329e-24
517	1.661e-21	687	1.166e-21	857	1.298e-22	1027	1.351e-24
518	1.691e-21	688	1.134e-21	858	1.243e-22	1028	1.402e-24
519	1.723e-21	689	1.101e-21	859	1.176e-22	1029	1.384e-24
520	1.761e-21	690	1.071e-21	860	1.091e-22	1030	1.420e-24
521	1.817e-21	691	1.040e-21	861	1.002e-22	1031	1.410e-24
522	1.877e-21	692	1.009e-21	862	9.202e-23	1032	1.463e-24
523	1.952e-21	693	9.845e-22	863	8.438e-23	1033	1.525e-24
524	2.041e-21	694	9.576e-22	864	7.720e-23	1034	1.564e-24
525	2.137e-21	695	9.294e-22	865	7.025e-23	1035	1.662e-24
526	2.233e-21	696	9.030e-22	866	6.419e-23	1036	1.764e-24
527	2.337e-21	697	8.758e-22	867	5.939e-23	1037	1.876e-24
528	2.434e-21	698	8.524e-22	868	5.533e-23	1038	1.944e-24
529	2.527e-21	699	8.303e-22	869	5.203e-23	1039	2.073e-24
530	2.609e-21	700	8.094e-22	870	4.958e-23	1040	2.282e-24

TABLE 12.1 (CONTINUED)
Ozone Absorption Coefficients (molecules⁻¹) @ the Given Wavelengths (nm)

Wave-length	Absorption coefficient	Wave-length	Absorption coefficient	Wave-length	Absorption coefficient	Wave-length	Absorption coefficient
531	2.678e-21	701	7.904e-22	871	4.745e-23	1041	2.512e-24
532	2.731e-21	702	7.757e-22	872	4.475e-23	1042	2.745e-24
533	2.764e-21	703	7.596e-22	873	4.239e-23	1043	3.104e-24
534	2.782e-21	704	7.449e-22	874	4.075e-23	1044	3.730e-24
535	2.784e-21	705	7.322e-22	875	3.960e-23	1045	5.270e-24
536	2.778e-21	706	7.205e-22	876	3.899e-23	1046	7.263e-24
537	2.777e-21	707	7.100e-22	877	3.852e-23	1047	5.683e-24
538	2.791e-21	708	7.000e-22	878	3.847e-23	1048	5.129e-24
539	2.820e-21	709	6.904e-22	879	3.879e-23	1049	5.166e-24
540	2.860e-21	710	6.820e-22	880	3.928e-23	1050	4.980e-24
541	2.908e-21	711	6.753e-22	881	3.991e-23	1051	4.406e-24
542	2.968e-21	712	6.690e-22	882	4.102e-23	1052	3.500e-24
543	3.029e-21	713	6.640e-22	883	4.246e-23	1053	3.188e-24
544	3.083e-21	714	6.626e-22	884	4.401e-23	1054	2.659e-24
545	3.135e-21	715	6.566e-22	885	4.587e-23	1055	2.647e-24
546	3.171e-21	716	6.487e-22	886	4.769e-23	1056	2.225e-24
547	3.209e-21	717	6.349e-22	887	4.887e-23	1057	2.262e-24
548	3.244e-21	718	6.186e-22	888	4.974e-23	1058	1.937e-24
549	3.268e-21	719	5.985e-22	889	5.038e-23	1059	1.926e-24
550	3.297e-21	720	5.775e-22	890	5.035e-23	1060	1.834e-24
551	3.329e-21	721	5.569e-22	891	4.985e-23	1061	1.705e-24
552	3.365e-21	722	5.378e-22	892	4.981e-23	1062	1.849e-24
553	3.406e-21	723	5.230e-22	893	5.100e-23	1063	1.852e-24
554	3.451e-21	724	5.070e-22	894	5.313e-23	1064	1.987e-24
555	3.505e-21	725	4.946e-22	895	5.597e-23	1065	2.347e-24
556	3.568e-21	726	4.801e-22	896	5.900e-23	1066	2.290e-24
557	3.641e-21	727	4.699e-22	897	6.117e-23	1067	1.889e-24
558	3.722e-21	728	4.567e-22	898	6.197e-23	1068	1.746e-24
559	3.818e-21	729	4.430e-22	899	6.143e-23	1069	1.879e-24
560	3.910e-21	730	4.307e-22	900	5.958e-23	1070	1.671e-24
561	4.010e-21	731	4.213e-22	901	5.685e-23	1071	1.422e-24
562	4.106e-21	732	4.141e-22	902	5.385e-23	1072	1.280e-24
563	4.195e-21	733	4.073e-22	903	5.068e-23	1073	1.116e-24
564	4.275e-21	734	4.035e-22	904	4.759e-23	1074	1.073e-24
565	4.343e-21	735	4.005e-22	905	4.410e-23	1075	1.104e-24
566	4.408e-21	736	3.988e-22	906	3.999e-23	1076	9.390e-25
567	4.464e-21	737	3.970e-22	907	3.670e-23	1077	1.014e-24
568	4.519e-21	738	3.948e-22	908	3.415e-23	1078	8.850e-25
569	4.579e-21	739	3.956e-22	909	3.168e-23	1079	8.400e-25
570	4.634e-21	740	3.966e-22	910	2.922e-23	1080	7.620e-25

(continued)

TABLE 12.1 (CONTINUED)

Ozone Absorption Coefficients (molecules^{-1}) @ the Given Wavelengths (nm)

Wave-length	Absorption coefficient	Wave-length	Absorption coefficient	Wave-length	Absorption coefficient	Wave-length	Absorption coefficient
571	4.683e-21	741	4.006e-22	911	2.683e-23	1081	6.970e-25
572	4.730e-21	742	4.063e-22	912	2.488e-23	1082	7.090e-25
573	4.756e-21	743	4.108e-22	913	2.329e-23	1083	7.230e-25
574	4.766e-21	744	4.176e-22	914	2.143e-23	1084	6.850e-25
575	4.763e-21	745	4.221e-22	915	2.014e-23	1085	6.620e-25
576	4.739e-21	746	4.265e-22	916	1.945e-23	1086	7.860e-25

12.3.4 TRANSMISSION OF THE ATMOSPHERE

Figure 12.3 illustrates the transmission of direct solar (beam) irradiance for the parameters given in the legend. Superimposed on the plot is the spectral irradiance (dot-dash line) for the given trace species with the sun in the zenith position at 1 AU The Rayleigh scattering is a smooth function of wavelength and the larger dips (and many of the smaller ones) that depart from this smooth function result from mixed gas absorption by O_2 and CO_2. Water vapor (light gray) and ozone (dark gray) absorption bands are plotted separately. Even a modest aerosol optical depth (dashed line) of 0.15 at 500 nm with a wavelength dependence given by

$$\tau = \beta\lambda^{-\alpha} \tag{12.8}$$

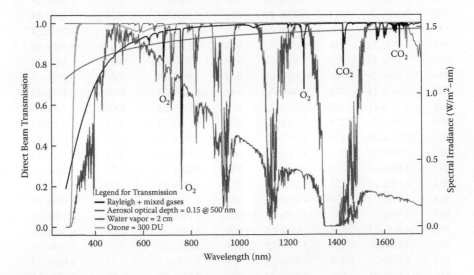

FIGURE 12.3 (See color insert.) Direct solar transmission of major components of the clear-sky atmosphere (left axis); superimposed on the transmission spectra is the spectral irradiance calculated for these conditions (right axis). (See color insert for clearer separation of curves.)

where $\beta = 0.06092$, and $\alpha = 1.3$ gives a reduced transmission that is typically the second most important contributor to solar radiation extinction after Rayleigh scattering. This is true for clear skies because the extinction from aerosols is typically large in the midvisible wavelengths where solar irradiance is highest. Note that some spectral absorption features in the spectral irradiance curve are not in the transmission spectra. Most of these extra absorption features are from the extraterrestrial solar spectrum and occur in the outer layers of the sun's atmosphere (see Figure 12.1).

12.4   BROADBAND FILTER RADIOMETRY

There are many reasons to measure isolated portions of the solar spectrum. For example, broadband measurements of red, green, and blue light are used for specifying color for both commercial and artistic reasons. Broadband measurements of the spectral distribution of sunlight can be used for calculating the efficiency of spectrally dependent photovoltaic devices (Michalsky and Kleckner, 1984). Although not the preferred method, broadband filters have been used with a pyrheliometer to approximately measure aerosol optical depths (Dutton and Christy, 1992). In this section we discuss some common types of broadband spectral measurements with solar applications.

12.4.1   PHOTOMETRY

Photometry is the measurement of light that the human eye can sense. Although there have been slight modifications to the definition, the Commission Internationale de l'Eclairage (CIE) defined the basic *luminosity function* for daytime (photopic) vision for the average human eye in 1924 (CIE, 1926). Figure 12.4 illustrates how the eye response varies with wavelength with a peak near 555 nm. *Scotopic response* is the wavelength response of the eye in darkened conditions. The response distribution with wavelength is similar but more skewed, and it shifts to shorter wavelengths with the peak response near 507 nm. Since we are mostly interested in daylight applications we will focus only on the photopic response that most photometers are designed to measure.

To convert any spectral irradiance, such as sunlight, into illuminance L, a very good approximation is

$$L = 683 \int_{380}^{770} \bar{y}(\lambda) I(\lambda) d\lambda \qquad (12.9)$$

where $\bar{y}$ is the photopic function plotted in Figure 12.3, $I(\lambda)$ is the spectral irradiance in $W/(m^2\text{-nm})$, and L has units of lumens/m^2.

Manufacturers of photometers try to match the photopic function as closely as possible using combinations of filters and detector responses. The success with which this is achieved varies; an example is given in Figure 12.5 from the photometry

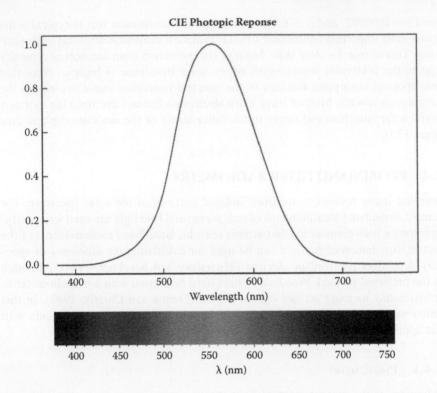

FIGURE 12.4   (See color insert.) The typical human eye's response for sunlit conditions as specified by the Commission International l'Eclairage (CIE); en.wikipedia.org/wiki/ File:Srgbspectrum.png is the source for the color diagram beneath the photopic response plot.

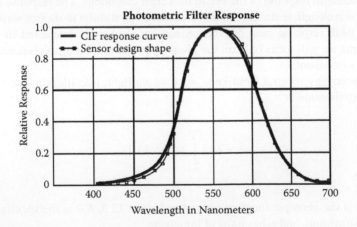

FIGURE 12.5   An example of one manufacturer's (Macam) attempt to match the CIE photopic response using filters. (From Austin, R., *The Guide to Photometry*, UDT Instruments, San Diego, CA, n.d. With permission.)

tutorial offered on UDT's website (http://www.udtinstruments.com/applications/photometric/photometry.shtml). When searching for a photometer, make sure that your search is for a photometer with a photopic response since the term *photometer* is used for many types of instruments that do not attempt to match the spectral response of the eye in daylight. If the photometer is to be used outdoors, make certain that it is designed to be waterproof and rugged. Although manufacturers try to match the CIE response curve, they all fail to some extent, so look for a reasonable photopic match. If the goal is to measure solar radiation for daylighting purposes, for example, performing the integral in Equation 12.9 using a modeled clear-sky irradiance first with the photopic function $\bar{y}$ and then using the manufacturer's estimate of the photopic function, say $\bar{y}'$, a bias can be calculated. Most daylight photometers will measure total horizontal illuminance. To do this well, a respectable cosine response for the instrument is required; that is, the signal should fall off approximately as the cosine of the angle of incidence. If the response is within 10% of the normalized response (actual response as a function of incident angle divided by cosine of the incident angle) out to about 80° incidence, then it can be considered acceptable. The left side of Figure 12.6 illustrates an acceptable cosine response for an actual instrument. Some manufacturers may illustrate cosine response as shown on the right side of Figure 12.6, but without the normalized cosine plotted for comparison. These plots may look approximately correct but should be plotted normalized as in the left half of Figure 12.6 for proper evaluation. Other factors to consider are instrument stability (responsivity changes are less than 3% per year) and absolute uncertainty (better than 5%). A final note on cost: It was possible to purchase good photopic photometers for outdoors use for around US$500 in 2010.

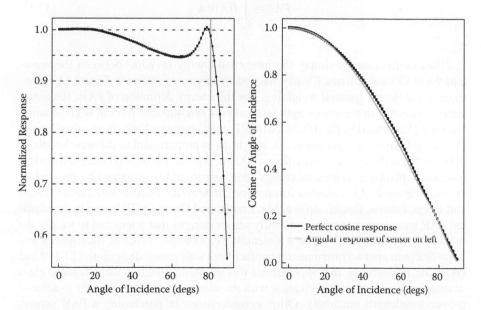

FIGURE 12.6 An example of the cosine response of a good receiver compared with an exact cosine response plotted normalized and unnormalized.

12.4.2 Photosynthetically Active Radiation (PAR)

Plants use PAR to transform water and carbon dioxide into sugars and oxygen through photosynthesis. PAR is measured with sensors that filter light to accept radiation between about 400 and 700 nm. Plants mostly use the light near either end of this wavelength range for photosynthesis and reflect more of the green light, thus giving plants their color. Photosynthesis is a quantum process. In other words, photons cause the reactions that occur in plants; therefore, it would seem logical to define PAR in terms of photons per unit time per unit area. If we know the spectral distribution of sunlight as a function of wavelength $I(\lambda)$, say in W/(m²-nm), and the energy of a photon is given by hc/λ in joules, then the photosynthetic photon flux density (PAR_{PPFD}) in units of photons/(sec-m²) can be calculated as

$$PAR_{PPFD} = \frac{1}{hc} \int_{400}^{700} \lambda I(\lambda) d\lambda \qquad (12.10)$$

where hc is the product of Planck's constant and the speed of light, respectively. To make the numbers reasonably convenient to work with, the photon number is usually expressed in units of micromoles 6.022×10^{17}. PAR with full sun (e.g., broadband DNI = 1000 W/m²) on a cloudless summer day would measure around 2000 µmoles/(m²-sec). PAR is sometimes defined by

$$PAR = \int_{400}^{700} I(\lambda) d\lambda \qquad (12.11)$$

PAR, in this case, is simply the integrated energy in W/m² between the wavelengths of 400 and 700 nm. Clearly, the two equations are different. Figure 12.7 illustrates the different spectral weighting. For the energy definition of PAR, the entire energy spectrum in the wavelength interval between 400 and 700 nm is given equal weighting illustrated by the thin black line. For the photosynthetic photon flux density (PPFD) definition in Equation 12.10, weighting is proportional to the wavelength. A 400 nm photon has more energy than a 700 nm photon, yet both produce a chemical reaction in plants; therefore, an energy spectrum should be weighted by the dashed curve in Figure 12.7 to calculate its photosynthetic effect. Federer and Tanner (1966) and Biggs, Edison, Eastin, Brown, Maranville, and Clegg (1971) began to standardize PAR measurements with some early sensor designs that attempted to weight the sensitivity proportionally to the wavelength. For example, LI-COR Biosciences and Kipp & Zonen are two instrument manufacturers with sensor designs, the LI-190 and PAR Lite, respectively, that approximate this weighting function using colored glass or interference filters in combination with the silicon photodiode detector to achieve proper wavelength sensitivity. Other considerations in purchasing a PAR sensor, besides the agreement between the theoretical and actual photosynthetic response (dashed line in Figure 12.6), are all-weather ruggedness, good cosine response, and

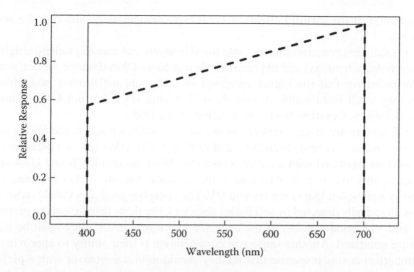

FIGURE 12.7 The weighting for PAR measured in energy units of W/m² (thin black line) and for weighting in photons/(sec-m²) (thick dashed line).

calibration stability. An annual stability of better than 3% and an absolute response of better that 5% are reasonable specifications to expect. Good PAR sensors could be found for about US$500 in 2010.

12.4.3 UVA AND UVB

Ultraviolet (UV) radiation has wavelengths below 400 nm and above 10 nm. The practical lower wavelength limit for the UV for solar radiation at the earth's surface is around 290 nm since ozone in the stratosphere absorbs virtually all of the radiation below this wavelength. High spectral resolution measurements of the solar UV spectrum require a spectrometer with two dispersive elements to minimize stray light from the stronger and longer wavelength portion of the solar spectrum if one is to achieve the best measurements at the very important shortest UV wavelengths. Because of the high cost of a double monochromator, it is more common to measure UV in broad wavelength bands. UVA covers the wavelengths between 315 nm and 400 nm and UVB covers the wavelengths between 290 nm and 315 nm. UVC includes wavelengths shorter than 290 nm but will not be discussed here because no significant amounts of UVC reach the earth's surface due to atmospheric absorption and scattering.

UVB is the most common UV measurement because of its deleterious effects on humans and material properties. Overexposure to UVB is associated with sunburn, eye disease (including cataracts), and suppression of the immune system. However, underexposure to UVB leads to lower vitamin D production, which has deleterious effects as well. Exposure to UVB at levels before sunburn may produce a net positive effect (Holick, 2005). The World Health Organization (WHO) has, however, taken the conservative position on its website (http://www.who.int/uv/faq/uvhealtfac/en/

index.html) that the harmful effects of UVB exposure generally outweigh the positive ones.

UVA radiation penetrates deeper into the skin layers and causes premature aging. It also produces hydroxyl and oxygen radicals that cause DNA damage. In fact, some scientists believe that the higher incidence of malignant melanomas in sunblock users may result from sunblock ingredients blocking UVB but not UVA (Autier, Dore, Schifflers, Cesarini, Bollaerts, Koelmel et al., 1995).

UVB sensors are usually tailored to match as closely as possible the Diffey erythemal response function (McKinlay and Diffey, 1987). UVA sensors measure the 315–400 nm passband with a uniform response. Most modern UVB and UVA sensors use a filter to define the passband of the radiation that one wishes to isolate followed by a phosphor that is sensitive to UV. The phosphor produces visible radiation that is more easily detected by solid-state detectors. For these instruments to perform well and with good stability at the expected low signal levels, they must be temperature stabilized. Another important consideration is their ability to approximate a Lambertian cosine response; that is, they should mimic a receiver with a perfect cosine angular response as described in the section on photometry. As far as absolute accuracy, UVB sensors that are better than 10% are likely the best that one can expect; relative calibration stability should be better than 5% per year. A list of UVB instrument manufacturers (see Appendix F) generally includes EKO Instruments, Kipp & Zonen, Solar Light, and Yankee Environmental Systems, Inc.

12.5 NARROW-BAND FILTER RADIOMETRY

Narrow-band filter radiometers are used for the measurement of atmospheric aerosol and molecular constituents, such as water vapor and ozone. Direct normal solar irradiance instruments are called sun photometers or, more generally, sun radiometers since they are used outside the photopic response band as well as within it. A sun radiometer measures in several narrow wavelength bands with responsivities defined by full widths (wavelength) at half-maximum (response) of 5–10 nm. The instrument should track the sun using a narrow field of view to observe the disk of the sun plus a small annulus around the sun that allows for tracking uncertainty and atmospheric refraction (e.g., 2° to 2.5° full view angle).

12.5.1 AEROSOL OPTICAL DEPTH

For a clear view of the sun with no clouds, one can use Beer's law to calculate the column aerosol optical depth using

$$I(\lambda) = I_o(\lambda)\exp(-\tau_{aerosol}(\lambda) \cdot m_{aerosol} - \tau_{ozone}(\lambda) \cdot m_{ozone} - \tau_{Rayleigh}(\lambda) \cdot m_{Rayleigh}) \quad (12.12)$$

where Equation 12.12 is Equation 12.1 expanded to explicitly show the optical depth components that are normally measured in the visible and near-infrared spectrum. For sun radiometers to measure aerosols, filter wavelengths and bandwidths are carefully chosen to avoid water vapor and molecular oxygen absorption bands. If $I(\lambda)$ is measured, the m's calculated using the sun's elevation, τ_{ozone} obtained from satellite or ground-based measurements, $\tau_{Rayleigh}$ calculated using Equation 12.2 and the atmospheric pressure is measured, then $\tau_{aerosol}$ can be calculated if $I_o(\lambda)$ is known. Determining of the extraterrestrial solar irradiance, $I_o(\lambda)$, is the crux of the matter when it comes to performing good sun radiometry and is the most demanding task in making valid aerosol optical depth measurements. For example, a 1% uncertainty in $I_o(\lambda)$ leads to an approximate uncertainty of 0.01 in aerosol optical depth.

The air masses for ozone, aerosol, and Rayleigh scattering differ because each is vertically distributed differently in the earth's atmosphere. However, an adequate approximation for many purposes is to assume that the air masses for ozone, aerosol, and Rayleigh scattering are the same, in which case Equation 12.13 reverts to Equation (12.1). If we take the natural logarithm of Equation 12.1 we get

$$\ln(I(\lambda)) = \ln(I_o(\lambda)) - \tau(\lambda) \cdot m \tag{12.13}$$

that has the form of a linear equation $y = y_o + b \cdot x$. If we plot $ln(I)$ versus air mass m, then for a perfectly stable atmosphere we should get a straight line whose intercept is the $ln(I_o)$ and whose slope is the negative of the total optical depth. If we now know $ln(I_o)$, then we can calculate or measure everything else in Equation 12.13 to derive total optical depth at any other time. Subtracting optical depths caused by Rayleigh scattering and ozone absorption yields the aerosol optical depth.

Although we have assumed that the air masses are the same for the different extinction components to illustrate this *Langley plot* technique, the more correct procedure for Langley analysis is to plot the left-hand side of the following equation versus aerosol air mass, that is,

$$\ln(I(\lambda)) + \tau_{ozone} \cdot m_{ozone} + \tau_{Rayleigh} \cdot m_{Rayleigh} = \ln(I_o(\lambda)) - \tau_{aerosol} \cdot m_{aerosol} \tag{12.14}$$

Using Equation 12.13 versus Equation 12.14 for the determination of the Langley intercept results is much less than 1% bias at visible and longer wavelengths, but it becomes crucial to use Equation 12.14 for studies in the ultraviolet because Rayleigh and ozone are generally the dominant contributions to the total optical depth.

The air mass relative to the zenith direction typically used in Equation 12.13 is for a pure Rayleigh atmosphere. Kasten and Young (1989) developed a revised approximation for a Rayleigh atmosphere, namely,

$$m_{Rayleigh} = \frac{1}{\cos(sza) + 0.50572 \cdot (96.07995 - sza)^{-1.6364}} \tag{12.15}$$

where the solar zenith angle sza on the right-hand side of the equation is in degrees. Almost always it is the case that more than 90% of the atmospheric ozone resides in the stratosphere between 20 and 40 km and varies with season. The ozone layer is lower in the winter than in the summer. The altitude also depends on the latitude with higher layers at the equator than at the poles. If one considers the geometry of an elevated layer, ozone air mass relative to the zenith direction is somewhat smaller than the Rayleigh air mass. Ozone air mass as calculated by the Dobson spectrophotometer community (Komhyr and Evans, 2008) is given by

$$m_{ozone} = \frac{R+h}{\sqrt{(R+h)^2 - (R+r)^2 \cdot \sin^2(sza)}} \tag{12.16}$$

where R is the mean radius of the earth 6371.299 km, r is the station height above mean sea level in km, and h is the mean height of the ozone layer in km. Although aerosol layers can be elevated, especially as a result of long-range transport in the free troposphere, they typically reside in the boundary layer near the surface. We therefore use the same air mass expression for aerosol that is used for water vapor (Kasten, 1966; based on Schnaidt, 1938), namely,

$$m_{aerosol} = \frac{1}{\cos(sza) + 0.0548 \cdot (92.65 - sza)^{-1.452}} \tag{12.17}$$

Two well-known locations for obtaining the clear, stable atmospheres needed for this straightforward determination of $ln(I_o)$ are Izaña Observatory on the island of Tenerife in the Canary Islands and at Mauna Loa Observatory on the island of Hawaii in the Hawaiian Islands. Even at these sites, one requires a number of Langley plots using only morning data since upslope winds can bring lower tropospheric air over the site. Examining multiple Langley plots (Langleys) allows one to identify measurement outliers and provides a robust estimate of the uncertainty of the calibration constant $ln(I_o)$.

Alternatively, indoor lamp calibrations, in principle, can be used to calibrate sun radiometers. A very careful study of lamp calibrations versus Langley calibrations from a high mountain site (Jungfraujoch in Switzerland) by Schmid and Wehrli (1995) demonstrated that the calibrations differ from as little as 0.2% to as much as 4.1% depending on the wavelength with most filter differences greater than 1.4%. A follow-on study (Schmid, Spyak, Biggar, Wehrli, Skeler, Ingold et al., 1998) suggested that typical differences between the two calibration techniques were at best 2–4% and became more variable than this if different published extraterrestrial spectra $(I_o(\lambda))$ were used for the extraterrestrial irradiance. Lamp calibration, however, may be a viable means for checking the stability of a calibration, but *Langley plots* are suggested for establishing the initial calibration of the radiometer. For example, if a good Langley calibration is associated with a near-concurrent lamp calibration, then future lamp calibrations could be used to track and correct for the degradation of the sun radiometer if needed for a site where good Langley plots occur infrequently or are impossible.

Michalsky, Schlemmer, Berkheiser, Berndt, and Harrison (2001) and Augustine, Cornwall, Hodges, Long, Medina, and Delusi (2003) describe two calibration methods that do not require high mountains or lamps. The fundamental idea is to use all possible Langleys to develop a robust and continuously updated calibration of the sun radiometer. These methods have demonstrated uncertainties around 1%.

AERONET and its affiliates form the largest aerosol optical depth network in the world (Holben, Eck, Slutsker, Tanre, Buis, Setzer et al., 1998). This network uses measurements on Mauna Loa to calibrate their standards using Langley plots over several good, stable days. These instruments are then transported to Goddard Space Flight Center in Maryland, where they are used to transfer calibrations to the network instruments via side-by-side comparisons. Recent comparisons between sun radiometers calibrated using this method and the method of Michalsky, Schlemmer et al. (2001) found biases in daily-averaged aerosol optical depths of less than 1% over a 3-year period (Michalsky, Denn et al. 2010).

12.5.2 Water Vapor

Using narrow-band filter radiometers to measure water vapor is more difficult than measuring aerosol optical depth. Thome, Herman, and Reagan (1992) reviewed the sun radiometric techniques, both empirical and theoretical, that had been used in the previous 80 years for water vapor column retrievals. The technique that has been used most often in the last 20 years to retrieve water vapor is based on the modified Langley method first introduced by Reagan, Thome, Herman, and Gall (1987). If measurements are made in the water vapor band, then Equation 12.1 can be modified; thus

$$I = I_o \exp(-\tau \cdot m) T_{H_2O} \qquad (12.18)$$

where τ and m are the optical depths associated with Rayleigh scattering and aerosol scattering and absorption. T_{H_2O} is the transmission in the water band due to water vapor absorption. Bruegge, Conel, Green, Margolis, Holm, and Toon (1992) modeled the transmission in the water band using

$$T_{H_2O} = \exp[-k(u \cdot m)^b] \qquad (12.19)$$

where k and b are constants. This accounts for the product of water vapor air mass m and column water u not behaving linearly in extinction. In other words, while doubling aerosol doubles aerosol optical depth, doubling water vapor increases the water vapor optical depth by about a factor of $\sqrt{2}$. Although Reagan et al. (1987) used the approximation that b was equal to 0.5, it should be calculated for the specific filter used for the measurements because the optical depth is measured over a finite filter width of, typically, 10 nm and includes water vapor and water vapor continuum contributions. A radiative transfer model is used to calculate the transmission over the expected range of $u \cdot m$. Each calculated transmission value $t(\lambda)$ for a given water vapor path is formed by convolving with the filter function $f(\lambda)$ according to

$$T = \int_\lambda t(\lambda) f(\lambda) d \bigg/ \int_\lambda f(\lambda) d\lambda \qquad (12.20)$$

Taking the natural logarithm of Equation 12.19 twice and rearranging terms gives

$$\ln(\ln(1/T_{H_2O})) = \ln(k) + b \cdot \ln(u \cdot m) \qquad (12.21)$$

A plot of the left-hand side versus $\ln(u \cdot m)$ should be linear, and a linear least-squares fit to the calculations yields the $ln(k)$ and b coefficients. Taking the natural logarithm of Equation 12.18 after substituting Equation 12.19 and rearranging terms yields

$$\ln(I) + \tau_{scat} \cdot m_{scat} = \ln(I_o) - k(u \cdot m)^b \tag{12.22}$$

If the left-hand side of Equation 12.22 is plotted versus m^b on clear days that have stable water vapor, the intercept of the linear plot is $ln(I_o)$, which can then be used to solve for water vapor for any subsequent clear-sky measurement. Langley plots for aerosol channels are less of a challenge than modified Langley plots for the water channel. Even seemingly clear days generally experience large water vapor changes; therefore, many more modified Langleys are required to obtain even an approximate calibration compared to the aerosol channels.

AERONET (Holben, Eck et al., 1998) implemented its version of this algorithm to produce a column water vapor product for its sites. Their latest methodology for water vapor retrieval is described in Smirnov, Holben, Lyapustin, Slutsker, and Eck (2004).

Michalsky, Min, Kiedron, Slater, and Barnard (2001) introduced an alternative technique for retrieving water vapor that depends on calibration with a stable spectral lamp and a trusted extraterrestrial spectral irradiance in the vicinity of the 940 nm water vapor band and the continuum near it. Besides the uncertainty in the extraterrestrial spectrum and the lamp calibration, the overall uncertainty of this procedure also depends on estimating the aerosol and Rayleigh scattering optical depths and on the accuracy of the water vapor transmission calculations, as was the case for the previously mentioned modified Langley approach.

If column water vapor is required, for example, for the spectral irradiance model calculations that are described in Section 12.6.2, and a sunradiometer with a water channel is not available, there are other reliable sources of column-integrated water vapor.

The National Weather Service launches radiosondes on balloons twice each day as part of a worldwide network of measurements that profile temperature, humidity, pressure, and wind speed and direction. The launches are coordinated with launch times centered on 0 and 12 GMT. The website http://www.wmo.int/pages/prog/www/ois/volume-a/vola-home.htm contains a list of stations launching radiosondes throughout the world. Although relative humidity is measured on these sondes, dew point temperature is the reported value. The actual vapor pressure e for any measurement in the profile can be calculated from the dew point temperature T_d using

$$e = 6.1121 \cdot \exp\left(\frac{17.502 \cdot T_d}{240.97 + T_d}\right) \tag{12.23}$$

The particulars of radiosondes measurements as practiced in the United States are explained in *Federal Meteorological Handbook No. 3* (1997, http://www.ofcm.gov/fmh3/text/default.htm). Column water vapor column can then be calculated using

$$W = \frac{1}{g} \int_{p_1}^{p_2} \frac{0.622 \cdot e}{p - e} dp \tag{12.24}$$

In Equation 12.24 from Fleagle and Businger (1963) W is the integrated water column, g is the acceleration due to gravity, and p is the pressure with the integration between pressure levels p_1 and p_2.

Alternatively, water vapor column can be derived from global positioning system (GPS) measurements (Bevis, Bussinger, Herring, Rocken, Anthes, and Ware, 1992; Businger, Chiswell, Bevis, Duan, Anthes, Rocken et al., 1996; Gutman, Chadwick, Wolfe, Simon, Van Hove, and Rocken, 1994; Gutman, Wolfe, and Simon, 1995; Kuo, Guo, and Westwater, 1993; Rocken, Ware, Van Hove, Solheim, Alber, Johnson et al., 1993; Yuan, Anthes, Ware, Rocken, Bonner, Bevis et al., 1993). These papers explain how column water vapor is derived, but the website http://gpsmet.noaa.gov/background.html provides the basics of the technique, which uses the delay time of the signal received from the average of multiple GPS satellite transmissions. The website http://gpsmet.noaa.gov can be used to find up-to-date column water vapor for many sites in the world but mostly in the United States.

12.5.3 SUN RADIOMETERS

Most sun photometers (or sun radiometers in this handbook) are made to point at the sun with a circular and narrow field of view, ideally just large enough to encompass the solar disk and allow for small tracking errors. Ideally, this is smaller than 2.5° in diameter. Some contain a moveable wheel that positions narrow-band (5–10 nm full width half-maximum) filters to permit sequential sampling over a fixed detector. Some contain channels with individual filter–detector combinations to ensure simultaneous measurements in all channels. Most must be mounted on a separate tracker, although some come with a tracker as part of the design. Some are temperature controlled, which is recommended, but others rely on temperature measurements made independently for correction of any temperature sensitivity of the sun radiometer detector or filter.

Rotating shadowband sun radiometers (Harrison, Michalsky, and Berndt, 1994) are different in that they measure the global horizontal irradiance and then shade the receiver from the direct solar and measure diffuse horizontal irradiance. By subtraction and division by the cosine of the solar zenith angle, they calculate direct normal irradiance; this must be further corrected for the departure of the receiver from a perfect cosine response. Since this instrument involves multiple measurements and a cosine correction, it is inherently somewhat less accurate than a pointing sun radiometer that makes a single measurement of direct spectral irradiance.

Sun radiometers use interference filters, which have improved but have trouble maintaining their transmission stability over time, often due to their hygroscopic nature. Ideally, the specifications for these instruments would indicate less than 1% response changes per year. The spectral bandpasses of these instruments are usually very reproducible even if the transmission changes. Temperature stability is desirable, but it is possible to use ancillary temperature measurements to correct instruments whose temperature response has been characterized.

Integrated, out-of-band light getting through the interference filters should be less than 0.1%. This is extremely important for good aerosol optical depth measurements. This can be tested using cut-on filters that pass light longward of the tail of the interference pass band. For example, for a 500 nm filter, a Schott glass filter OG 570 should not pass any radiation if placed over a good 500 nm filter with a 10 nm pass band. Measurements of the direct sun on a clear day around solar noon should be made with a completely dark cover to determine zero offset, then with full sun, followed by full sun with the OG 570 filter covering the radiometer; two or three measurements with this sequence should be averaged. After subtracting the offsets from the measurements, the ratio of the OG 570 measurement average to the uncovered measurement average should be less than 0.001.

12.6 SPECTROMETRY

Spectrometers separate light into wavelengths typically using a diffraction grating or a prism. The fundamental layout of a spectrometer is illustrated in Figure 12.8: There is an entrance slit on which the source of light is incident; a collimating lens, whose focal length is the distance from the slit, that causes the light exiting the slit to become parallel (collimated) before it reaches the dispersing element; either a prism or a diffraction grating, that may be rotated or held stationary; and after the dispersing element is another lens with a focal length that allows the light to come to a focus either on an exit slit or on a detector array. If the last element is a slit, then a detector is placed after the slit, and the dispersing element is rotated to scan different wavelengths of radiation. Of course, in practice there are other complications that will not be discussed here; for example, optical aberrations and order sorting for grating instruments.

Spectrometers can be calibrated applying standard procedures and using absolute spectral irradiance sources available from the national standards laboratories of many countries, such as the National Institute of Standards and Technology (NIST) in the United States, the Physikalisch-Technische Bundesanstalt (PTB) in Germany, and the National Physical Laboratory (NPL) in the United Kingdom. These sources are typically 1000 W FEL lamps operated at prescribed electrical currents at a fixed

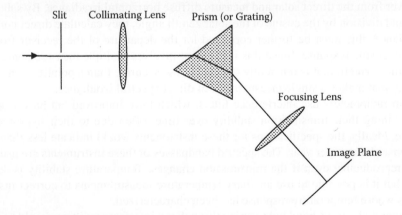

FIGURE 12.8 Schematic of the key components of a spectrometer.

distance between source and spectrometer (e.g., Walker, Saunders, Jackson, and McSparron, 1987).

12.6.1 SPECTROMETERS

Manufacturers of spectrometers are numerous; spectrometers that operate unattended in an outdoor environment are few. LI-COR Biosciences offered a weatherproofed scanning spectrometer for many years. The LI-1800 scanned solar irradiance between 300–1100 nm with typically about 6 nm spectral resolution although higher resolution using narrower slits was possible. The instrument is no longer manufactured, but it is still supported if a used one can be located; however, the support ends in 2012. EKO Instruments Co., Ltd. offers two all-weather spectroradiometers that cover the UV between 300 and 400 nm and the visible near-infrared between 350 and 1050 nm. These are still offered on its website. There are many companies offering diode array and CCD spectrometers but not generally for continuous outdoor use. Custom-built outdoor instruments based on these devices are being developed, such as Herman, Cede, Spinei, Mount, Tzortziou, and Abuhassan (2009), but not as a commercial product at this time.

12.6.2 SPECTRAL MODELS

Mathematical models for calculating the spectral distribution of solar radiation can be useful for estimating responses of broadband and spectrally selective devices, for example, photovoltaic modules. These models do very well at predicting the clear-sky spectral irradiance if the inputs are well specified. Some models work for cloudy skies, but specifying the cloud properties is problematic. For clear skies, the critical inputs are the aerosol optical depth, for at least two wavelengths, water vapor, ozone, and the spectral surface reflectivity at a number of wavelengths. For concentrating systems, the surface albedo is irrelevant because the direct normal irradiance (DNI) is the only irradiance component of interest. Time, latitude, and longitude are required to calculate the solar position. Two models in the public domain that are widely used for solar calculations are the Simple Model of the Atmospheric Radiative Transfer of Sunshine (SMARTS; Gueymard, 1995, 2001) and the Santa Barbara DIScreet Ordinate Radiative Transfer (SBDART; Richiazzi, Yang, Gautier, and Sowle, 1998).

The solar energy community is the primary user of SMARTS for clear-sky calculations. It is useful for calculating DNI radiation and global tilted irradiance (GTI) falling on a surface of any orientation. It is useful, therefore, for calculating the radiation that would be incident on a rooftop photovoltaic (PV) panel or a solar hot water installation, for example. It also has useful built-in functions to calculate, for example, PAR, photometric flux, and various UV action spectra. The software and instruction manual are available from the National Renewable Energy Laboratory's Renewable Resource Data Center at http://www.nrel.gov/rredc/smarts/.

SBDART (Richiazzi, Yang et al., 1998) was written to calculate radiative transfer in cloudy as well as clear atmospheres. This code is widely used in the atmospheric radiation community. It is accessible online and as downloadable code. An introduction and explanation was available at http://arm.mrcsb.com/sbdart/html/sbdart-intro.html, although at this time it is no longer supported. Both SMARTS and SBDART take a little effort to understand the input structure, but once this is learned many useful calculations can be performed.

For calculations that focus on the UV spectrum, a very useful, well-tested, and straightforward code is available from the National Center for Atmospheric Research (http://cprm.acd.ucar.edu/Models/TUV/). Madronich (1993) and Madronich and Flocke (1999) developed this model for biological and atmospheric chemistry applications, respectively.

QUESTIONS

1. What is the range of the spectrally integrated solar irradiance at the top of the earth's atmosphere for an average separation of the sun and earth that is in dispute?
2. Explain why the clear winter sky is blue. Explain why the setting sun is red.
3. What is the term that is used to quantify aerosol absorption? What does it mean?
4. What is the most significant gas absorber in the solar spectrum?
5. What wavelength is the easiest for the eye to see? What color would you say this is? What term is used for the eye's response to daylight?
6. In general the shorter the ultraviolet light's wavelength, the more dangerous it is; why do we not worry about UVC when we go outside?
7. If one has a number of measurements of the direct solar radiation on a clear day through a narrowband filter centered at 500 nm, how could one estimate the response of that instrument at the top of the atmosphere?
8. What is the ozone optical depth at 555 nm if the OMI satellite measures 308 DU of ozone?
9. Calculate the Rayleigh air mass if the sun is at a solar zenith angle of 81 degrees?
10. What are the two types of sunradiometers? How do they differ?
11. What are two ways to separate colors in a spectrometer?
12. What is the biggest impediment to performing a Langley calibration for a sunradiometer with a filter in a water vapor band?
13. Name two ways to get information on the water vapor column.

REFERENCES

Augustine, J.A., C.R. Cornwall, G.B. Hodges, C.N. Long, C. I. Medina, J.J. DeLuisi. 2003. An automated method of MFRSR calibration for aerosol optical depth analysis with application to an Asian dust outbreak over the United States. *Journal of Applied Meteorology* 42: 266–278.

Autier P., J. F. Dore, E. Schifflers, J. P. Cesarini, A. Bollaerts, K. F. Koelmel, O. Gefeller, A. Liabeuf, F. Lejeune, D. Lienard, M. Joarlette, P. Chemaly, and U. R. Kleeburg. 1995.

Melanoma and use of sunscreens: An EORTC case control study in Germany, Belgium and France. *International Journal of Cancer* 61:749–755.

Bevis, M., S. Businger, T. A. Herring, C. Rocken, R. A. Anthes, and R. H. Ware. 1992. GPS meteorology: Remote sensing of atmospheric water vapor using the global positioning system. *Journal of Geophysical Research* 97:15787–15801.

Biggs, W., A. R. Edison, J. D. Eastin, K. W. Brown, J. W. Maranville, and M. D. Clegg. 1971. Photosynthesis light sensor and meter. *Ecology* 52:125–131.

Bodhaine, B. A., N. B. Wood, E. G. Dutton, and J. R. Slusser. 1999. On Rayleigh optical depth calculations. *Journal of Atmospheric and Oceanic Technology* 16:1854–1861.

Bruegge, C. J., J. E. Conel, R. O. Green, J. S. Margolis, R. G. Holm, and G. Toon.1992. Water vapor column abundance retrievals during FIFE. *Journal of Geophysical Research* 97:18759–18768.

Businger, S., S. R. Chiswell, M. Bevis, J. Duan, R. A. Anthes, C. Rocken, R. H. Ware, M. Exner, T. Van Hove, and F. S. Solheim. 1996. The promise of GPS in atmospheric monitoring. *Bulletin of the American Meteorological Society* 77:5–18.

CIE. 1926. *Commission Internationale de l'Éclairage Proceedings, 1924*. Cambridge, UK: Cambridge University Press.

Dutton, E. G. and J. R. Christy. 1992. Solar radiative forcing at selected locations and evidence for global lower tropospheric cooling following the eruptions of El Chichon and Pinatubo. *Geophysical Research Letters* 19:2313–2316.

Eberhard, W. L. 2010. Correct equations and common approximations for calculating Rayleigh scatter in pure gases and mixtures and evaluation of differences. *Applied Optics* 49:1116–1130.

Federer, C. A. and C. B. Tanner. 1966. Sensors for measuring light available for photosynthesis. *Ecology* 47:654–657.

Fleagle, R. G. and J. A. Businger. 1963. *An introduction to atmospheric physics.* New York: Academic Press.

Fröhlich, C. 2006. Solar irradiance variability since 1978: Revision of the PMOD Composite during Solar Cycle 21. *Space Science Reviews* 125:53–65.

Gueymard, C. 1995. SMARTS2, Simple model of the atmospheric radiative transfer of sunshine: Algorithms and performance assessment. Rep. FSEC-PF-270-95, Florida Solar Energy Center, Cocoa Beach, FL.

Gueymard, C. 2001. Parameterized transmittance model for direct beam and circumsolar spectral irradiance. *Solar Energy* 71:325–346.

Gueymard, C. A. 2004. The sun's total and spectral irradiance for solar energy applications and solar radiation models. *Solar Energy* 76: 423–453.

Gutman, S. I., R. B. Chadwick, D. E. Wolfe, A. M. Simon, T. Van Hove, and C. Rocken. 1994, September. Toward an operational water vapor remote sensing system using GPS. FSL Forum, Forecast Systems Laboratory, Boulder, CO, 13–19.

Gutman, S. I., D. E. Wolfe, and A. M. Simon. 1995, December. Development of an operational water vapor remote sensing system using GPS: A progress report. FSL Forum, Forecast Systems Laboratory, Boulder, CO, 21–32.

Hansen, J. E. and L. D. Travis. 1974. Light scattering in planetary atmospheres. *Space Science Reviews* 16:527–610.

Harder, J, W., J. M. Fontenla, P. Pilewskie, E. C. Richard, and T. N. Woods. 2009. Trends in solar spectral irradiance variability in the visible and infrared. *Geophysical Research Letter* 36:L07801, 5 pp. doi:10.1029/2008GL036797

Harrison, L., J. Michalsky, and J. Berndt. 1994. Automated multifilter rotating shadow-band radiometer: An instrument for optical depth and radiation measurements. *Applied Optics* 33:5118–5125.

Henyey, L. and J. Greenstein. 1941. Diffuse radiation in the galaxy. *The Astrophysical Journal* 93:70–83.

Herman, J., A. Cede, E. Spinei, G. Mount, M. Tzortziou, and N. Abuhassan. 2009. NO_2 column amounts from ground-based Pandora and MFDOAS spectrometers using the direct-sun DOAS technique: Intercomparisons and application to OMI validation. *Journal of Geophysical Research* 114: D13307, 20 pp. doi:10.1029/2009JD011848

Holben, B. N., T. F. Eck, I. Slutsker, D. Tanre, J. P. Bluis, A. Setzer, E. Vermote, J. A. Reagan, Y. J. Kaufman, T. Nakajima, F. Lavenu, I. Jankowiak, and A. Smirnov. 1998. AERONET—A federated instrument network and data archive for aerosol characterization. *Remote Sensing of Environment* 66:1–16. doi:10.1016/S0034-4257(98)00031-5

Holick, M. F. 2005. The vitamin D epidemic and its health consequences. *Journal of Nutrition* 135:2739S–2748S.

Lord Rayleigh. 1871. On the light from the sky, its polarization and color. *Philosophical Magazine* 41:107–120, 274–279.

Kasten, F. 1966. A new table and approximate formula for relative optical air mass. *Archives for Meteorology, Geophysics, and Bioclimatology–Series B: Climatology, Environmental Meterology, Radiation Research* 14:206–223.

Kasten, F. and A. T. Young. 1989. Revised optical air mass tables and approximation formula. *Applied Optics* 28:4735–4738.

Komhyr, W. D. (with revisions by R. D. Evans). 2008. *Operations handbook—ozone observations with a Dobson spectrophotometer.* GAW No. 193, WMO/TD-No. 1469.

Kuo, Y., Y. Guo, and E. R. Westwater. 1993. Assimilation of precipitable water vapor measurements into a mesoscale numerical model. *Monthly Weather Review* 121:1215–1238.

Liou, K. N. 1992. *Radiation and cloud processes in the atmosphere.* New York: Oxford University Press

Madronich, S. 1993. UV radiation in the natural and perturbed atmosphere. In *Environmental effects of UV (ultraviolet) radiation*, ed. M. Tevini, 17–69. Boca Raton, FL: Lewis Publishers.

Madronich, S. and S. Flocke. 1999. The role of solar radiation in atmospheric chemistry. In *Handbook of environmental chemistry*, ed. P. Boule, 1–26. Heidelberg, Germany: Springer-Verlag.

McKinlay, A. F. and B. L. Diffey. 1987. A reference action spectrum for ultraviolet induced erythema in human skin. In *Human exposure to ultraviolet radiation: Risks and regulations*, ed. W. F. Passchier and B. F. M. Bosnajakovic, 83–87. Amsterdam: Elsevier.

Michalsky, J., F. Denn, C. Flynn, G. B. Hodges, P. Kiedron, A. Kountz, J. Schlemmer, and S. E. Schwartz. 2010. Climatology of aerosol optical depth in north central Oklahoma: 1992–2008. *Journal of Geophysical Research* 115:1–16. doi:10.1029/2009JD012197

Michalsky, J. J. and E. W. Kleckner. 1984. Estimation of continuous solar spectral distributions from discrete filter measurements. *Solar Energy* 33:57–64.

Michalsky, J. J., Q. Min, P. W. Kiedron, D. W. Slater, and J. C. Barnard. 2001. A differential technique to retrieve column water vapor using sun radiometry. *Journal of Geophysical Research* 106:17433–17442.

Michalsky, J., J. Schlemmer, W. Berkheiser III, J. L. Berndt, and L. C. Harrison. 2001. Multiyear measurements of aerosol optical depth in the atmospheric radiation measurement and quantitative links programs. *Journal of Geophysical Research* 106:12099–12107.

Reagan, J. A., K. Thome, B. Herman, and R. Gall. 1987. Water vapor measurement in the 0.94 micron absorption band: Calibration, measurement, and data applications. *Proceedings of the International Geosciences and Remote Sensing Symposium* IEEE 87CH2434-9, 63–67.

Ricchiazzi, P., S. Yang, C. Gautier, and D. Sowle. 1998. SBDART: A research and teaching software tool for plane-parallel radiative transfer in the earth's atmosphere. *Bulletin of the American Meteorological Society* 79:2101–2114.

Rocken, C., R. H. Ware, T. Van Hove, F. Solheim, C. Alber, J. Johnson, and M. G. Bevis. 1993. Sensing atmospheric water vapor with the global positioning system. *Geophysical Research Letters* 20:2631–2634.

Rottman, G. 2005. The SORCE mission. *Solar Physics* 203:7–25.

Schmid, B. and C. Wehrli. 1995. Comparison of Sun photometer calibration by use of the Langley technique and the standard lamp. *Applied Optics* 34:4500–4512.

Schmid, B., P. R. Spyak, S. F. Biggar, C. Wehrli, J. Sekler, T. Ingold, C. Maetzler, and N. Kaempfer. 1998. Evaluation of the applicability of solar and lamp radiometric calibrations of a precision sun photometer operating between 300 and 1025 nm. *Applied Optics* 37:3923–3941.

Schnaidt, F. 1938. Berechnung der relativen Schnichtdicken des Wasserdampfes in der Atmosphäre. *Meteorologische Zeitschrift* 55:296–299.

Shettle, E. P. and S. Anderson. 1995. New visible and near IR ozone absorption cross-sections for MODTRAN. In *Proceedings of the 17th Annual Review Conference on Atmospheric Transmission Models*, ed. G. P. Anderson, R. H. Picard, and J. H. Chetwynd, 335–345. PL-TR-95-2060, Phillips Laboratory, Hanscom AFB, MA.

Smirnov, A., B. N. Holben, A. Lyapustin, I. Slutsker, and T. F. Eck. 2004. AERONET processing algorithms refinement, AERONET Workshop, El Arenosillo, Spain.

Stamnes, K., S. C. Tsay, W. Wiscombe, and K. Jayaweera. 1988. Numerically stable algorithm for discrete-ordinate-method radiative transfer in multiple scattering and emitting layered media. *Applied Optics* 27:2502–2509.

Thome, K. J., B. M. Herman, and J. A. Reagan. 1992. Determination of precipitable water from solar transmission. *Journal of Applied Meteorology* 31:157–165.

Walker, J. H., R. D. Saunders, J. K. Jackson, and D. A. McSparron. 1987. *NBS Measurement Services: Spectral Irradiance Calibrations*. National Bureau of Standards Special Publication 250-20.

Yuan, L. L., R. A. Anthes, R. H. Ware, C. Rocken, W. D. Bonner, M. G. Bevis, and S. Bussinger. 1993. Sensing climate change using the global positioning system. *Journal of Geophysical Research* 98:14925–14937.

13 Meteorological Measurements

But unless we patiently and accurately carry on solar observing throughout this generation, our successors will be as much at a loss as we are...

Charles Greeley Abbot
(1872–1973)

13.1 INTRODUCTION

The World Meteorological Organization (WMO) is working to bring standards to the measurements affecting weather, climate, and water. In 2008, WMO issued its seventh edition of the Commission for Instruments and Methods of Observation (CIMO) guide to meteorological measurements (http://www.wmo.int/pages/prog/www/IMOP/publications/CIMO-Guide/CIMO_Guide-7th_Edition-2008.html). The CIMO guide is written to encourage standardized meteorological measurements around the world.

This chapter covers the basic elements of the CIMO guide for the measurement of ambient temperature, relative humidity, wind speed and direction, and pressure. In addition, advantages and disadvantages of typical practices are discussed.

13.2 AMBIENT TEMPERATURE

Temperature is related to the average kinetic energy of molecules of a substance. The fundamental temperature scale is the Kelvin scale based on the average kinetic energy of an "ideal" gas. An "ideal" gas is composed of noninteracting point particles and is used because idealized gases can be described and analyzed with great precision using statistical mechanics. The basic physical properties of gases are most easily understood using an ideal gas as the example, and most gases behave in a manner very similar to an ideal gas under most circumstances.

The triple point of water has a defined kelvin temperature of 273.16 K (Preston-Thomas, 1990). The triple point of a substance is the state in which gas, solid, and liquid forms can coexist in thermodynamic equilibrium. A unit kelvin is 1/273.16 of the thermodynamic temperature of the triple point of water.

Historically, the centigrade scale is 1/100 the difference between the temperature of melting ice and boiling water under standard atmospheric pressure (101,325 Pascal). Centigrade is now called the Celsius scale after Celsius, who worked on developing the scale. The Celsius scale is now defined so that 0°C = 237.15 K and a change of 1°C is equal to a change of 1 K. Note that when referring to degrees kelvin the "°" symbol is not used because it is for the Fahrenheit and Celsius scales.

As with the precise definitions for the temperature scale, it is necessary to clearly define how air temperature is measured to have consistency in measurements. Important considerations are airflow over the sensor, height of the sensor above the ground, environment in which the temperature is measured, and the shielding of the sensor from direct sunlight. These details are also important for most meteorological and solar and infrared radiation measurements.

13.2.1 TYPES OF TEMPERATURE SENSORS

There are two common ways to measure temperature: with (1) a thermocouple; or (2) a resistive temperature detector (RTD), or a thermistor. A thermocouple is made of two dissimilar metals that generate a voltage proportional to the temperature difference between the junctions of the two metal junctions. This is the same device used for thermopiles in pyranometers. Thermocouples are very versatile, operate over a wide range of temperature with varying linearity, and have a fast response time but also require a measuring device that provides a stable reference and linearization. Thermocouples are difficult to recalibrate because their response is dependent on the environment in which they are used.

RTD detectors depend on the known change of resistance of the material as a function of temperature. RTDs are stable over time, are more accurate over their given range, and are easier to recalibrate. RTDs work over a limited temperature range compared with thermocouples and can be fragile in harsh industrial environments.

The most common types of thermometers used in automated weather stations are platinum resistance thermometer (PRT) detectors. They are considered the most accurate of thermometers and operate accurately over most ambient temperature ranges.

Specifications for a good temperature detector include an operating range of –30 to 45°C, with an accuracy of better than 0.2°C with a deviation of less than 0.1°C over any 10°C range. Initial calibrations should be done at accredited laboratories with periodic performance checks using reference PRTs.

13.2.2 RESPONSE TIMES

Air temperature can fluctuate up or down a degree K or two over a few seconds. However, the reported value is the average ambient temperature, so it is not necessary to have thermometers that react on timescales of seconds. The WMO guidelines recommend that a thermometer register 63.2% (or more exactly $1-e^{-1}$) of a step change in air temperature in 20 seconds.

13.2.3 MEASURING TEMPERATURE

For meteorological measurements, ambient air temperature (also called "dry bulb" temperature) should be measured between 1.2 and 2 meters above the ground. For the best temperature measurements, the ground should be level with little obstruction from surrounding objects such as trees or structures. Temperature measurements on the tops of building are affected by the vertical temperature gradient and the effect of the building itself.

FIGURE 13.1 Cotton Region Shelter use for temperature measurements. Temperature and relative humidity sensors are inside this shelter.

Solar radiation directly from the sun or the sky dome or reflected from the ground and surrounding objects can heat the temperature sensor and give false readings of ambient air temperature. Therefore, it is necessary to shield the thermometer from such radiation. Radiation shields also help protect the detector from rain and fog, which can artificially depress the temperature. Radiation shields can provide naturally aspirated or forced air measurement environments. For example, the U.S. Weather Bureau (now the National Weather Service) adopted the Cotton Region Shelter for air temperature measurements beginning in the nineteenth century (Köppen, 1915) (see Figure 13.1). The current Automated Surface Observing System (ASOS) uses a ventilated instrument housing, or aspirator, to move air across the senor (see Figure 13.2). Regardless of the mounting design, the thermometer should be thermally isolated from the radiation shield to prevent conductive heating or cooling, especially if the thermometer is not ventilated.

Temperature measurements should have a steady airflow over the detector, and, if it is ventilated, the airflow should be between 2.5 and 10 m/s. High-efficiency fan motors should be used because the motor can slightly warm the air passing over the motor, especially if the flow rate is low enough. An example of a meteorological tower is shown in Figure 13.3 with the temperature and relative humidity sensors in a radiation shield and a tipping bucket rain gauge on the side.

FIGURE 13.2 Image of an ASOS weather station in Milford, Utah. (Photograph courtesy of the Western Regional Climate Center.)

As with any measurement it is useful to fully describe the surroundings in which the measurements are made and to record and date any changes that are made to the surroundings.

Information on temperature measurements made with glass or other types of thermometers, minimum–maximum temperature measurements, and wet bulb measurements can be found in the WMO Guide to Meteorological Instruments and Methods of Observation. This guide is recommended reading before purchasing meteorological equipment and setting up a station.

13.3 WIND SPEED AND WIND DIRECTION

Individual air molecules move at a variety of speeds and directions. One can describe the motion as a vector in three-dimensional space with the speed assigned to the magnitude of the vector. As the vector evolves over time, the generalized motion (or flow) of the air is obtained. Often with wind measurements, one is dealing with surface flow and measurements are limited to wind speed and wind direction in a plane 10 m above the ground level. A more complete characterization of wind flow would require a measurement of the wind in the vertical direction as well as in the horizontal plane. Various wind profilers have been designed based on remote-sensing techniques. Light detection and ranging (LIDAR) and sound detection and ranging (SODAR) technologies are two such commercially available technologies.

In addition, there is considerable interest in the gustiness of the wind (i.e., rapid fluctuations). Therefore, peak wind speed and standard deviations of the wind speed and direction are also useful. Often for meteorological purposes wind speeds and directions are averaged over 10 minutes and are used for forecasting. Wind speed measured over a 1-minute interval is often termed a long gust.

FIGURE 13.3 Picture of a met tower with the temperature and relative humidity sensor in a radiation shield mounted halfway up the unistrut beam. The lightning finial is on the top of the beam. The rain gauge is on another unistrut beam on the right.

13.3.1 Sensor Terminology

The time response of an instrument is the time it takes to detect and indicate a change. More specifically, given an input step change, the response time to register 63.2% of the step change is defined at the response time. For anemometers, it is more appropriate to use a representative response length to characterize the instrument. For example, if the wind speed is 5 m/s and the response time is 1 sec, then the response length would be 5 meters. For many anemometers, the higher the wind speed, the quicker the response time. Therefore, the instrument response to changes in wind speed is characterized as response length. Response length is approximately the passage of wind (in meters) required for the output of a wind-speed sensor to indicate 63.2% of a step-function change of the input speed. For most cup and propeller anemometers, the response time for acceleration is faster than for deceleration. This leads to an overestimation of actual average wind speed.

An important topic of wind vanes is the critical damping factor. When there is a change in wind direction, there is a tendency for the vane to overshoot and oscillate until it is pointing in the right direction. The critical damping factor is the damping value for a step change in wind direction without overshooting the value. The damping ratio is the ratio of the actual damping factor to the critical damping factor.

13.3.2 ANEMOMETER

There are two commonly used anemometers: (1) the rotary cup and (2) the propeller type. Pitot tubes measuring static and dynamic air pressures can be used and produce satisfactory results but are not commonly deployed. Advanced technologies, such as sonic anemometers, are being used in some new research tools and are now being incorporated in some monitoring systems for routine measurements.

Both the cup and propeller type anemometers consist of two components: a rotor and a signal generator. The rotors of both the propeller and cup instruments turn at a speed proportional to the wind speed. The responsivity of the instruments to changes in wind speed is characterized by a response length that is related to the moment of inertia of the rotor and other geometrical considerations (Busch and Kristensen, 1976; Coppin, 1982).

13.3.3 CUP ANEMOMETERS

The first anemometer on record was invented by Leon Battista Alberti, an Italian architect, in 1450. This consisted of a disk placed perpendicular to the airflow, and its rate of rotation depended on the wind speed. The hemispherical cup anemometer was invented in 1846 by Irish astronomer John Thomas Romney Robinson. The original cup anemometer had four cups, but work by Patterson (1926) showed that three-cup anemometers responded faster to changes in wind speed. Patterson's work also showed that the larger the ratio of the cup radius to the arm length the better the linearity of the response.

The response to increases in wind speed is typically faster than the response to decreases in wind speed, and this asymmetry in response results in overestimates of the wind speed (Kristensen, 1993, 1999). In addition, the cup instruments are sensitive to the vertical motion of the air while the propeller instruments are insensitive to the vertical component. The overestimate of the average wind speed can be as much as 10% depending on the anemometer type and characteristics.

The rate of rotation of the rotor is roughly the difference between the wind speed and the initial wind speed necessary to rotate the rotor divided by the length of the column of air necessary to cause the rotor to rotate 1 radian. Therefore, the wind speed (ws) is obtained by multiplying the rotation frequency (rf) times a gain (g) plus the initial wind speed ($inws$) necessary to start the rotation (usually around 0.2 m/s). The gain is the length of the column of air necessary to cause the rotor to rotate 1 radian.

$$ws \approx rf \cdot g + inws \qquad (13.1)$$

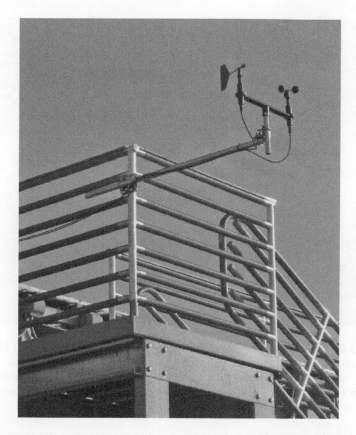

FIGURE 13.4 A cup anemometer and a wind vane are mounted from a pole several feet from the railing of a deck.

The friction that prevents the anemometer from rotating at very low wind speeds introduces a nonlinear response. Friction also varies with temperature. In practice, the anemometer bearings have a burn-in period before the instrument runs at a steady rate. This is sometimes done at the factory. After a year or two of use, the bearing friction starts to affect the anemometer and the bearings should be replaced. Cup anemometers also have a slight dependence on temperature and pressure. A thorough discussion of cup anemometers can be found in the IEA Wind Task 11 report on wind speed measurements and cup anemometers.

An example of a cup anemometer and a wind vane are shown in Figure 13.4.

13.3.4 PROPELLER ANEMOMETERS

H. W. Dines first published information on the propeller anemometer in 1887. In the early 1940s the Bendix Friez Company developed the "Aerovane" anemometer under contract with the U.S. Navy. Modern helicoid (propeller) anemometers are based on the Aerovane design (propeller anemometers are shown in Figure 13.5). The helicoid propeller is mounted on a rod with a vane at one end to pivot the

FIGURE 13.5   (See color insert.) Wind anemometers—sonic and vane. Pacific Northwest National Laboratory has developed a special filtering technique to enhance data gathered by an inexpensive cup anemometer. With the technique, cup anemometer measurements are surprisingly close to those made with a costly and sophisticated sonic anemometer (center of photo). (Courtesy of Warren Gretz, NREL staff photographer.)

propeller into the wind. Above the wind speed that overcomes the friction of the bearing, the rate of rotation of the propeller is proportional to the wind speed.

The propeller responds mainly to the airflow parallel to the axis of rotation. By mounting three propeller anemometers in an orthogonal array, the three-dimensional airflow can be measured. These instruments have been used to study turbulence. To achieve accurate results, corrections must be applied to account for the instruments' cosine response with respect to wind direction and the interaction of the wind with the support arms.

Disadvantages of propeller anemometers are that they need to point into the wind and there is a lag time or they may overshoot the direction. Therefore, the propeller is not always pointed directly into the direction of wind flow. This is particularly a problem with low wind speed or if there are eddies in the wind flow. Airflows vertical to the propeller and parallel to the plane of the vane can also cause problems.

For fixed-propeller anemometers, response decreases as the angle from the direction of airflow increases. For example when the angle reaches 85 degrees from the direction of airflow, the amount of air needed to push the propeller through 1 radian increases by a factor of 3.

In theory, the instrument should not require calibration except at low wind speeds. In practice, calibrations should be performed to obtain accurate measurements at low wind speeds and to check for excessive drag as the bearings begin to wear.

Overspeed problems should be less than with cup anemometers because the propeller is pointed into the wind.

13.3.5 SONIC ANEMOMETERS

Sonic waves travel at different speeds depending on the direction of the path with respect to the flow of the wind. Measuring the time for the pulse to travel between the source and a receiver can then be used to calculate the wind speed. A sonic anemometer is shown in Figure 13.5. The typical frequency of a sonic anemometer is about 100 kHz.

$$W_x = \Delta t \cdot kRT \Big/ 2d \tag{13.2}$$

where the wind speed in the "x" direction (W_x) is equal to the time it takes to travel the distance (Δt) times the velocity of sound squared (or equivalently kRT, where k is the ratio of specific heats, R is the gas constant, and T is the air temperature) divided by twice the path length (d). Some sonic anemometers can measure with a resolution as low as 0.005 m/s.

Sonic anemometers are ideally suited for measuring the structure of turbulence because they have high resolution and precision. However, the presence of precipitation can affect the operation of the instrument. In addition, the geometry of the sensing heads introduces distortions that can affect wind speed measurements.

13.3.6 INSTALLING ANEMOMETERS

When mounting anemometers on a meteorological tower, it is important to minimize distortions caused by the tower and cabling. It is best to locate the anemometer at the top of the tower. Any lightning rod should be far enough away from the anemometer to minimize its effect. The location of the meteorological tower should also avoid local obstructions. The standard measurement exposure for wind speed and direction is 10 m above the ground over level, open terrain (WMO, 2008). Because wakes from trees and buildings can easily extend downwind to 12 or 15 times the obstacles height, a properly located measurement system should be at least 10 obstruction heights away for proper "wind fetch."

For boom-mounted anemometers, the anemometer should be mounted above the cabling. For cylindrical wind towers, the boom supporting the anemometer should be at least 6 diameters from the support structure. If the wind comes from a prevalent direction, the minimum distortion occurs when the boom is mounted 45° to the prevailing direction of the wind.

For a lattice mast, the boom should be mounted perpendicular to the prevalent wind direction. To reduce distortion of the measurements, the anemometer should be mounted 12 to 15 times the boom diameter above the boom (Hunter, 2003). To reduce distortions, cylindrical booms should be used. In addition, angular structures should be avoided along with untidy cabling. A good distance also should be kept from tower guys and lightning finials. An example of a wind tower is shown in Figure 13.6. The tower is a narrow pole with wind measurements at two heights. The tower is well guyed for stability, and a lightning finial is at the top of the pole

FIGURE 13.6 A wind tower with guys. Wind sensors are located at two levels with guy wires attached below the wind sensors. The tower is a pole with the wires running to the wind sensors securely attached to the pole. A lightning finial is mounted on top of the tower.

and away from the wind sensors. It is important to firmly guy the wind tower to the ground. The energy in a gust of wind is proportional to the cube of the wind speed. Even wind gusts of 25 m/s (roughly 60 miles per hour) carry a lot of energy.

Note that some anemometers can freeze in position during cold periods, especially if there is little airflow. This happens especially if there is freezing rain. Some anemometers have heaters to prevent freezing. An additional source of information on wind monitoring is Schreck, Lundquist, and Shaw (2008).

13.3.7 WIND VANES

Wind vanes have been around since at least 450 BCE and probably decorated homes and public buildings much earlier. Besides providing decorative art to the building, wind vanes let people know which way the wind blows. This is important information that is added to anemometer measurements, and wind direction measurements often are taken alongside wind-speed measurements.

Wind direction is measured in a clockwise direction from true north (geographic north, not local magnetic north). It is very important to align the wind vane due north when it is installed. Any alignment error will be reflected in the measurement. When wind speeds are less than 1 knot or 0.514 ms^{-1}, calm is reported, and direction is often given as 0. The abbreviation used for knot is kn. Typically, it takes a wind speed of about 1 knot flowing at a 45° angle to move the wind vane. This will vary by model. Accuracies are typically reported for steady winds above a certain speed, such as 5 ms^{-1}. With fluctuating wind speeds and directions, the wind vane can over-shoot an angle and oscillate about the true direction. These oscillations are usually damped exponentially.

There are many different ways to sense the direction the vane is pointing, ranging from potentiometers to rotary switches. Wind vanes with potentiometers will be discussed here. Wind vanes that use a potentiometer require a voltage to function. About one-half the excitation voltage should be recorded when the vane is pointed at 180°.

The shaft of the wind vane should be vertical. Deviations by more than a degree or two will impart systematic errors into the readings. Resistors, hence potentiometers, are dependent on temperature, and this temperature dependence can affect readings in extreme temperatures. Like the anemometer, a heater is sometimes an additional component that can be purchased for the wind vane. The heater should be powered by a separate unit from the one that powers the potentiometer.

It is useful to calibrate the wind vane periodically to test for problems. One can often start by making measurements at 180° and then subsequent measurements at 15° increments. Calibrations that are out of spec require replacement. The manufacturer should be contacted to make sure the component is replaced correctly.

Wind speed and wind direction information are often plotted in a figure called a wind rose (see Figure 13.7). The wind rose is a polar plot with the wind direction going from 0° to 360°. The distance from the center of the plot is the frequency that the wind is from a given direction. Often the length of the frequency is divided into wind-speed ranges.

13.4 RELATIVE HUMIDITY

The water vapor content of air can vary considerably over the day and affects many aspects of our life from comfort level to the performance of electronic devices. There are several ways to measure the water vapor content of our atmosphere, including most commonly the dew-point temperature and relative humidity measurements. Relative humidity, U, is the percent ratio of the observed vapor pressure to the saturated vapor pressure of water at the temperature and pressure at the time of observation. Relative humidity measurements will be discussed since one can calculate the water vapor content of the atmosphere given temperature, pressure, and any measurement of water vapor and since relative humidity measurements can be made automatically.

A chilled-mirror hygrometer is a field and reference instrument used to measure dew-point temperature. The hygrometer works by cooling a surface until dew or ice

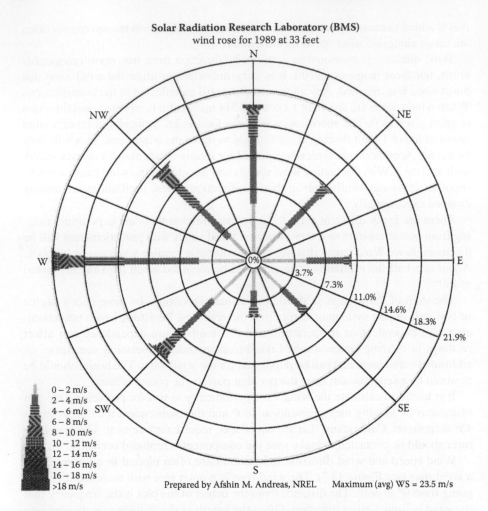

Solar Radiation Research Laboratory (BMS)
wind rose for 1989 at 33 feet

0 – 2 m/s
2 – 4 m/s
4 – 6 m/s
6 – 8 m/s
8 – 10 m/s
10 – 12 m/s
12 – 14 m/s
14 – 16 m/s
16 – 18 m/s
>18 m/s

Prepared by Afshin M. Andreas, NREL   Maximum (avg) WS = 23.5 m/s

FIGURE 13.7 Example of a wind rose plot from the NREL MIDC website. Data are for all of 1989 for an anemometer 10 meters above the ground. Wind rose plots can represent wind data from days to years, depending on the time period selected. The wind directions are divided into eight groups, one group every 45°. Each circle represents 3.7% of the data. The different segments represent different wind speed regimes. For example, the plot shows that approximately 3% of the time, the wind blowing from the southwest is between 4 m/s and 6 m/s. (Note that the bars run horizontally.)

forms on the surface. The chilled-mirror hygrometer has a solid, nonhygroscopic surface, which does not absorb or release water vapor. The dew-point temperature (T_d), or frost-point temperature (T_f), if it is cold enough, is the temperature at which dew, or ice, forms on the surface. Most chilled-mirror hygrometers use a polished metal mirror that is heated electronically by a Peltier-effect heater. An optical detector determines when moisture starts to condense on the mirror, and the temperature of the surface is then recorded. A pressure measurement is needed to calculate the relative humidity.

Still in use today, the sling psychrometer is used to measure relative humidity. The sling psychrometer measures both the ambient temperature and dew-point temperature, and the relative humidity can be determined using psychrometric charts. Many hygrometers are used to measure relative humidity such as the hair hygrometer, the chilled-mirror dew-point hygrometer, lithium chloride heated condensation hygrometer (dew cell), hygrometers using absorption of electromagnetic radiation, and electrical resistive and capacitive hygrometers. Only the electrical resistive and capacitive hygrometers will be discussed in this book because they lend themselves to automated measurements. For information on the other methods of measuring relative humidity see the CIMO guide.

Electrical resistance hygrometers have a plastic surface with a conductive surface on top of a nonconductive substrate. Water molecules adhere to the surface, changing the conductivity. A water-attracting (hygroscopic) electrolyte is used, and the conductivity is a function of the amount of water adsorbed. Alternating current is used to avoid polarizing the electrolyte.

Hygrometers based on electrical capacitance use solid hygroscopic material and measure the dielectric properties of the material. Water binds to the polymer, and the large dipole moment of water alters the dielectric property of the material. A capacitor is created by sandwiching the polymer foil between two electrodes. The capacitance is usually small, on the order of a few hundred picofarads. Therefore, it is necessary to have the electrical interface near the probe to minimize the electrical and capacitive effects. The effect of temperature on the probe must also be taken into consideration.

Electrical resistive and capacitive hygrometers usually are packaged with PRT temperature sensors. Enclosures that block the sun also prevent rain from wetting the probes. These probes may be damaged if they come in contact with liquid water.

The range of measurement for these probes can vary from 1 to 100%. These hygrometers have a temperature dependence of around 0.05%/°C. The response time is measured is seconds (roughly 10–20 seconds) and depends on the temperature. Typically, the accuracy of these probes in the field is 2 to 3% with the larger uncertainty above 90%. Stability of these hygrometers is about 1% per year. Of course, manufacturer specifications should be checked.

The response of these instruments to water vapor is nonlinear, and formulas are often supplied by the manufacturer that render the measured values into percent relative humidity. Relative humidity probes should be calibrated annually, either at the factory or by use of salt solutions that have known water vapor pressures. A good hair hygrometer can be used to spot check the calibration of these hygrometers.

13.5 PRESSURE

Atmospheric pressure is the force per unit area exerted by the weight of the atmosphere. The unit of pressure is the Pascal (Pa) or Newton per square meter. Typically hectopascals are the units used for pressure. 1 hPa = 100 Pa as 1 hPa is equivalent to 1 millibar (mbar), the old scale of measurement. 1013.250 hPa is equivalent to 760 mm of mercury (Hg), or one standard atmosphere of pressure.

The invention of the mercury barometer has been credited to Evangelista Torticelli in 1643, although there were discussions of similar barometers around the same time or earlier. The mercury barometer compares the pressure of a column of mercury against air pressure. The height of the column of Hg is proportional to the atmospheric pressure. Temperature, gravity, and the meniscus are among the most prevalent sources of error that can affect the reading of a mercury barometer.

Several types of electronic barometers are used in automated weather stations. These barometers use materials that respond to pressure and use transducers to translate these changes into electrical signals. Some of these sensors are triply redundant to improve accuracy and have temperature sensors mounted internally to provide automatic temperature compensation.

13.5.1 ANEROID DISPLACEMENT TRANSDUCERS

Aneroid displacement transducers use materials that expand or contract in response to changes in pressure. Typically, a metal alloy of beryllium and copper are used for the instrument and expansion or contraction is measured by levers that are connected to a device to display the pressure. Alternatively, designs can be created to generate electronic output.

13.5.2 PIEZORESISTIVE BAROMETERS

Piezoresistive pressure sensors work by running a current through a crystal that changes resistance resulting from deformations in the structure. The current then goes through temperature compensation circuitry and is amplified. Piezoelectricity is the charge that occurs when pressure or strain is applied to certain solid materials. Piezoresistivity and piezocapacitance are the changes in resistivity and capacitance that occur when strain is applied. Piezoresistive barometers use crystalline quartz as the sensor material because it has well-defined and stable piezoelectric properties. Strain changes interatomic spacing, and this affects the bandgap, making it easier or harder for electrons to be raised into the conduction band. The strain results in a change in resistivity of the semiconductor that is proportional to the strain being applied. A frequency control circuit is connected to the piezoelectric oscillator, and alternating current (AC) resistance is proportional to the pressure being applied. Again, crystalline quartz is usually chosen because frequency characteristics are precisely reproducible and the dependence on temperature is relatively small.

The gage factor, the change in resistance per change in length, is typically about 2 for metals, 10 to 35 for poly silicon, and 50 to 150 for single crystalline silicon (Schubert, Jenschke, Uhlig, and Schmidt, 1987; Smith, 1954; Tufte and Stelzer, 1963).

Pressure at an inlet port moves a flexible bellows upward and applies an upward axial force that creates a compressive force on the quartz crystal. The quartz crystal diaphragm is very rigid, and the pressure creates minimal deflections. Strain gages are attached around the circumference of the diaphragm or diffused into the silicon. A silicon chip with diffused strain gauges can act as the diaphragm itself.

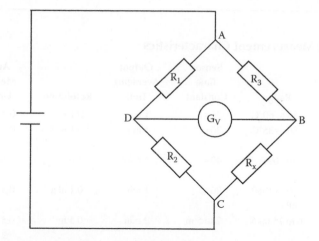

FIGURE 13.8 Generic drawing of a Wheatstone bridge used to determine an unknown resistance. A gauge, G_V, measures the voltage between points D and B. Resistor R_X is unknown resistance.

A silicon wafer can be fabricated to serve as the pressure sensor. A thick silicon rim supports a monocrystalline silicon diaphragm. Etching away the bulk of the silicon creates the diaphragm. Two to four piezoresistors are then diffused into the diaphragm. The silicon is doped with n-type atoms that provide an extra conduction electron to the silicon. Diffusion of p-type atoms, atoms that seek an electron to complete a pair bond, is applied to small squares to create piezoresistors. The piezoresistors are arranged in the Wheatstone bridge configuration to achieve higher voltage sensitivity and low temperature sensitivity.

A generic Wheatstone bridge (Figure 13.8) has three known resistances and one unknown. When the bridge is balanced, there is no current flow from point B to point D in Figure 13.8. When the bridge is unbalanced by the change in resistance of the strain gages, a current is created and this current is proportional to the change in resistance and, hence, pressure. Some piezoresistive barometers have thermostats to eliminate temperature effects.

Modern piezoresistive barometers measure two resonance frequencies of the piezoelectric element. By calculating a linear function of these frequencies, and with an appropriate set of variables obtained after calibration, a pressure is calculated by a microprocessor that is independent of the temperature of the sensor. Accuracies of piezoresistance barometers can reach 0.01%.

13.5.3 Piezocapacitance Barometers

In a piezocapacitance barometer, a metal or silicon diaphragm acts as one terminal of the capacitor. The other terminal is a metal substrate on a ceramic or glass substrate. Pressure deflects the diaphragm, which in turn changes the gap spacing and the capacitance. In a differential piezocapacitance barometer, the silicon diaphragm is between the two electrodes and the pressure deforms the diaphragm and

TABLE 13.1

Operational Measurement Characteristics

Variable	Range	Sensor Time Constant	Output Averaging Time	Resolution	Achievable Measurement Uncertainty
Air temperature	−80 to 60°C	20 s	1 min	0.1 K	0.2 K
Dew-point temperature	−80 to 35°C	20 s	1 min	0.1 K	0.5 K
Relative humidity	0 to 100%	40 s	1 min	1%	3%
Pressure	500 to 1080 hPa	20 s	1 min	0.1 hPa	0.3 hPa
Wind speed	0 to 75 ms^{-1}	2 to 5 m	2 min	0.5 m^{-1}	0.5 ms^{-1} for ≤ 5 ms^{-1} 10% for > 5 ms^{-1}
Wind direction	0 to 360°	1 s	2 min	1°	5°

increases the capacitance on one side and decreases the capacitance on the other. Piezocapacitance barometers can have an accuracy of 0.1%.

13.6 RECOMMENDED MINIMUM ACCURACIES FOR OPERATIONAL INSTRUMENTS

When planning for meteorological measurements, it is important to get a feeling for the resolution, data recording intervals, uncertainties, time constants, and achievable measurement uncertainties of meteorological variables. The Guide to Meteorological Instruments and Methods of Observation (2008) is a good reference for this type of information. Table 13.1 shows some of the information for the measurements discussed in this chapter.

QUESTIONS

1. What international organization sets the standards for meteorological measurements?
2. If the temperature is 300 K, what is the temperature in Celsius? Fahrenheit?
3. Name two types of temperature sensors.
4. What is an aspirator, and how does it help provide a more accurate temperature measurement?
5. What is dew-point temperature, and how is it measured?
6. Name three types of anemometers.
7. What is an anemometer's response length?
8. A zero reading for wind direction means the wind is blowing from what direction?
9. What are some of the problems that affect wind measurements?

10. What is a wind rose?
11. What does a sling psychrometer measure, and how does it work?
12. Relative humidity sensors do not respond linearly to changes in atmospheric moisture. How is this nonlinear behavior corrected?
13. Name two types of pressure sensors, and describe how each one works.
14. What is the difference between resolution and measurement uncertainty? Which is larger?
15. What are the international units for measuring atmospheric pressure?

REFERENCES

Busch, N. E. and L. Kristensen. 1976. Cup anemometer overspeeding. *Journal of Applied Meteorology*,15, 1328–1332.

Coppin, P.A., 1982. An examination of cup anemometer overspeeding. *Meteorologische Rundschau*, 35, 1–11.

Hunter, R. S. (Ed.). 2003. *Expert group study on recommended practices for wind turbine testing and evaluation - 11. Wind speed measurement and use of cup anemometry* (2d ed.). International Energy Agency Wind Productions.

Köppen, V. 1915. A uniform thermometer exposure at meteorological stations for determining air temperature and atmospheric humidity. *Monthly Weather Review* 42:389–395.

Kristensen, L. 1993. *The cup anemometer and other exciting instruments*. Technical Report R-615(EN), Risø National Laboratory, Roskilde, Denmark.

Kristensen, L., 1998. Cup anemometer behavior in turbulent environments. *Journal of Atmospheric and Oceanic Technology*. 15:5–17.

Kristensen, L. 1999. The perennial cup anemometer. *Wind Energy* 2:59–75.

Kristensen, L. 2005. Fragments of the cup anemometer history. Available at: http://www.cup-anemometer.com/technical/The%20Cup%20Anemometer%20History.pdf

Patterson, J. 1926. The cup anemometer. *Transactions of the Royal Society of Canada*, Ser. III 20:1–54.

Preston-Thomas, H. 1990. The International Temperature Scale of 1990 (ITS-90). *Metrologia* 1990(27):3-10, 107.

Schreck, S., J. Lundquist, and W. Shaw. 2008. *U.S. Department of Energy workshop report: Research needs for wind resource characterization*, June. NREL/TP-500-43521.

Schubert, D., W. Jenschke T. Uhlig, and F. M. Schmidt, 1987. Piezoresistive properties of polycrystalline and crystalline silicon films. *Sensors and Actuators* 11(2):145–155.

Singh R., L. L. Ngo, H. S. Seng, and F. N. C. Mok. 2002. A silicon piezoresistive pressure sensor. Proceedings of the First IEEE International Workshop on Electronic Design, Test and Applications (DELTA 2002). Christchurch, New Zealand.

Smith, C. S. 1954. Piezoresistance effect in germanium and silicon. *Physics Review* 94:42–49.

Tufte, O. N. and E. L. Stelzer. 1963. Piezoresistive properties of silicon diffused layers. *Journal of Applied Physics* 34:313–318.

WMO. 2008. WMO guide to meteorological instruments and methods of observation, WMO-No. 8 (7th ed.). Chapter 8. Available at: http://www.wmo.int/pages/prog/www/IMOP/publications/CIMO-Guide/CIMO_Guide-7th_Edition-2008.html

COLOR FIGURE 3.23 The WRR is determined by a group of absolute cavity radiometers named the World Standard Group. At the moment, the WSG is composed of six instruments: PMO-2, PMO-5, CROM-2L, PACRAD-III, TMI-67814, and HF-18748.

COLOR FIGURE 5.3 Picture of a copper-constantan thermopile manufactured by Eppley Laboratory, Inc. Constantan wire is wrapped around a solid core and then dipped into a copper bath. (Photograph by Warren Gretz, NREL staff photographer.)

COLOR FIGURE 5.4 Picture of Kipp & Zonen CM 22 pyranometers in ventilators at SRRL in Colorado.

COLOR FIGURE 5.12 Three ventilators mounted on a Sci-Tech tracker. Going from right to left, the ventilator on the left is a Swiss PMOD ventilator holding a Kipp & Zonen CM 22 pyranometer. The ventilator is mounted on an Eppley ventilator base so the height of the dome matches the height of the domes of the other instruments. PMOD also makes a base for leveling. The middle ventilator is a Schenk ventilator for Schenk Star pyranometers. The ventilator on the left is an Eppley ventilator holding an Eppley precision infrared radiometer or pyrgeometer.

COLOR FIGURE 6.3   The Eppley model SBS shadowband blocks direct sunlight on a clear summer day in Eugene, Oregon. Note that the detector surface is well centered beneath the shadowband. The shadowband blocks a significant part of the brightest region of the clear-day skylight.

COLOR FIGURE 6.4   An example of a shaded-disk measurement of DHI. The arm holds disks that shade each of three black-and-white pyranometers.

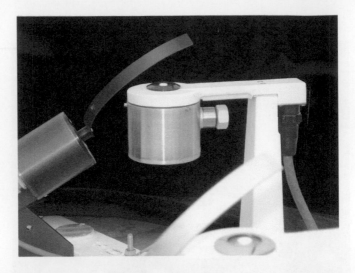

COLOR FIGURE 7.2 The Yankee Environmental Systems, Inc. seven-channel multifilter rotating shadowband radiometer model MFR-7.

COLOR FIGURE 9.8 Photograph of the SURFRAD albedo tower near Sioux City, South Dakota. Note the shields that ensure the direct sunlight will not reach the upwelling detector even for minor horizontal leveling errors.

COLOR FIGURE 10.5 Three Eppley PIR pyrgeometers operated in the optimum manner; the instruments are continuously shaded from direct sunlight using an automatic tracker and are placed in heated ventilators to minimize dew formation and dust deposition on the instrument domes.

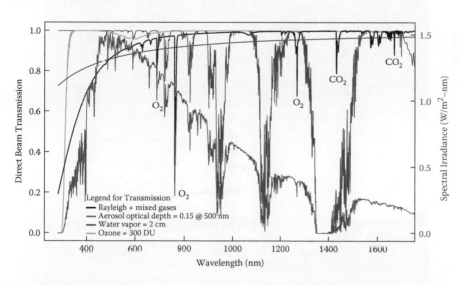

COLOR FIGURE 12.3 Direct solar transmission of major components of the clear-sky atmosphere (left axis); superimposed on the transmission spectra is the spectral irradiance calculated for these conditions (right axis).

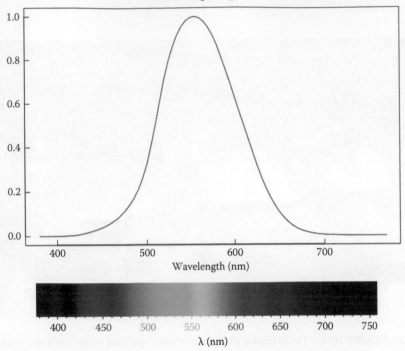

COLOR FIGURE 12.4 The typical human eye's response for sunlit conditions as specified by the Commission International l'Eclairage (CIE); en.wikipedia.org/wiki/File:Srgbspectrum. png is the source for the color diagram beneath the photopic response plot.

COLOR FIGURE 13.5 Wind anemometers—sonic and vane. Pacific Northwest National Laboratory has developed a special filtering technique to enhance data gathered by an inexpensive cup anemometer. With the technique, cup anemometer measurements are surprisingly close to those made with a costly and sophisticated sonic anemometer (center of photo). (Courtesy of Warren Gretz, NREL staff photographer.)

COLOR FIGURE 14.10 Photograph of the SURFRAD station near Sioux Falls, South Dakota.

COLOR FIGURE 14.11 Photograph of the SRRL station in Golden, Colorado, looking north. Sample of instruments in continuous use (foreground) and under test at SRRL. Note platform on which the instruments are mounted is a grate that helps to prevent snow from building up around instruments.

COLOR FIGURE 14.13 Photograph of the University of Oregon Solar Radiation Monitoring Laboratory monitoring station in Eugene, Oregon. Note that most tilted measurements use a shield to block some of the ground-reflected irradiance.

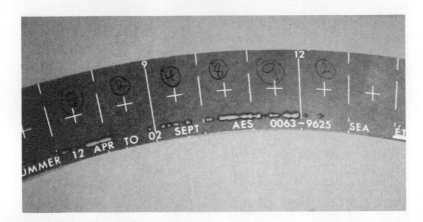

COLOR FIGURE H.1 Northern hemisphere summertime Campbell–Stokes measurement of sunshine duration. Fractional hours with bright sunshine are estimated by an analyst and hand recorded.

14 Setting Up a Solar Monitoring Station

The question always arises as to the true reliability of any measurement performed with an instrument under different conditions than those of the calibration.

John R. Hickey
(1970)

14.1 INTRODUCTION

Considerable planning and forethought should go into the design and layout of a solar monitoring station. Initial considerations should identify project goals and data accuracy requirements for the project. Once the basic parameters for the solar monitoring station are established, it becomes easier to select between various design options. The optimum site specifications and location, instrument selection, data acquisition system functions, and the operation and maintenance requirements for the measurement equipment all directly affect the quality of the data that will be produced. Budget constraints and continuity of funding for station operation and maintenance are also important. Specifically, high-performance (expensive) instrumentation should not be considered if the project resources are insufficient to support the maintenance and calibration schedule recommended by the sensor manufacturer and to purchase auxiliary supporting equipment necessary to obtain the accuracy specified. A well-chosen, designed, and constructed solar monitoring station will enhance the quality of collected data and help minimize the time and effort necessary to service the station and validate the data. Three factors usually dictate the location of a solar monitoring station: (1) a clear field of view (adequate solar access); (2) proximity to a site operator who will maintain the station; and (3) most importantly, the requirements of the station sponsor who will rely on the data.

While the main focus of this book is to characterize the instruments used in measuring solar and infrared irradiance, the resulting data are only as good as the station installation, operation, and maintenance allows. To expand on the limited number of publications available on this topic (McArthur, 2005; Stoffel et al., 2010), this chapter discusses many of the details necessary to locate, construct, and maintain a quality solar and infrared radiation monitoring station. Specifically, this chapter discusses factors to consider in choosing a site and determining the requirements for the data-logging system. Maintenance of the station is described along with best practices for cleaning the instruments and creating standard procedures to record station maintenance. The importance of viewing the data on a regular basis is discussed along with other quality control measures. Next, suggestions for field calibrations of the radiometers are presented. Finally, examples of solar monitoring stations are shown.

14.2   CHOOSING A SITE

Often the general location of the solar station is selected to fill certain requirements, such as providing data for a solar electric facility or characterizing the climate for a specific ecosystem. However, just as important, and maybe more so, is the ease of access and maintenance. If the station is located close to the person responsible for servicing the instrumentation and the measurement system, the maintenance can be performed on a more regular basis. Access to electricity and to remote communication is important, although photovoltaic panels, cellular modems, and satellite data links can provide sufficient power and communications to even the remotest stations.

Before selecting a location for the monitoring station, it is important to identify the person or group who will maintain the station. Working with the station maintenance personnel, the potential locations with the best field of view as well as optimum maintenance access and security should be identified. It should be noted that finding a location that is totally unobstructed as the sun moves across the sky throughout the year is difficult. Trees, buildings, and power poles are among the many structures that present problems. To identify the significance of the problem and to compare potential locations, it may be useful to obtain a sun path diagram for the site under consideration and take the diagram out to the location under consideration. Sun path diagrams are discussed in Chapter 2 and can be created using the Sun Chart Program from the University of Oregon (http://solardata.uoregon.edu/SunChartProgram.php). Sun path charts for every 10° of latitude are provided in Appendix C. Use a compass to locate magnetic north, and make the magnetic declination corrections so that the sun path diagram can be aligned facing the equator (south in the northern hemisphere and north in the southern hemisphere). A horizontal plot of the obstructions can be made with the compass and clinometer to mark the location and angular height of the obstructions. There are tools available that can take photographs of a location and superimpose sun path charts on the panoramic image of the site. These tools are designed for siting solar systems, and, while they can be used to quickly identify obstructions, they do not necessarily produce accurate enough obstruction height information near the horizon (Figure 14.1).

To simplify descriptions for the rest of this chapter, it will be assumed that the station is located in the northern hemisphere. For stations in the southern hemisphere, switch the terms north and south. During the summer you will notice that the sun will rise and set somewhat behind you if you are facing the equator. The farther away from the equator the site is located, the farther behind one's back the sun will rise or set during the summer months.

A site with no obstructions higher than 5° above the horizon is ideal, but sites with obstructions up to 10° are acceptable, especially if the obstructions do not block the sun's path. Sometimes obstructions cannot be avoided, such as nearby hills or trees. If at all possible, keep obstructions out of the sun's path and as far back (north) as possible from the east–west axis. If the obstructions are behind the site, they will not block the sun's path during the winter when the sun is lowest in the sky. If there is an obstruction that is not too tall (e.g., less than 15° above the horizon) or wide, then it may be possible to locate the station so that the obstruction is directly south of the station. Study the sun path chart, and make sure that the sun passes over the

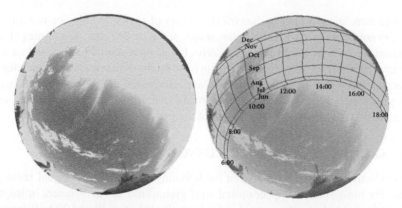

FIGURE 14.1   Image of Ashland, Oregon, station using the Solmetric Suneye. The Suneye first takes a fisheye view of the station, shown on the left-hand side of the image, and then places a grid of the path followed by the sun during different times of year, shown on the right-hand side of the image. Local standard time is given in the image. Objects on the horizon that block the sun's path are then analyzed, and the percent of the shaded area is then calculated.

obstruction during the winter months. If the structure is too tall, it may be necessary to move the station farther east or west to avoid shading as much as possible.

Trees can be a particular problem for solar-monitoring sites. Trees grow, and while they may not block the sun path at first this may not be the case several years later. Therefore, it is prudent to know future plans for development near the site. Sometimes the station can be mounted on a raised platform or a rooftop to avoid blockage by some local obstruction. If the station is located on the roof of a building, adequate and safe access to the equipment should be incorporated into the design.

Some solar stations are located near meteorological towers that are used to monitor the wind resource. It is important to locate the station to the south of such towers and to make sure the tower is not highly reflective. In fact, any object that is likely to reflect light toward the pyranometers should be removed or painted with a flat black, gray, or nonreflective coating.

Some stations are located in areas that are subject to snow and snow drifts. The solar-monitoring stations should be located high enough above ground to avoid direct reflections from the snow and should be constructed so that minimal snow accumulates on the platforms around the pyranometers and other instruments. Nonconductive gratings used for walkways can make useful platforms. If the temperature gets cold enough, enclosures containing the data logger or other equipment may need electrical heaters controlled by thermostats.

The equipment in the boxes should be mounted so that moisture from condensation cannot drip on electric connections. If appropriate, desiccant packs can be used to help keep sealed boxes dry. Sometimes "Weep Holes" at the base of the enclosure are used to let the condensate drain freely. Electrical connections to the boxes should be laid in conduit or contain a drip loop so that water will not run down the cables and into the electronics box. Signal cables and power cables should be separated as much as possible to avoid electrical interference with the instrument signals.

While most snow is easy to dust off the domes of pyranometers, ice is a different story. Do not chip the ice off the pyranometer dome or diffuser. Sometimes alcohol can be used to melt the ice from a glass dome, but alcohol should *not* be used on white (acrylic) diffuser disks. Other times, a warm hand on the dome or disk will melt the ice. A sponge soaked in *warm* water can also help. Hot water should be avoided because too large a temperature differential might break seals, crack the glass, or cause other damage to this important optical element of the pyranometer or pyrheliometer.

14.3   GROUNDING AND SHIELDING

A solid electrical ground is important for lightning protection and signal noise protection. For most areas, a copper-coated steel ground rod 1.5 to 3 meters in length is sufficient for grounding. The ground rod is typically between 10 and 20 millimeters in diameter. To drive the ground rod into dirt, a heavy metal pipe with an end cap can be used. For areas of frequent lightning strikes, a more robust grounding strategy might be needed. For more detailed information on grounding and shielding see Morrison (1998).

All masts and metal poles should be grounded to allow an easy path for lightning to follow to ground. A lightning finial is often recommended for areas subject to more frequent lightning strikes. Surge protectors such as metal oxide varistors or avalanche diodes can be used for lightning protection. These surge protectors degrade over time, and manufacturers' recommendations should be followed when determining when these surge protectors need to be replaced.

Sensitive equipment usually has a ground lug that can be used for grounding. A 10 or 12 gauge wire (AWG units) with 2.6 mm to 2.0 mm diameters, respectively, should be used for connecting the equipment to the central grounding point.

Twisted pair shielded cable should be used for low-voltage signals and should be connected as a differential measurement at the data logger. By isolating the signal low from ground, some of the noise can be avoided. The shield should be connected at one end to the ground and should be left unattached at the other end to prevent ground loops. Signal cables should also be physically isolated from power cables, and if they have to cross they should cross at right angles to minimize any induced signals.

14.4   DATA LOGGER AND COMMUNICATIONS

Many data-logging systems are available; however, it is important to have data loggers that can withstand field environmental conditions for many years without changing their electrical specifications and maintaining their specifications over the range of temperatures experienced at the station. Providing a temperature-controlled enclosure should help to reduce the subtle effects caused by temperature drift, but this increases the electrical power requirements for the system. A good outdoor data logger should be able to maintain specifications for the temperature ranges expected at the measurement site.

The clock on the data logger should have minimal drift. Some data loggers now have auxiliary global positioning systems (GPSs) that automatically maintain the clock accuracy. This should be considered for any new station, especially those that sample with high frequency such as every second. Some data loggers come with software that will automatically synchronize the data-logger clock with the computer clock. If that is the case, then one must ensure that the computer clock is correct and doesn't experience drifts. Coordinated Universal Time (UTC) and local time in the United States can be found at http://www.time.gov, and either is recommended for data logging. Daylight saving time is not recommended. Sometimes it is better to monitor the data logger time and change it only if it has drifted more than one or two seconds since the previous check.

The data logger should be selected to have enough analog inputs, voltage excitation sources, and control ports to handle the number of sensors being installed at the site. Plan to have room for additional instruments. Over time, requirements change, and it is helpful if data loggers don't have to be changed to accommodate additional measurements. Some data loggers come with auxiliary multiplexers to handle additional channels of data.

The amount of storage in the data logger is an important consideration. The more frequent the sampling and the greater the number of instruments, the greater the amount of storage required to ensure continuous data reporting. The data storage should be sufficient to account for the longest time between data transfers by a central computer.

Data loggers should be able to handle a variety of inputs. Some anemometers generate pulses that need to be counted, temperature sensors require a reference voltage, and many pyranometers need data loggers that can measure with μV precision. Make sure the data logger can handle the signals from the sensors being chosen and from sensors that might be needed in the future.

Accuracy and linearity of the data logger is important and should be better than the sensors being installed. Data loggers with a 0.5% accuracy or better are needed to avoid adding to the measurement uncertainty of the data.

It is important to be able to communicate with and download data from the data logger. It is very helpful if the data logger can be queried remotely and the data logging routines can be changed from a central location. Remote communication enables troubleshooting from the central office where technical expertise is available. Communications can be through a phone modem, Internet, cellular phone, satellite phone, or use of special communication networks. Over the long-term, communication costs can add up, and it is important to be able to change to new methods as they are developed. For ease of access, Internet communication is best, but simple phone modems are very reliable if a phone line is available. One should calibrate the data logger periodically, just as the radiometers should be calibrated. This helps ensure the quality of the data and also helps spot potential problems.

Power failures should be expected in the field, even if the system is connected to the utility grid. The data logger should have a battery backup system and should be able to automatically recover data and programs if the power outage drains the battery. It is very helpful if the data logger draws only small amounts of power. This minimizes the size of the backup power needed and provides for the opportunity to use photovoltaic (PV) panels for the power source.

Software packages that come with the data logger are useful, but the software should be able to output data from the logger in a format that is readable to a wide variety of computer programs or comes with a translator that can export the data to an ASCII file.

With solar irradiance measurements, the ideal would be to have the sensor read continuously over the time interval. This would yield an integrated average of the sensor being read. The practical alternative is to have the sensor read at very short intervals of a second or two and these readings averaged over the time interval of interest. If one is interested in the variance of the irradiance over the time interval, it would be necessary to have a data logger that could record the standard deviation of the measurement over the time the measurement is averaged. Similarly, recording the maximum and minimum signal during the averaging period can provide useful information.

Other factors to consider are the complexity of programming the data logger, readable and informative instruction manuals, and a quick response to requests for assistance with problems. The data logger is a key component, and reliability and backup are needed to make the data set as complete as possible.

14.5 MEASUREMENT INTERVAL

Choice of time interval for the measurements is often specified by the requirements of the user of the data. For example, many electric utilities use 15-minute readings for their power monitoring. However, irradiance data has many uses, and where one time interval is preferred by one user another time interval may be preferred by another. With shorter interval data, one can always sum or average into longer time intervals, but one cannot take a 15-minute data set and divide it into a 5-minute data set without introducing additional uncertainties.

Lately, there has been interest in 1-minute or shorter time interval data. With short time intervals one must look at the sensor's response time, and unless one is examining the response time of a sensor the time intervals chosen should be long enough so that the sensor can provide an accurate measurement integrated over the time period (McArthur, 2004). Storing data in 10-minute intervals should be avoided because the data will not divide evenly into the utility standard of 15 minutes.

Note that most satellite-derived data come from images that are analyzed each half-hour, and the timing of the images does not necessarily fall on the hour or half-hour mark. The older GOES East geosynchronous orbit weather satellite produces images of the northern hemisphere 15 and 45 minutes after the hour. The GOES West geosynchronous orbit weather satellite produces images of the northern hemisphere on the hour and 30 minutes after the hour. To validate models that use satellite images to estimate solar data, the time sampling interval for ground-based measurements should be sufficiently short that it captures the data at the point in time when the satellite image is taken (Wilcox and Marion, 2008). Ground-based solar data integrated over 5-minute intervals around the time the satellite images were taken would provide a better comparison to the solar data derived from the satellite images than ground-based data taken at 15-minute or hourly intervals.

Some data loggers and computers will automatically change from daylight saving time to local standard time (LST) and vice versa. However, it is best to record data in either local standard time or UTC. Keeping the same time year round means that the data files can be continuous and don't have the jumps associated with daylight time. The matter of consistent data record time stamping is important for accurate solar position calculations needed for data quality assessment.

UTC relates global time to an atomic clock standard. GPS, computer time, and other equipment times are related to UTC time. The advantage of UTC is that it is the same throughout the world; the disadvantage is that the 24-hour UTC day divides the local solar day over two-day boundaries at many locations. LST is best for small networks and local data, especially if the data are to be accessed by the general public. UTC time will be found with larger databases or with satellite-derived data sets.

The granularity of the data (time interval) affects the precision to which changes in the solar irradiance can be perceived. Short time intervals can help in spotting problems with the data such as reductions in the output when the instrument is cleaned. Data applications dictate the maximum time interval needed for recording. A sample plot of 1-minute, 5-minute, 15-minute, and hourly data is given in Figures 14.2a–14.2d. Note that as the time interval increases, the information on the variability in the data becomes more and more obscured.

14.6 CLEANING AND MAINTENANCE

High-quality pyranometers require minimal maintenance other than cleaning the dome on a regular basis and making sure the desiccant does not become saturated with moisture. It should be months to a year between desiccant changes for a properly assembled pyranometer. If the desiccant needs to be changed more often, that is usually an indication that a seal has developed a leak (typically at the desiccant chamber/body interface). Gel desiccants that change color when they absorb moisture are recommended because they do not leave dust that is associated with some desiccants (calcium chloride should be avoided). The dust from a desiccant can travel to the surface of the sensor or to the inside surface of the dome, thus affecting the responsivity of the pyranometer. Occasionally, the rubber O-rings sealing the desiccant holder will need to be greased to keep from drying out and cracking. Grease that is used for O-rings on vacuum pumps works well as it does not outgas when heated and leave a residue inside the pyranometer.

Moisture, frost, or snow also blocks or scatters incident solar radiation. While dust may build up slowly over time and regular cleaning can minimize the effects of dust, frost or snow can affect the instrument immediately. A ventilator will keep most frost and light snow off a pyranometer while minimizing the accumulation of dust. Of course, heavy snow can accumulate on the shield or around the ventilator, so the platform should be constructed so that snow cannot accumulate around the instruments, for example, by mounting instruments on a grate.

Pyrheliometers are more susceptible to the effects of dust or moisture. A little dust or a drop of moisture on the instrument window can scatter the incident irradiance so that it does not reach the detector. For the most accurate results, the window of the pyranometer should be cleaned on a daily basis, or no less often than twice each week. This depends, of course, on the environmental setting of the measurements and time of year. It is important to note on a log sheet when the instruments were cleaned.

To clean the instrument windows or domes, distilled water is the preferred cleaning agent. Ethyl alcohol (ethanol) may be used in icing conditions, although if the temperature is above freezing during part of the day it is better to wait and clean with distilled water. Overuse of alcohol can cause the seals to dry out and crack. Cracked seals can allow moisture to enter the instrument and degrade the measurements. If alcohol is used, 95% alcohol should be used because the other 5% is water. With 99% alcohol, other chemicals are present that enable the purity to reach 99%. These

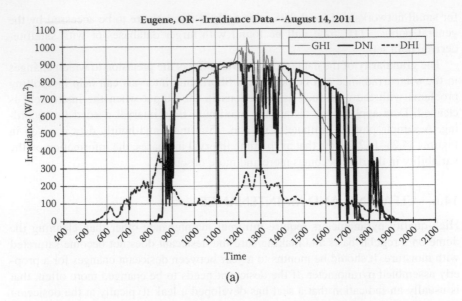

(a)

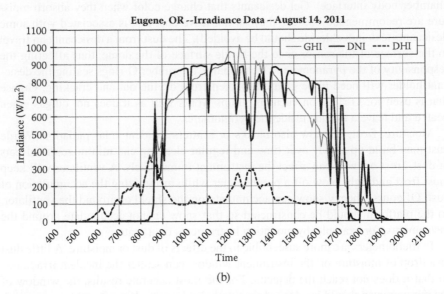

(b)

FIGURE 14.2 Plots of GHI, DNI, and DHI on August 14, 2011, at time intervals from (a) 1 minute, (b) 5 minutes, (c) 15 minutes, and (d) 1 hour. Information on the variability of the irradiance data becomes more and more obscured as the averaging time interval of the data increases.

chemicals may leave a residue. A lint-free cotton cloth or paper products designed for cleaning optical surfaces should be used to clean the radiometers. Any lint or substance left behind affects the performance of the instrument. If a cotton cloth is used, it should be laundered frequently to prevent grime from accumulating. It is also possible to use a can of pressurized air to dry the dome to avoid the microscopic

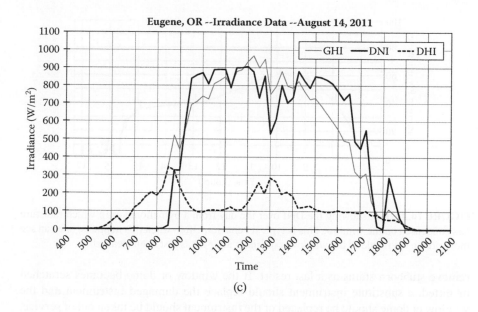

(c)

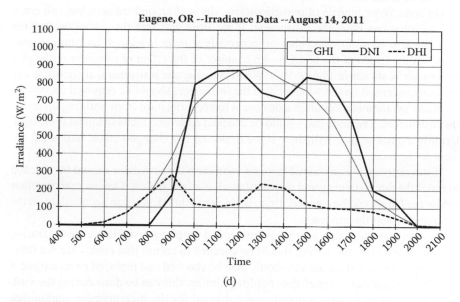

(d)

FIGURE 14.2 (Continued.)

scratching from rubbing the dome with a dry cloth. Be sure that the spray can does not have an oil-based propellant or other chemicals that can coat the surface of the window or dome.

Occasionally, chemicals from rainwater or treated water form a film on the instrument that cannot be cleaned by water or alcohol. A little vinegar should remove these water stains. A little cesium oxide that is used for polishing mirrors can be used to

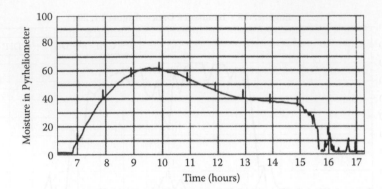

FIGURE 14.3 Plot showing the DNI over the day from a pyrheliometer in which moisture has condensed on the window. This is typical behavior. Often when this happens, one can see moisture on the inside of the window, especially after it has just been cleaned.

remove stubborn stains as a last resort. If the window or dome becomes scratched or pitted, a substitute instrument should replace the damaged instrument and the window or dome should be replaced or the instrument should be taken out of service.

On some older models of pyrheliometers, the seal around the window can crack and moisture can enter the instrument. If moisture gets into the pyrheliometer, the data begin to show strange behavior, especially if there is enough moisture to condense on the inside of the window. An example of the effect of moisture in the pyrheliometer is shown in Figure 14.3. When cleaning the instrument, if moisture can be seen condensed on the window then the instrument should be removed for repair.

Many pyranometers contain a desiccant to keep moisture from building up inside the instrument. The desiccant should be checked each time the instrument is cleaned. Usually a quick glance will show if the desiccant has changed color, indicating the need for replacement.

If the instrument is mounted inside a ventilator, the ventilator should be checked to see if the fan is running and the ventilator has a free airflow. Once a year, the filter below the fan should be cleaned. Dust, mold, and other debris can accumulate on the filter and inhibit the airflow.

At least one site visit a year is recommended for preventive maintenance. During an annual site visit, the wiring should be checked to ensure that cables are not deteriorating. Batteries at the station should also be checked and replaced or recharged if necessary. If cabinets or platforms require painting, this can be done during the visit.

It is important to have a maintenance manual for the measurement station (see Appendix I for troubleshooting hints). This will help people who tend the station know what needs to be done, what to look out for in the way of problems, and how to contact the appropriate person should a problem occur. These manuals can also help in training and can be use to refresh one's memory if a procedure has not been performed in a while. The manual should contain simple information like how to clean a pyranometer to more complex tasks like setting the clock on an automatic tracker.

Putting together the manual while planning the station is most helpful. Potential issues that might have been overlooked in the initial station design can arise when one is describing how to operate and maintain the station. This helps to document why certain decisions were made and is very useful when other stations are established. Make an electronic version of the manual so that it can be changed and modified as experience is gained with operating the station.

14.7 RECORD KEEPING

Keeping records of maintenance, site visits, instrument serial numbers, instrument calibration, and all activities associated with the station is essential. If there are questions about the data, the records can be examined to see if there were any activities that might explain the questioned data. Carefully maintained records enable one to go back and see when instruments were changed and what calibration numbers were used during particular time periods.

Log sheets should be available for those maintaining the site. They are useful reminders of the daily tasks and help record what maintenance was performed. An example log sheet is shown in Figure 14.4. Table 14.1 lists some information that might be included on the log sheet. NREL is working on an application for a table that would allow the technician to record work with automatic time stamping of each action and photograph the noteworthy observations at the time.

Room on the log sheet should be reserved to record the beginning and end times when the maintenance was performed. In addition, it is helpful to have a list of the solar declination for each day of the year. This is necessary if a fixed shadowband has to be adjusted or a manual tracker is used at the site. It is also useful information when checking the data. It is important to leave room on the log sheets for notes. Unusual events such as birds roosting or flying insects swarming near instruments should be recorded. A digital form of the log sheet should be available and the information on the paper log sheet should be transferred to the electronic database. The NREL Measurement & Instrumentation Data Center (MIDC) online maintenance log can be view at http://www.nrel.gov/midc/apps/maint.pl?BMS. The resulting metadata records of maintenance findings for each instrument are available as links from the data pages. For example, the maintenance records for the measurement of GHI using a CM22 pyrameter are accessible from http://www.nrel.gov/midc/apps/fmpxml.pl?type=maint;find=Global%20CM22;site=BMS.

14.8 IMPORTANCE OF REVIEWING DATA

It is helpful to review the data daily. To make viewing the data easier, a routine should be written to plot the data in time series. Data intervals of 15 minutes or less are best for visual inspection of the data. The trained eye can be more precise than computer programs and can usually spot a wider variety of problems. Viewing plots of the data routinely will help spot problems quickly and minimize the amount of bad data that needs to be flagged and possibly eliminated from the database. For example, if there are problems with the alignment of the pyrheliometer, a quick scan of the data will show the problem. Dirt can build on the dome of the pyranometer, and while it may not be easily visible to the eye it will show up as a depression on the plot of the calculated diffuse irradiance (see Figure 14.5). Misalignments of shadow balls or disks

Solar Measurement Station Maintenance Check List and Weather Log

Eugene, OR   From 1/23/2011 to 1/29/2011

	Sun	Mon	Tue	Wed	Thu	Fri	Sat
Date	Jan 23	Jan 24	Jan 25	Jan 26	Jan 27	Jan 28	Jan 29
Local Standard Time							
Observer's initials							

Key: ✓ = good condition   X = bad condition   O = corrected

Global horizontal PSP

Dome condition							
Ventilator							
Sensor level							
Desiccant							

Diffuse horizontal PSP

Dome condition							
Ventilator							
Sensor level							
Desiccant							
Shadow alignment							

Diffuse PSP shadow

Direct normal

Window condition							
Tracking alignment							
Adjusted tracker time							

NIP target

Shadowband

Declination	-19.9	-19.1	-18.9	-18.6	-18.4	-18.1	-17.9
Adjusted shadowband							

Cloud key: O = clear (<10%)   ◑ = scattered (10-50%)   ◑ = broken (50-90%)   ⊕ = overcast (>90%)

Cloud cover							
Temperature (°C)							
Time temperature read							

Solar cells washed

Comments (e.g.; ice or frost on pyranometer; additional site visit; change of instrument, chart paper, pen, etc.)							

FIGURE 14.4 Sample page from the log sheets used at the high-quality stations in the University of Oregon Solar Radiation Monitoring Network. It is important to note when the person was maintaining the instruments and who did the maintenance. It is also useful to leave room for comments. It is easier to remember all the tasks if the log sheet has a space for each task that should be performed.

TABLE 14.1
Types of Information on a Daily Log Sheet

Task	Preferred Performance Schedule	Status
Clean the instrument	Daily	Status before and after cleaning (note unusual build-up)
Check if desiccant is OK	Weekly	Color OK
Check if ventilator is working	Daily	Air flow, no air flow
Pyrheliometer alignment	Daily	Show alignment status
Weather	Daily	Show type of weather
Shade ball shadow covering pyranometer (DHI)	Daily	Yes or no or partially; adjust shadowing

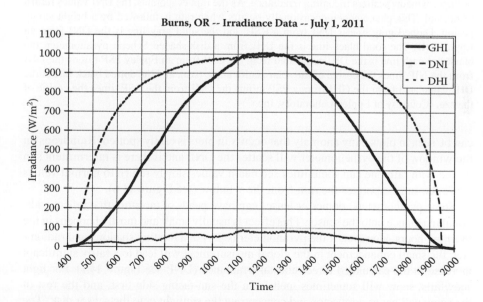

FIGURE 14.5 Plot of GHI, DNI, and calculated DHI on July 1, 2011, in Burns, Oregon. There is dirt on the pyranometer that causes a slight decrease in the GHI values. When subtracting the DNI times the cosine of the sza, the calculated DHI shows bumps. Because the DNI is relatively smooth, the bumpiness of the DHI indicates there are problems with the GHI. The pattern of the bumps for several days confirms the problem with the GHI data. Alcohol was required to clean the thin film that caused this problem.

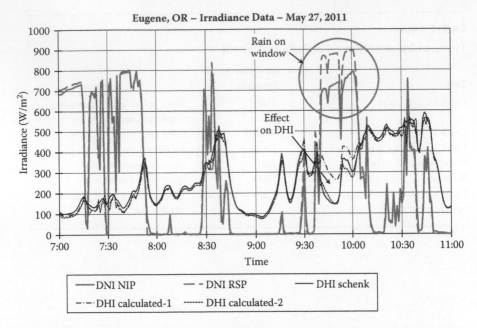

FIGURE 14.6 Plot of DNI and DHI on May 27, 2011. Raindrops on the front of the pyrheliometer window scatters incoming irradiance. As the rain evaporates, the DNI values return to normal. This phenomenon occasionally occurs when rain is followed by a bright sunny period. Dashed gray line is DNI from a RSP, and the solid gray line is the DNI from an Eppley NIP. The solid black line is the DHI from a disk-shaded Schenk pyranometer, the black dot-dash line is the DHI calculated from the GHI from an Eppley PSP minus the DNI from the NIP times the cosine of the solar zenith angle (sza), and the dotted black line is the DHI calculated from the GHI from a PSP minus the DNI from the RSP times the cosine of the sza. (Courtesy of Eppley Laboratory, Inc.)

can be seen in plots. Any anomaly that occurs in plots is easily spotted. Raindrops on the window of the pyrheliometer will scatter the DNI, and if there is intermittent sun mixed with rain one can sometimes see beam values slowly return to normal values while the sun evaporates the rain from the window (see Figure 14.6).

Some pyranometers' domes or lenses can have beads of moisture that reflect additional irradiance onto the sensor. This effect is usually small and mostly happens in the morning hours. However, if the pyranometer is part of a rotating shadowband radiometer, the small enhancement in the global measurement will lead to a fairly significant increase in the calculated direct normal irradiance (DNI; see Figure 14.7). For light snowfalls, snow will sometimes melt from the sun-facing side first, and the rest of the snow will act as a reflector and concentrate the sunlight onto the sensor disk. The enhanced global horizontal irradiance (GHI) can show up clearly when plotted.

Obstructions blocking the sun also show up on clear days. If a dip shows up in the same spot over several clear days, it is likely to be caused by an obstruction. In summary, there are many ways of checking the data using plots that can be useful, and while it does take a little time, a better, more complete data set will result.

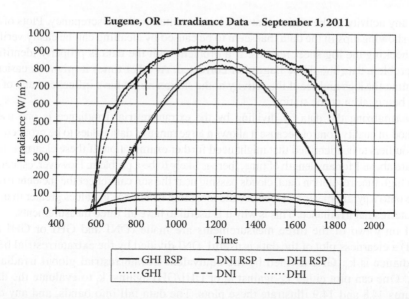

FIGURE 14.7 Plot of 1-minute RSP data along with reference data. There is dew on the diffuser of the LI-COR LI-200 pyranometer in the morning between 6:00 and 7:00. The GHI increases slightly while there is a large bump on the DNI, which is calculated by subtracting DHI from GHI and dividing by the cosine of the solar zenith angle. This RSP in this plot was calibrated against the DNI values and no correction algorithms have been applied to the DHI data.

14.9 QUALITY CONTROL OF DATA

Quality control of data is very important. Quality control can spot problems as well as give confidence in the data. Since irradiance data can vary rapidly and significantly, it is useful to have more than one instrument measuring the solar irradiance. The best quality control method is to have three measurements: the DNI, GHI, and diffuse horizontal irradiance (DHI). The GHI is equal to the DNI times the cosine of the solar zenith angle plus the DHI. If the values agree to within a few percent (e.g., ±3%), all three measurements are likely to be correct. If raindrops on the window of the pyrheliometer scatter some of the incident irradiance, the calculated global value may differ markedly from the measured GHI. Therefore, these checks can identify problems in the data.

It is important to have a procedure to ensure the quality of the data and to identify any data that are bad. For example, when an instrument is cleaned, a spike (either an increase or decrease) is often observed in the reported irradiance. This is especially true when 1 minute or shorter time-interval data are obtained. The data can be flagged or even corrected as long as the corrected data are labeled. The procedure for flagging the data or correcting the data should be documented. With data intervals longer than 5 minutes, the ease of observing problems decreases along with the accuracy to which corrections can be made.

One method of identifying problem data is to plot the data and compare them with notes on the log sheets. If an instrument's data start to deviate from the data of similar instruments at the station, then it is worthwhile to examine log sheets to see if there was any activity at the station that may have resulted in this discrepancy. Plots of the data showing dips in the data that seem to be caused by cleaning can be easily verified by checking the log sheet records. If a problem with the data is properly identified, the problem data should be flagged or corrected. At other times it might be easier to examine the logs and see if the activity at the station might have influenced any of the data being taken at the station.

Normalizing the data by dividing by the extraterrestrial irradiance is one useful method in quality control because it allows a large range of data values to be grouped and any outliers to be identified that might need further examination. If these clearness indices are above 1 for any length of time, there is likely to be a problem. However, especially with high time resolution data, clouds passing near the sun can reflect the incident irradiance onto a pyranometer and increase the incoming irradiance to values greater than the extraterrestrial irradiance. This is why it is useful to have all three measurements.

If only two of the three measurements are made (DNI and GHI or GHI and DHI) a clearness plot of the data is useful. DNI divided by the extraterrestrial beam irradiance is k_b; GHI divided by the equivalent extraterrestrial global irradiance is k_t. One can plot either k_b against k_t or DHI/GHI against k_t to evaluate the data. Figures 14.8 and 14.9 illustrate these plots. The data fall into bands, and any data

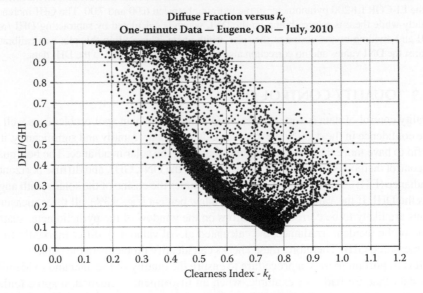

FIGURE 14.8 Plot of the diffuse fraction plotted against clearness index k_t for 1-minute data. Most of the data fall along the lower left-hand side of the plot. Those data points represent clear periods. A spread of data points to the left represents data when clouds are present. When there is no DNI, the diffuse fraction is 1, and GHI = DHI. When there is some sun present, the clouds can act as reflectors, and a wide variety of GHI values can occur for any diffuse fraction. Note the number of clearness index values near 1 or greater than 1. These GHI values are enhanced by sunlight reflecting off clouds and into the sensor. As the averaging time interval increases, the number of data points with k_t values near or greater than 1 significantly decreases. When plotting hourly average values, there are very few data points with k_t greater than 0.8.

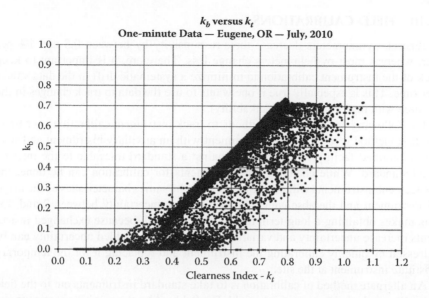

FIGURE 14.9   Plot of k_b versus k_t for 1-minute data. As with the diffuse fraction relationship, shown in Figure A.1, the majority of points during clear periods fall on the left-hand side of the plot. Values with the largest clearness index are not values with the largest beam clearness index values. This results from sunlight reflecting off clouds and into the global pyranometer. This reflected light is not seen by the pyrheliometer that measures the DNI values. Note that there are very few data points with k_b values above a maximum of 0.8. The maximum k_b values are somewhat dependent on altitude because higher altitudes have a lower air mass through which sunlight has to traverse.

outside the bands should be examined in more detail. Certain values, such as k_b greater than k_t, are not generally possible. For partially cloudy periods, GHI can be greater than the extraterrestrial radiation that results from reflections off of clouds adding to the GHI. This leads to clearness values greater than 1 and represents real data occurrences. These periods when the GHI are greater than the extraterrestrial radiation are short-lived and are not seen in 10-minute, 15-minute, or hourly data files (see k_b vs. k_t plots in Appendix A). Offsets or fluctuations when GHI values are small, say less than 10 Wm^{-2}, can lead to spurious data points. If the spread of the data is too great or if there are a lot of points outside the main grouping, then there may be problems with the data. Common problems are dirt on the pyranometer, misalignment of the pyrheliometer, misalignment of the shadowband, systematic errors with the instrument, or poor calibrations for the instrument.

For DNI and GHI measurements, the National Renewable Energy Laboratory (NREL, 1993) developed the SERI QC model that defines the region where most data points should fall. The more the data points fall outside the region, the larger the SERI QC flags. The SERI QC flags are found in the National Solar Radiation Data Base. With analysis of a lot of data, programs such as SERI QC can isolate possible problem data without visual inspection of every data point. SERI QC does not spot all problems, but it does spot the more egregious ones.

14.10   FIELD CALIBRATIONS

In general, pyranometers decrease their responsivity by between 0.5 and 1% per year, whereas most pyrheliometers change less. Therefore, it is important to keep track of the instrument calibration to minimize a systematic drift in the data values over time. This is especially true if one wants to use the data to track changes in the resource with high confidence (bankability).

The following are two of the methods to track instrument calibration over time. The first method is to replace the instrument with an auxiliary instrument and send it to a qualified facility for calibration against a standard traceable to the international standard. While changing out instruments for calibration can be done, the replacement instruments will not have precisely the same characteristics as the original instrument and the absolute calibration has an uncertainty between 2 and 3%. This makes obtaining a long-term record more difficult because exchanged instruments increase uncertainty and variation in the data. This added uncertainty can be addressed by quickly calibrating the instrument and replacing it for the temporary substitute instrument at the site.

An alternate method of calibration is to take standard instruments out to the field and calibrate the instruments in the field. For field calibrations, it is important to use the same type of instrument for calibration because each model of instrument deviates from a perfect response in its own characteristic manner. Therefore, one would not want to calibrate a photodiode pyranometer from one manufacturer with a photodiode pyranometer from another manufacturer or with a thermopile pyranometer. The best side-by-side calibrations are done by comparing the same model instruments from the same manufacturer. The reference instrument should be calibrated yearly against standards traceable to the international standards as discussed in the calibration sections of Chapters 4 and 5.

The station should be constructed so that it is easy to place the comparison instrument next to the instrument being calibrated. The instruments should be on the same level. If the field instrument is in a ventilator and the reference instrument is not using a ventilator, then the reference instrument needs a stand to hold it at the same level. If the calibration is being done on an automatic tracker, be sure that the instrument is secured to the platform. Cables from these temporary instruments should have enough play so that movement of the tracker will not stress the cable and be placed or safety flagged so that they will not present tripping hazards while working around the instruments.

It is also helpful to run the calibration overnight, especially for instruments that are subject to thermal offsets. An estimate of the thermal offset can be made looking at the nighttime values. In addition, any offsets or problems with the data logger or cabling stand out during the nighttime when there is no solar radiation input to the irradiance sensor.

Often other site maintenance tasks are performed when the instruments are being calibrated. Make sure that these tasks do not interfere with the calibrations, or if they do note the time and the questionable data can be removed.

Calibration numbers usually have a ±2–3% uncertainty. Less variation is introduced into the data set if the calibration number is not changed each time a new one is obtained provided the new calibration number is within the uncertainty of the last calibration. However, calibration values typically change less than 1% per year, and this is within the uncertainty of the calibration measurement. Plot the change in calibration over time. Once a consistent trend is observed, then that calibration trend can be used to adjust the data to the more accurate value.

Calibration numbers often depend to some degree on the ambient temperature. Note the temperature and irradiance level during the calibration. This information may be useful at a future date if the responsivity of the reference instrument is correlated with temperature.

Always keep good records during field calibrations. When the data are examined at a later date, it will be easier to address questions about questionable data values.

14.11   PHYSICAL LAYOUT OF A SOLAR-MONITORING STATION

When setting up a solar-monitoring station, it is important to keep a few principles in mind:

1. The instruments at a solar-monitoring station should have minimal shading, especially by other instruments at the station.
2. The instruments should be easy to access for cleaning and maintenance.
3. Wiring should be shielded from the environment and animals. Exposed wiring should be kept to a minimum.
4. Access to the station should be limited to those who need to maintain or calibrate the instruments.
5. Growth around the station should be manageable.
6. Station layout should allow for side-by-side calibrations when possible.

An example of a Surface Radiation (SURFRAD) Network station is shown in Figure 14.10. The various monitoring equipment are well separated to minimize shading. The meteorological tower is to the north of the solar-monitoring equipment. Most wiring is in conduit, and power cables are buried and protected. The equipment is on sturdy concrete pads, and the equipment is tied down securely to the pads. The vegetation is low growing and has been cleared around the instruments.

Two examples of the SRRL station in Golden, Colorado, are shown in Figures 14.11 and 14.12. A wide variety of instruments are used and tested at SRRL. The tower allows further testing to be done from tilted irradiance to wind speed, temperature, and relative humidity. Ground-reflected longwave and shortwave radiation are also measured along with net radiation. Some of these instruments are not easy to reach, such as the tilted pyranometers, and hence are not as likely to receive as much cleaning as instruments that are within easy reach. Of course, this tower is located near the SRRL main station, and there are people there to provide maintenance on a regular basis. The reference solar monitoring station for UO SRML network in Eugene, Oregon, is shown in Figure 14.13.

FIGURE 14.10   (See color insert.) Photograph of the SURFRAD station near Sioux Falls, South Dakota.

FIGURE 14.11   (See color insert.) Photograph of the SRRL station in Golden, Colorado, looking north. Sample of instruments in continuous use (foreground) and under test at SRRL. Note platform on which the instruments are mounted is a grate that helps to prevent snow from building up around instruments.

FIGURE 14.12 Photograph of the SRRL Baseline Measurement System's radiometer tower in Golden, Colorado. Located away from the elevated instrumentation platform shown in Figure 14.11, this auxiliary tower provides for a wide variety of radiometers at various orientations for continous measurements (http://www.nrel.gov/midc/srrl_bms).

FIGURE 14.13 (See color insert.) Photograph of the University of Oregon Solar Radiation Monitoring Laboratory monitoring station in Eugene, Oregon. Note that most tilted measurements use a shield to block some of the ground-reflected irradiance.

QUESTIONS

1. What are a few issues to consider when selecting a solar monitoring station?
2. How will sun path charts help in this process?
3. Why are site logs useful?
4. Why are field calibrations preferred if it is possible?
5. What is the advantage of having all three primary solar measurements, GHI, DNI, and DHI?
6. What is the primary cleaning fluid for solar instruments?
7. Explain why the calculated DHI is larger than the measured DHI in Figure 14.6.

REFERENCES

Hickey, J. R. 1970. Laboratory methods of experimental radiometry including data analysis. In *Advances in Geophysics* (vol. 14), eds. H. E. Landsberg and J. Van Mieghem; *Precision Radiometry*, ed. A. J. Drummond. New York: Academic Press.

Long, C. N. and Y. Shi. 2008. An automated quality assessment and control algorithm for surface radiation measurements. *Open Atmospheric Science Journal* 2:23–37.

McArthur, L. J. B. 2004. Baseline Surface Radiation Network (BSRN). Operations Manual, Version 2.1. WMO/TD-No. 1274, World Climate Research Programme, World Meteorological Organization, Geneva, Switzerland, WCRP/WMO.

Morrison, R. 1998. *Grounding and shielding techniques*, 4th ed. New York: John Wiley & Sons.

NREL. 1993. *User's manual for SERI_QC software- assessing the quality of solar radiation data.* NREL/TP-463-5608. Golden, CO: National Renewable Energy Laboratory.

Stoffel, T., D. Renne, D. Myers, S. Wilcox, M. Sengupta, R. George, and C. Turchi. 2010. Concentrating solar power: Best practices handbook for the collection and use of solar resource data (CSP). NREL Report No. TP-550-474651, 46 pp. http:// www.nrel.gov/docs/fy10osti/47465.pdf.

Wilcox, S. and W. Marion. 2008. User's manual for TMY3 data sets. Technical Report NREL/TP-581-43156, revised May 2008.

Appendix A: Modeling Solar Radiation

A.1 INTRODUCTION

Before incident solar radiation is measured, it's important to understand the characteristics and magnitude of the incoming irradiance and how the radiation is changed as it passes through the atmosphere. Solar radiation basics are discussed in Chapter 2, and spectral characteristics are covered in Chapter 12. From the existing data and with knowledge of the characteristics of solar irradiance, a wide variety of models have been created to estimate irradiance from meteorological parameters, atmospheric constituents, and a limited set of irradiance measurements. Some models exist to estimate solar radiation on clear days, and other models have been created to estimate direct horizontal irradiance (DHI) or direct normal irradiance (DNI) from global horizontal irradiance (GHI) data under all weather conditions. This appendix contains an overview of a variety of solar radiation models used to estimate irradiance values when there are limited or no solar radiation measurements at a site.

The first models covered in this appendix are the clear-sky models that estimate the DNI irradiance when clouds are not present. Clear-sky models are also referred to as clear-day models in some literature. The clear-sky modeling section is followed by the definition and initial discussion of the clearness index parameters, K_t, K_{df}, and K_b. The clearness index parameters are used with many of the other models. Next, models that estimate the diffuse fraction are described followed by a discussion of the beam–global (DNI–GHI) correlations. The diffuse fraction is the ratio of DHI divided by GHI. The diffuse irradiance is then obtained by multiplying the diffuse fraction by GHI. Estimating the irradiance on tilted surfaces will then be demonstrated. A summary of the International Energy Agency (IEA) report on model comparisons will then be given along with references to a variety of models.

A.2 CLEAR-SKY MODELS

As sunlight passes through the earth's atmosphere, the radiation is scattered (reflected), absorbed, and transmitted in varying amounts by the atmosphere before reaching the surface. At the surface, the amount of sunlight coming directly from the sun is called the direct normal irradiance (DNI). If the exact constituents of the atmosphere between the sun and the observer are known, the DNI can be calculated to a high degree of accuracy knowing the spectral composition of the extraterrestrial irradiance, the earth–sun distance at the time of interest, and the length of the path through the atmosphere. The physics behind this calculation is described by the Beer–Lambert Law.

In general

$$\text{DNI} = DNI_o \exp(-m^*(\tau_a + \tau_r + \tau_g + \tau_{NO_2} + \tau_w + \tau_{O_3}))$$ (A.1)

where DNI_o is the extraterrestrial beam irradiance, m is the air mass, τ_a is the *optical depth* due to aerosol absorption and scattering, τ_r is the optical depth due to Rayleigh scattering from atmospheric molecules, τ_g is the optical depth due to absorption by gases other than those specified such as carbon dioxide (CO_2) and oxygen (O_2) and nitrogen dioxide (NO_2), τ_{NO_2} is the optical depth due to absorption by nitrogen dioxide, τ_w is the optical depth due to water vapor absorption, and τ_{O_3} is the optical depth due to absorption by the ozone.

Exact calculations are complicated by the wavelength dependence of the absorption and scattering, by the nonuniform distribution of the atmospheric constituents, and by the lack of information on the exact amount of the constituents at the time of interest.

Clear-sky models were developed to be simple to use and to give an approximate estimate of DNI on clear days. Since the total solar irradiance (TSI) from the sun is constant to within a few tenths of a percent (the currently accepted value of 1,366 Wm^{-2} is used in this book) and the earth–sun distance is known with a considerable degree of accuracy, the incident irradiance outside the earth's atmosphere can be accurately calculated. The extraterrestrial direct normal irradiance is

$$DNI_o = 1366 Wm^{-2} \times (R/R_o)^2$$ (A.2)

where R is the distance from the earth to the sun for the day of interest, and R_o is the mean earth–sun distance. The mean earth–sun distance is called 1 AU (astronomical unit). Because of the elliptical orbit, the distance between the earth and the sun varies by about $\pm 1.6\%$ over the year. Since the sun's radiance can be treated as a point source, the irradiance is inversely proportional to the square of the separation distance; the incident solar radiation at the top of the atmosphere varies by approximately 6.6% from the closest approach (perihelion) around January 3 to the farthest distance (aphelion) around July 4.

On clear days, there are no clouds to interact with the DNI. However, other constituents of the atmosphere absorb and scatter the irradiance as it passes. Since the clear-sky model is an attempt to simplify a complex process, there is no exact prescription for obtaining the simplified model. In fact, over a dozen clear-sky models attempt to estimate the incoming irradiance as the DNI passes through the atmosphere (Atwater and Ball, 1978, 1979; Bemporad, 1904; Bird and Hulstrom, 1979, 1980; Davies and Hay, 1979; Hoyt, 1978; Kasten, 1964; Kondratyev, 1969; Lacis and Hansen, 1974; Mahaptra, 1973; McDonald, 1960; Moon, 1940; Vignola and McDaniels, 1988a; Watt, 1978). These models use the measured or average values of atmospheric elements that could interact with the sunlight to model the scattering or absorbance of the irradiance. Specifically, the amount of scattering depends on the amount and types of the aerosols in the atmosphere and the amount of water vapor, ozone, and other gases. Since it is difficult to know the exact mixture of these elements in the atmosphere and since the models were developed at different sites and with varying quality of instruments, it is not surprising that there are a variety of model estimates given the same initial conditions. Bird and Hulstrom (1981) provide a good survey of the clear-sky models.

A.3   CLEARNESS INDICES K_T, K_{DF}, AND K_B

Normalization of the incident radiation by dividing by the equivalent extraterrestrial component is very useful for irradiance modeling and data analysis. This helps to compare data from different times of year or different times of day. Normalization is especially useful when developing generalized correlations that hold true over the year or over the day. The averaging can be done over a minute, an hour, a day, or a month, depending on the available data and information required.

The standard formula for extraterrestrial DNI_o is given in the previous section for clear-sky modeling. Since DNI_o is used over various time periods, it is usual to obtain the value by averaging over the time period.

$$\overline{DNI_o} = \frac{24}{2\pi} \int_{ha_1}^{ha_2} DNI_o dha = \frac{24}{2\pi} DNI_o \cdot (ha_2 - ha_1) \tag{A.3}$$

where ha is the hour angle defined in Section 2.5. The factor $24/(2\pi)$ changes from hour angles to radians. The hour angles should also be expressed in radians. Since DNI_o changes by less than 0.2% per day, it is generally assumed to be constant over the day.

The global extraterrestrial value, GHI_o, is equal to DNI_o projected onto a horizontal surface by multiplying by the cosine of the solar zenith angle (computed from site location, day of year, and time of day as described in Appendix D).

$$\overline{GHI_o} = \frac{24}{2\pi} \int_{ha_1}^{ha_2} DNI_o \cdot (\sin(lat) \cdot \sin(dec) + \cos(lat) \cdot \cos(dec) \cdot \cos(ha)) \cdot dha \tag{A.4}$$

or

$$\overline{GHI_o} = \frac{24}{2\pi} DNI_o \cdot (\sin(lat) \cdot \sin(dec) \cdot (ha_2 - ha_1) + \cos(lat) \cdot \cos(dec) \\ \cdot (\sin(ha_2) - \sin(ha_1))) \tag{A.5}$$

The hour angles in Equation A.5 are in radians and not degrees. For daily $\overline{GHI_o}$, the sunrise and sunset hours as determined in Equation 2.11 can be used. For monthly averaged $\overline{GHI_o}$ the daily $\overline{GHI_o}$ should be averaged over the month. For many months, the middle of the month produces an adequate monthly averaged $\overline{GHI_o}$. However, for both June and December, the average $\overline{GHI_o}$ comes later or earlier, respectively, because the solar declination reaches a maximum or minimum during the month and the declination at the middle of the month is not typical of the average monthly declination (Klein, 1977; Klein and Theilacker, 1981). The extraterrestrial DHI_o is essentially zero as there are few atoms, molecules, or particles in space to scatter the DNI_o irradiance. To differentiate between short interval, daily, and monthly average clearness values, the following convention is used. For the global clearness index, k_t is used for hourly or shorter time intervals, K_t is used for the daily clearness index, and $\overline{K_t}$ is used for the monthly average index. The DHI and DNI clearness indices have a similar structure with t replaced by df or b, respectively.

A.4   IRRADIANCE ON A TILTED SURFACE

To calculate irradiance on a tilted surface, irradiance models sum the contributions from the DNI, the DHI, and the ground-reflected radiation. Instantaneously, the DNI contribution is equal to DNI times the cosine of the incident angle to the surface. The formula for the cosine of the incident angle to the surface is given by Equation 2.13 in Chapter 2. When using longer time interval data, the product must be integrated over time. Because the average beam irradiance (DNI) varies systematically over the day, the same systematic weighting of the beam component should be used. This is not always done, and for hourly values the weighting is significant only in the morning and evening hours when the atmospheric scattering and absorption start to significantly attenuate the incoming DNI (Vignola and McDaniels, 1988b).

The projection of the DHI onto the tilted surface is complex as the diffuse radiance varies across the sky (is anisotropic), depending on the relative position of clouds and sun. In general, diffuse radiation can be broken down into four components: the forward-scattered radiation around the sun (aureole), the diffuse radiance brightening near the horizon, the minimum diffuse radiance 90° from the sun (i.e., anti-solar point), and the diffuse radiance from the rest of the sky. Mathematically, it can be shown that if the diffuse radiance around the sun is evenly distributed, then it can be projected onto a surface much like DNI. Modeling other diffuse components can be complex. One of the more rudimentary models treats the diffuse radiance as distributed evenly across the sky (i.e., isotropic), and DHI_T on a tilted surface is simply

$$DHI_T = DHI \times (1 + \cos(\theta_T))/2 \qquad (A.6)$$

where θ_T is the tilt of the surface from the horizontal. This is called the isotropic diffuse model and is the simplest method (Kondratyev, 1977; Kondratyev and Manolova, 1960; Liu and Jordan, 1963). More comprehensive models that take into account the various diffuse components such as brightening around the sun or horizon brightening perform better than the simple isotropic diffuse models (ASHRAE, 1971, 1976; Brink, 1982; Bugler, 1977; Cohen and Zerpa, 1982; Gueymard, 1983; Hay, 1979; Hay and Davies, 1980; Hay and McKay, 1985; Hay, Perez, and McKay, 1986; Ineichen, 1983; Klucher, 1979; Lawrence Berkeley Laboratories, 1982; Oegema, 1971; Page, 1978; Perez, Scott, and Stewart, 1983; Puri, Jiminez, and Menzer, 1980; Rogers, Page, and Souster, 1979; Skartveit and Olseth, 1986; Temps and Coulson, 1977; Threkheld, 1962).

A tilted surface also receives solar radiation reflected from the ground. Ground-reflected radiation is very difficult to accurately model because the ground is not a uniform reflector and can scatter light in a specular fashion. The radiative character-istics of this scattering are wavelength dependent. In addition, many objects shade the ground from direct sunlight such as vegetation and structures, further complicat-ing any modeling. Therefore, most models that calculate irradiance on tilted surfaces use a simple isotropic ground-reflecting model to estimate the light reflected from the ground.

$$Ground\_reflected = GHI \cdot \rho \cdot (1 - \cos(\theta_T))/2 \qquad (A.7)$$

where θ_T is the tilt angle of the surface from the horizontal, and ρ is the albedo of the ground, usually a value of 0.2 for vegetated surfaces, but the albedo can vary from 0.1 for a dark surface to 0.9 for a surface covered with snow (see Chapter 7 for more information about albedo). Errors associated with modeling ground-reflected light are usually not significant. For example, given the most extreme case when the surface is perpendicular to the ground (tilted 90°) and the albedo is 0.2, the ground-reflected irradiance is just 10% of the GHI. Therefore, a 20% error in the model will produce only ~2% error in the calculated tilted irradiance, well within the measurement uncer-tainty of most tilted surface irradiance measurements. Of course, if the surface faces a snowfield with an albedo between 0.7 and 0.9 or faces a highly reflective surface, there can be a large contribution from the ground-reflected light.

The IEA Task 9 compared many of the models that estimate the irradiance on tilted surfaces (Hay and Mckay, 1988). The study divided the models into three types. The first type used hourly or short time interval data to estimate diffuse irradiance on tilted surfaces. Twenty-one models were compared (ASHRAE, 1971, 1976; Brink, 1982; Bugler, 1977; Cohen and Zerpa, 1982; Gueymard, 1983; Hay, 1979; Hay and Davies, 1980; Hay and McKay, 1985; Hay et al., 1986; Ineichen, 1983; Klucher, 1979; Lawrence Berkeley Laboratories, 1982; Oegema, 1971; Page, 1978; Perez et al., 1983; Puri et al., 1980; Rogers et al., 1979; Skartveit and Olseth, 1986; Temps and Coulson, 1977), and the Perez, Gueymard, and Hay models were rated the best when compared with data from 27 data sets from around the world. These models varied from using isotropic distributions to the sky distributions of Moon and Spencer (1942) to detailed diffuse sky models like Perez et al. (1983).

Seven models were tested that used daily GHI values to determine the daily GTI data values. These models were used to estimate the DNI and then calculated the direct component on the tilted surface (Bremer, 1983; Desnica et al., 1986; Jones, 1980; Klein and Theilacker, 1981; Liu and Jordan, 1960; Page, 1961; Revfeim, 1976, 1979, 1982a, 1982b). To perform these calculations, some model of the diurnal variation has to be included. Identical diffuse values were used for all comparisons (Collares-Pereira and Rabl, 1979a, 1979b, 1979c). The Page (1961) model gave the best results for the direct component on the horizontal surface. The models did have problems when estimating irradiance data for surface orientations away from due south. Any asymmetry in the daily irradiance, more irradiance in the morning or afternoon, would lead to such results (Vignola and McDaniels, 1988b, 1989).

Estimation of daily diffuse irradiance on tilted surfaces from daily GHI values was also studied. Either an isotropic diffuse model (Kondratyev and Manolova, 1960; Liu and Jordan, 1963) or a model to estimate the diurnal variation over the day (Collares-Pereira and Rabl, 1979a, 1979b, 1979c) can be used, and then one of the hourly diffuse models can be used to estimate the diffuse on a tilted surface. This was the approach used by Gueymard (1985, 1986a, 1986b) and McFarland (1983). All these approaches work better than the isotropic models.

A.5   DIFFUSE FRACTION

All models that calculate irradiance on a tilted surface require the use of DNI and DHI values rather than starting with simply the GHI value. Since GHI data are more often available, models were developed to estimate the DHI and DNI values from GHI values. The relationship between DHI and GHI is double-valued and not linear. Low values of DHI occur when GHI values are low and the sky is very overcast and on very clear days when GHI is high.

Diffuse modelers address this problem by dividing DHI by GHI, creating the diffuse fraction. When the diffuse fraction is plotted against k_t, a functional relationship results. Modelers use this relationship to estimate diffuse irradiance from GHI measurements. A sample plot of the diffuse fraction plotted against k_t is shown in Figure A.1 for hourly data. The densest area of data points represents clear periods

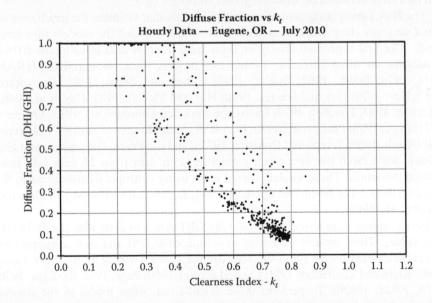

FIGURE A.1   Plot of the diffuse fraction plotted against clearness index k_t for hourly data. Most of the data fall along the lower left-hand side of the plot. Those data points represent clear periods. There is a spread of data points to the left that represents data when clouds are present. When there is no DNI, the diffuse fraction is one and GHI = DHI. When there is some sun present, the clouds can act as reflectors and a wide range of GHI values can occur for any diffuse fraction.

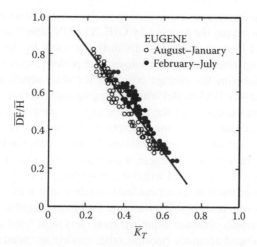

FIGURE A.2   Plot of monthly average diffuse fraction (DHI/GHI or in this figure DF/H) versus monthly average clearness index $\overline{K_t}$. Data are plotted in two groups, February through July and August through January. There is a clear seasonal dependence of the diffuse fraction in relation to the monthly average clearness index.

with little or no clouds. Clouds give the broad scatter to the possible GHI values because there are numerous ways for clouds to reflect sunlight onto the pyranometer and add to the measured GHI and DHI values. Gaps in the clouds that let some or most of the DNI through lead to the widest scatter in the data points in Figure A.1. A general relationship between the diffuse fraction and clearness index can be used for short interval, daily, or monthly average correlations. The hourly, daily, and monthly relationships will differ. As the time period increases over which the diffuse fraction is analyzed, the scatter in the data is greatly reduced. For hourly data (Figure A.1), the spread of the data is reduced by about one-half from the scatter of 1-minute data (see Figure 14.8). For daily or monthly averaged values, the scatter of the data is even further reduced. In addition to the reduction of data scatter, the average relationship between the diffuse fraction and clearness index also changes. Therefore, the relationship for daily data is not the same as the relationship for monthly average data. Because of the reduction in the scatter of the data, a seasonal relationship for daily and monthly average relationships becomes evident with higher diffuse fractions in the spring than fall as shown in Figure A.2 (Vignola and McDaniels, 1984b). This seasonal dependence is probably related to the varying aerosol optical depth determined by the dust loading in the atmosphere.

Scattering from dust and most aerosols is Mie-type scattering with a majority of the scattered light going in the forward direction. With higher levels of aerosols and dust, the DHI component increases, whereas the GHI decreases only slightly because the scattering is mostly in the forward direction. Therefore, it is expected the relationship between the diffuse fraction and the clearness index will vary at locations with different seasonal aerosol loading or during periods affected by volcanic eruptions.

When the diffuse fraction is determined from the functional relationship, the DHI is estimated by multiplying the fraction by GHI. The DNI can then be calculated by subtracting the DHI from the GHI and dividing the result by the average value of the cosine of the solar zenith angle. For longer time periods, such as an hour, a more accurate way to determine the average cosine of the solar zenith angle is to weight the cosine by a clear-sky DNI model when averaging (see Section A.7 for details).

For the daily data the cosine of the incident angle is integrated over the day, and for monthly averaged data the cosine is averaged for the day in the month that is closest to the average extraterrestrial irradiance for the month. For daily and monthly averaged cosine of the solar zenith angle, a constant DNI value is assumed because the existing models were developed without a weighting factor.

Most of the diffuse fraction, clearness index studies have used DHI and GHI values that come from instruments that have large uncertainties (±5% or more). Models derived from DHI values obtained using pyranometers with fixed shadowbands (see Figure 6.3) have reduced accuracy because other models are needed to estimate the shadowband correction factor to account for blocking the sky as well as the solar disk. When the DHI values are obtained by subtracting DNI times the cosine of the solar zenith angle from the GHI data, the systematic errors in the GHI and DNI add significant uncertainties to the DHI values. However, it remains to be seen if new studies with better data will be able to improve the results, especially considering the natural variability in the data and the variation of the relationship over the year and from site to site. The studies by Gueymard (2009) and Evseev and Kudish (2009) are examples of model testing and illustrate some of the problems associated with verification of any model. Some of the models tested that were not referenced earlier are Willmott (1982), Iqbal (1983), Perez, Ineichen, Seals, Mickalsky, and Stewart (1990), Reindl, Beckman, and Duffie (1990), Gueymard (1987), Muneer (2004), and ASHRAE (2005).

A.6   BEAM–GLOBAL CORRELATIONS

A more direct way to obtain the DNI value is by using a beam–global correlation. Using the results of a beam–global correlation, the DNI value can be calculated directly without having to calculate the DHI first and then divide the difference between the GHI and DHI values by the cosine of the solar zenith incident angle. It is important to remember that, at large zenith angles, small errors in either the diffuse or global values can lead to large errors in the beam values because the cosine of large zenith angles is a small number and this small number divides into the difference between GHI and DHI.

As with the diffuse fraction–global relationships, the beam–global relationships have a seasonal dependence that reflects the seasonal change in aerosols and other atmospheric constituents. The contrast is greatest between spring and fall where the solar angles are the same, but the aerosol content of the atmosphere generally differs.

There are fewer beam–global models than the diffuse fraction–global models because historically the number of DHI measurements greatly outnumbers DNI measurements. (For example, see Randall and Biddle, 1981.) Diffuse measurements with fixed shadowbands were easier to make and less expensive. DNI measurements that required manual trackers that were difficult to keep in alignment and were more expensive than shadowbands. Interest in deriving DNI directly from GHI started to

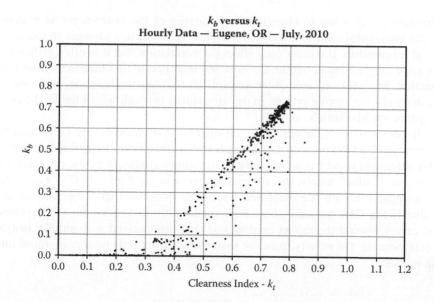

k_b versus k_t
Hourly Data — Eugene, OR — July, 2010

FIGURE A.3 Plot of k_b versus k_t for hourly data. As with the diffuse fraction relationship, shown in Figure A.1, the majority of points during clear periods fall on the left-hand side of the plot. This results from sunlight reflecting off clouds and into the global pyranometer. This reflected light is not seen by the pyrheliometer that measures the DNI values. Note that there are very few data points with k_b values above a maximum of 0.8. The maximum k_b values are somewhat dependent on altitude because higher altitudes have a lower air mass through which sunlight has to traverse.

increase when DNI values became of primary interest, when more DNI data became available, and when it became clear that DNI measurements were more accurate than DHI measurements. With a number of stations now using automatic trackers that enable both high-quality DNI and DHI measurements, the relationship between GHI, DNI, and DHI is once again being studied with hopes of improving on older models (Vignola and McDaniels, 1988b, Vignola, 2003).

One of the main problems facing beam–global models is that the rise of the normalized DNI (k_b) as a function of k_t is fairly steep, resulting in a given value of k_t having a wide range of possible k_b values. This is because many of the aerosols preferentially scatter the incident radiation in a forward direction, and this does not significantly change the GHI, but it does affect the DNI. A typical plot of k_b versus k_t is given in Figure A.3. The data obtained during clear skies fall into a concentrated group on the left side of the plot. When clouds are present, reflected light can greatly enhance the GHI values. As opposed to one minute data (Figure 14.9), hourly data produces very few k_t values that are greater than 0.8 and doesn't result in any data with values of k_t greater than 1.0 or greater than the extraterrestrial irradiance.

The more accurate models for the relationship between k_b and k_t use the variability of GHI over the day, the time of day, and other parameters that help separate partially cloudy from clear periods. For hourly and other short time interval correlations, other meteorological variables can also affect the correlation. These extra parameters may relate to changing responsivities of the instruments as affected by the meteorological parameters and might not reflect true changes in the functional relationship. Haziness does affect the correlation, but it is often difficult to get good information about the aerosols and incorporate that information into the function. Seasonal variance of the aerosols also affects the function, as do volcanic aerosols that can be present in the air column through which the DNI passes (Vignola and McDaniels, 1984b).

Of course, once k_b is obtained, the diffuse irradiance can be calculated by subtracting the DNI projected on a horizontal surface from the global irradiance. DHI values obtained from disk-shaded pyranometers should be the reference for validation. Unfortunately, the systematic errors in the GHI and DNI can affect the comparison with the measured DHI. One question that sometimes occurs is that if the GHI measurements were made with an instrument with systematic errors, should the model used be one that was derived with similar instruments because the offsets from the systematic errors will be incorporated into the model?

A.7 DETERMINING THE AVERAGE COSINE OF THE SOLAR ZENITH ANGLE

For short time periods, the average of the cosine of the solar zenith angle can be calculated to an acceptable level of accuracy at the midpoint of the time in the interval. For 15-minute and especially hourly intervals, choosing the midpoint of the interval may result in systematic errors in the estimated average cosine value. This is especially true when one is using the cosine of the incident angle to project DNI onto a horizontal (or tilted) surface during morning or evening hours. For many models, the main reason to calculate the average cosine of the incident angle is to determine the direct component on the surface. For example, restating Equation 5.1 as an average over a time interval results in

$$GHI = \frac{\int_{ha_1}^{ha_2} DNI \cdot \cos(sza)\,dha}{\int_{ha_1}^{ha_2} dha} + \frac{\int_{ha_1}^{ha_2} DHI\,dha}{\int_{ha_1}^{ha_2} dha} \tag{A.8}$$

where ha_1 is the start time hour angle and ha_2 is the end time hour angle. The second term on the right-hand side of Equation A.8 is just the average value of $\overline{DHI}$ for the time period. The first term of the right-hand side of Equation A.8 is not equal to the average $\overline{DNI}$ times the average value of the cosine of the solar zenith angle, $\overline{\cos(sza)}$, because DNI varies in a systematic way over the interval. (See how the DNI varies on a clear day in Figure 2.12.) One method of weighting the cosine of the solar zenith angle is to assume that DNI is equal to a constant times a clear-sky function of DNI, CS(DNI). For a clear day, the constant would equal 1. For cloudy periods, the envelope of the clear day still constrains the DNI values, and on average the DNI can be approximated by a constant times CS (DNI) (Vignola and McDaniels, 1988b).

$$\overline{DNI} \approx a \cdot \int_{ha_1}^{ha_2} CS(DNI)\,dha \qquad (A.9)$$

Substituting Equation A.9 into Equation A.8 and replacing a by the ratio $\overline{DNI}$ divided by the integral of the clear-sky function over the period, a modified formula for $\overline{GHI}$ as a function of $\overline{DNI}$ and $\overline{DHI}$ is obtained:

$$\overline{GHI} = \overline{DNI} \cdot \overline{w\cos(sza)} + \overline{DHI} \qquad (A.10)$$

where the weighted average of the cosine of the solar zenith angle is

$$\overline{w\cos(sza)} = \left. \int_{ha_1}^{ha_2} CS(DNI) \cdot \cos(sza)\,dha \middle/ \int_{ha_1}^{ha_2} CS(DNI)\,dha \right. \qquad (A.11)$$

The DNI contribution to the GHI can then be calculated by multiplying the average value of the hour by the weighted average of the cosine of the solar zenith angle.

For accurate hourly calculations, this weighting of the cosine of the solar zenith angle becomes important especially during morning and evening hours (Vignola and McDaniels, 1988b).

When calculating DNI from the difference between GHI and DHI, using just the average cosine of the zenith angle can lead to systematic errors. The weight removes the systematic error but there will still be some variability because clouds affect the DNI during the period. This variability is mostly random, and weighting with an appropriate clear-sky model results in a more accurate estimate of the average of the cosine of the solar zenith angle.

QUESTIONS

1. What is meant by clearness index?
2. Why is the diffuse fraction used in correlations with GHI instead of DHI?
3. What is a clear-sky model?
4. Approximately what is the maximum value of hourly k_t and k_b values?
5. Should the model used to estimate a variable be derived with similar instruments used to make the GHI measurement? - Discuss.
6. When the average value of the cosine of the solar zenith angle is determine, would it be appropriate to weigh the average with a clear-sky model?

REFERENCES

ASHRAE. 1971. *Procedure for determining heating and cooling loads for computerized energy calculations—algorithms for building heat transfer subroutines.* American Society of Heating, Refrigerating and Air Conditioning Engineers, New York.

ASHRAE. 1976. *Procedures for determining heating and cooling loads for computerizing energy calculations.* T. Kusuda (ed.), American Society of Heating, Refrigerating and Air Conditioning Engineers, New York.

ASHRAE. 2005. *Handbook of fundamentals, SI edition.* American Society of Heating, Refrigerating, and Air-Conditioning Engineers, Atlanta, GA.

Atwater, M. A. and J. T. Ball. 1978. A numerical solar radiation model based on standard meteorological observations. *Solar Energy* 21:163–170.

Atwater, M. A. and J. T. Ball. 1979. Erratam. *Solar Energy* 23:275.

Bemporad, A. 1904. Zur Theorie der Extinktion des Lichtes in der Erdatmosphare. *Mitteilungen der Grossherzoglichen Sternmwarte Zu Heidelberg.* No. 4, 1–78.

Bird, R. E. and R. L. Hulstrom. 1979. *Application of Monte Carlo techniques to insolation characterization and prediction.* SERI/RR36-306. Golden, CO: Solar Energy Research Institute.

Bird, R. E. and R. L. Hulstrom. 1980. *Direct insolation models.* SERI/TR-335-344. Golden, CO: Solar Energy Research Institute.

Bird, R. E. and R. L. Hulstrom. 1981. Simplified clear sky model for direct and diffuse insolation on horizontal surfaces, Technical Report SERI/TR-642-761. Golden, CO: Solar Energy Research Institute.

Bremer, P. 1983. *Modeles de rayonnement siples testes sous Ie soleil suisse.* Quatrieme Symposium Recherche et Developpementen Energie Solaire en Suisse, October 17–18, EPFL-GRES EDS, 1015 Lausanne, Switzerland.

Brink, G. J. van den. 1982. *Climatology of solar irradiance on inclined surfaces, Pt II: Validation of calculation models.* Report 803.229 IV-2 (ESF-006-80 NL B), Technisch Physische Dienst TNO-TH, Delft, the Netherlands, 48 pp.

Bugler, J. W. 1977. The determination of hourly insolation on an inclined plane using a diffuse irradiance model based on hourly measured global horizontal insolation. *Solar Energy* 19:477–491.

Cohen, B. and N. Zerpa. 1982. Spatial transformations of insolation measurements: Preliminary results. *Solar Energy* 28:75–76.

Collares-Pereira, M. and A. Rabl. 1979a. Derivation of a method for predicting long term average energy delivery of solar collectors. *Solar Energy* 23:223–233.

Collares-Pereira, M. and A. Rabl. 1979b. Simple procedures for predicting long term average performance of nonconcentrating and of concentrating solar collectors. *Solar Energy* 23:235–253.

Collares-Pereira, M. and A. Rabl. 1979c. The average distribution of solar radiation: Correlations between diffuse and hemispherical and between daily and hourly insolation values. *Solar Energy* 22:155–164.

Davies, J. A. and J. E. Hay. 1979. Calculation of the solar radiation incident on a horizontal surface. In *Proceedings of the First Canadian Solar Radiation Data Workshop, April 17–19.* Canadian Atmospheric Environment Service.

Evseev, E. G. and A. I. Kudish. 2009. The assessment of different models to predict the global solar radiation on a surface tilted to the south. *Solar Energy* 83:377–388.

Gueymard, C. 1983. *Utilisation des donnees meteorologiques horaires pour Ie calcul du rayonnement solaire sur des batiments solaires passifs.* Ph.D. thesis, University of Montreal.

Gueymard, C. 1985. *Radiation on tilted planes: A physical model adaptable to any computational time-step.* Presented at INTERSOL 85, Congress of the International Solar Energy Society, June, Montreal.

Gueymard, C. 1986a. Mean daily averages of beam radiation received by tilted surfaces as affected by the atmosphere. *Solar Energy* 37:261–267.

Gueymard, C. 1986b. Monthly averages of the daily effective optical air mass and solar related angles for horizontal and inclined surfaces. *Journal of Solar Energy Engineering* 108:320–324.

Gueymard, C. A. 1987. An anisotropic solar irradiance for tilted surfaces and its comparison with selected engineering algorithms. *Solar Energy* 38:367–386.

Gueymard, C. A. 2009. Direct and indirect uncertainties in the prediction of tilted irradiance for solar engineering applications, *Solar Energy* 83:432–444.

Hay, J. E. 1979. *A study of shortwave radiation on non-horizontal surfaces.* Final report, Contract Serial Number OS878 00053, Atmospheric Environment Service, Downsview, Ontario, Canada, 140 pp.

Hay, J. E. 1984. *Calculation of solar irradiances for inclined surfaces: Verification of models which use hourly and daily data.* Interim Report, Contract 02SE.KM147-3-3029 (DSS I AES), June 22, Canadian Atmospheric Environment Service, Downsview, Ontario, Canada, 55 pp.

Hay, J. E. 1993. Calculating solar radiation for inclined surfaces: Practical approaches. *Renewable Energy* 3(4–5):373–380.

Hay, J. E. and J. A. Davies. 1980. Calculation of the solar radiation incident on an inclined surface. In *Proceedings of the First Canadian Solar Radiation Data Workshop,* ed. J. E. Hay and T. K. Won, Atmospheric Environment Service, Downsview, Ontario, Canada, 59–72.

Hay, J. E. and D. C. McKay. 1985. Estimating solar irradiance on inclined surfaces: A review and assessment of methodologies. *International Journal of Solar Energy* 3:203–240.

Hay, J. E. and D. C. McKay. 1988. Final Report IEA Task IX-Calculation of solar irradiances for inclined surfaces: Verification of models which use hourly and daily data. *International Energy Agency Solar Heating and Cooling Programme.*

Hay, J. E., R. Perez, and D. C. McKay. 1986. Addendum and errata to the paper, estimating solar irradiance on inclined surfaces: A review and assessment of methodologies. *International Journal of Solar Energy* 4:320–324.

Hoyt, D. V. 1978. A model for the calculation of solar global insolation. *Solar Energy* 21:27–35.

Ineichen, P. 1983. *Quatre annees de mesures d'ensoleillement a Geneve 1978–1982.* Thesis 2089, Geneva University, Geneva Switzerland, 221 pp.

Iqbal, M. 1983. *An introduction to solar radiation.* Canada: Academic Press.

Jones, R. E. 1980. Effects of overhang shading of windows having arbitrary azimuth. *Solar Energy* 24:305–312.

Kasten, F., 1964. A new table and approximation formula for the relative optical air mass. Technical Report 136. Hanover, NH, U.S. Army Material Command, Cold Region Research and Engineering Laboratory.

Klein, S. A. 1977. Calculation of monthly average insolation on tilted surfaces. *Solar Energy* 19:325–329.

Klein, S. A. and J. C. Theilacker. 1981. An algorithm for calculating monthly-averaged radiation on inclined surfaces. *Journal of Solar Energy Engineering* 103:29–33.

Klucher, T. M. 1979. Evaluation of models to predict insolation on tilted surfaces. *Solar Energy* 23:111–114.

Kondratyev, K. Y. 1969. *Radiation in the atmosphere.* New York: Academic Press.

Kondratyev, K. Y. 1977. *Radiation regime on inclined surfaces.* Technical Note 152, World Meteorological Organization, WMO-No. 467, Geneva, Switzerland, 82 pp.

Kondratyev, K. Y. and M. P. Manolova. 1960. The radiation balance of slopes. *Solar Energy* 4:14–19.

Lacis, A. L. and J. E. Hansen. 1974. A parameterization for absorption of solar radiation in the earth's atmosphere. *Journal of Atmospheric Science* 31:118–133.

Lawrence Berkeley Laboratories. 1982. DOE-2 engineers manual, version 2.1 A, LBL11353 (LA-8520-M), (DE83004575), Technical Information Center, United States Department of Energy.

Liu, B. Y. H. and R. C. Jordan. 1960. The interrelationship and characteristic distribution of direct, diffuse and total solar radiation. *Solar Energy* 4:1–9.

Liu, B. Y. H. and R. C. Jordan. 1962. Daily insolation on surfaces tilted toward the equator. *Transactions, American Society of Heating, Refrigerating and Air Conditioning Engineers* 67:526–541.

Liu, B. Y. H. and R. C. Jordan. 1963. The long-term average performance of flat-plate solar energy collectors. *Solar Energy* 7:53–74.

Ma, C. C. Y. and M. Iqbal. 1983. Statistical comparison of models for estimating solar radiation on inclined surfaces. *Solar Energy* 31:313–317.

Mahaptra, A. K. 1973. *An evaluation of a spectroradiometer for the visible-ultraviolet and near-ultraviolet.* Ph.D. thesis, University of Missouri, Columbia, 121 pp. (University Microfilms 74-9964).

McDonald, J. E. 1960. Direct absorption of solar radiation by atmospheric water vapor. *J. Meteorology* 17:319–328.

McFarland, R. D. 1983. Solar radiation approximations. Appendix E: Passive solar design handbook, Vol. 3: Passive solar design analysis and supplement, ed. R. W. Jones. Boulder, CO: American Solar Energy Society, pp. 293–302.

Moon, P. 1940. Proposed standard solar-radiation curves for engineering use. *Journal of the Franklin Institute* 230:583–617.

Moon, P. and D. E. Spencer. 1942. Illumination from a non-uniform sky. *Transactions of the Illumination Engineering Society* 37:707–726.

Muneer, T. (ed.). 2004. *Solar radiation and daylight models*, 2d ed. New York: Elsevier.

Oegema, S. W. T. M. 1971. *De digitale berekening van de zonnestraling.* Report 332-111, Technisch Physische Dienst TNO-TH, Delft, the Netherlands.

Page, J. K. 1961. The estimation of monthly mean values of total daily solar short-wave radiation on vertical and inclined surfaces from sunshine records for latitudes 40N-40S. In *Proceedings of the United Nations Conference on New Sources of Energy,* Paper 35/5/98, 378–389.

Page, J. K. 1978. *Methods for the estimation of solar energy on vertical and inclined surfaces.* BS 46, Department of Building Science, University of Sheffield, UK, 61 pp.

Perez, R., P. Ineichen, R. Seals, J. J. Michalsky, and R. Stewart. 1990. Modeling daylight availability and irradiance components from direct and global irradiance. *Solar Energy* 44:271–289.

Perez, R. R., J. T. Scott, and R. Stewart. 1983. An anisotropic model for diffuse radiation incident on slopes of different orientations and possible applications to CPC's. *Progress in Solar Energy* 25:85–90.

Puri, V. M., R. Jiminez, and M. Menzer. 1980. Total and non-isotropic diffuse insolation on tilted surfaces. *Solar Energy* 25:85–90.

Randall, C. M. and J. M. Biddle. 1981, September. Hourly estimates of direct insolation: Computer code ADIPA user's guide. The Aerospace Corporation, AIR-81 (7878)-1.

Reindl, D. T., W. A. Beckman, and J. A. Duffie. 1990. Evaluation of hourly tilted surface radiation models. *Solar Energy* 45:9–17.

Revfeim, K. J. A. 1976. Solar radiation at a site of known orientation on the earth's surface. *Journal of Applied Meteorology* 15:651–656.

Revfeim, K. J. A. 1979. Maximization of global radiation on sloping surfaces. *New Zealand Journal of Science* 22:293–297.

Revfeim, K. J. A. 1982a. Estimating global solar radiation on sloping surfaces. *New Zealand Journal of Agricultural Research* 25:281–283.

Revfeim, K. J. A. 1982b. Simplified relationships for estimating solar radiation incident on any flat surface. *Solar Energy* 2:509–517.

Rogers, G. G., J. K. Page, and C. G. Souster. 1979. Mathematical models for estimating the irradiance falling on inclined surfaces for clear, overcast and average conditions. *In Proceedings of the Meteorology for Solar Energy Applications, Conference (C18) at the Royal Institution, United Kingdom International Solar Energy Society,* pp. 48–62.

Skartveit, A. and J. A. Olseth. 1986. Modeling slope irradiances at high latitudes. *Solar Energy* 36(4):333–344.

Temps, R. C. and K. L. Coulson. 1977. Solar radiation incident upon slopes of different orientations. *Solar Energy* 19:179–184.

Threkheld, J. L. 1962. Solar radiation of surfaces on clear days. *ASHRAE Journal* 69:43–54.

Vignola, F. 2003. Beam-tilted correlations. In *Proceedings of the Solar 2003 American Solar Energy Society Conference,* Austin, TX.

Vignola, F. and D. K. McDaniels, 1984a. Correlations between diffuse and global insolation for the Pacific Northwest. *Solar Energy* 32:161–168.

Vignola, F. and D. K. McDaniels. 1984b. Diffuse-global correlation: Seasonal variations. *Solar Energy* 33:397–402.

Vignola, F. and D. K. McDaniels. 1988a. Beam-global correlations in the Pacific Northwest. *Solar Energy* 36:409–418.

Vignola, F. and D. K. McDaniels. 1988b. Direct beam radiation: Projection on tilted surfaces. *Solar Energy* 40:237–247.

Vignola, F. and D. K. McDaniels. 1989. Direct radiation: Ratio between horizontal and tilted surfaces. *Solar Energy* 43:183–190.

Watt, D. 1978. *On the nature and distribution of solar radiation.* HCP/T2552-0l. U.S. Department of Energy.

Willmott, C. J. 1982. On the climatic optimization of the tilt and azimuth of flat-plate solar collectors. *Solar Energy* 28:205–216.

Appendix B: Solar Radiation Estimates Derived from Satellite Measurements

B.1 INTRODUCTION

Satellite-derived irradiance data augment ground-based measurements as satellites survey large areas and provide continuous information for long-term studies of the solar resource. Satellite data and images are used to generate time- and site-specific irradiance data and high-resolution (10 km × 10 km or smaller) maps of solar radiation. In fact, if nearby high-quality ground-based irradiance measurements are not available, the best characterization of the solar resource comes from satellite-derived irradiance data.

Satellite-derived solar resource surveys are used to characterize the long-term variability of the solar resource and climatologies areas for the best location for solar facilities, while ground-based monitoring stations are essential for accurately quantifying the solar irradiance at a specific site, measuring short-term variability of the solar resource, and providing ground truth for the satellite-derived values. However, when extrapolating the solar resource to nearby locations, satellite-derived hourly values become more accurate beyond 25 km from a ground station (Zelenka, Perez, Seals, and Renné, 1999).

Geostationary satellites are most suitable for modeling solar irradiance as they monitor the state of the atmosphere and the earth's cloud cover with a spatial resolution near 1 km in the visible range and with a 30-minute time resolution. However, the view of the earth from these satellites limits their effectiveness at high latitudes (see Figure B.1). Polar satellites are closer to the earth's surface and provide a variety of measurements that can be transformed into surface solar irradiance values, but they pass over a particular area only once during the day, which limits their temporal coverage at the lower latitudes.

In this appendix, the satellites used to generate the measurements for the irradiance models will be discussed, along with a brief description of the Perez, Ineichen, Moore, Kmiecik, Chain, George, and Vignola (2002) satellite irradiance model.

B.2 GEOSTATIONARY SATELLITES

Geostationary satellites are located 35,880 km (22,300 miles) above the equator in a geosynchronous orbit. The United States has two satellites, GOES-West (135° W) and GOES-East (75° W), that cover North and South America. The European Union operates two satellites, Meteosat-9 (0°) and Meteosat-7 (57.5° E), that cover Europe, Africa, and the Middle East. The Japanese operate MTSAT (140° E), which covers

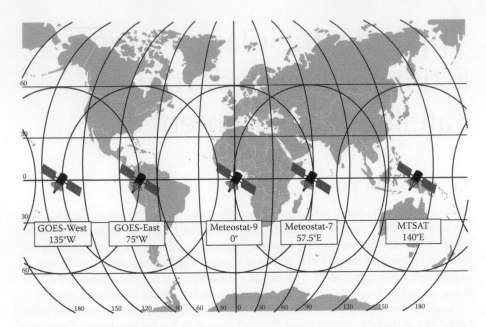

FIGURE B.1 Geostationary weather satellites cover most of the globe from latitude 60°S to 60°N. This figure shows the coverage of five geostationary satellites. (Illustration courtesy of Angie Skovira of Luminale, LLC.)

Asia and Australia (Figure B.1). The curvature of the earth limits the useable images to between −66° and +66° latitude, so there is some redundancy in the satellite images and this allows some coverage even if one satellite has problems. Russia (GOMS), China (FY-2 series), and India (InSat and KALPANA) also have geosynchronous satellites that provide meteorological data and images.

Geosynchronous satellites have to be replaced periodically because they become outdated or run out of fuel for stabilization. For example, GOES-West and GOES-East are the twelfth and thirteenth satellites in the series. Historic meteorological satellite data from the United States, the European Union, and Japan are more complete and readily available than satellite data from China, India, and Russia.

B.3   DERIVING IRRADIANCE FROM SATELLITES

Many models exist to derive irradiance values from satellite data. An overview of satellite models will be presented in this section. For a more complete summary of satellite models see Chapter 4 of the NREL Best Practices Handbook (Stoffel et al., 2010). The empirical satellite models derive a cloud index (*CI*) from the satellite visible-channel measurements of surface radiance and use this index to modulate a clear-sky global irradiance model of the solar resource. Physical models use satellite radiance data and other atmospheric information to calculate the irradiance as it passes through the atmosphere by accounting for the radiative transfer processes. Empirical models estimate irradiance by using correlations to bypass some of the more cumbersome and time-consuming atmospheric modeling procedures.

Physical models can take considerable computer time and require knowledge of the distribution of the gases, aerosols, and particles that make up the atmosphere in addition to understanding how each constituent affects the incoming irradiance. Physical models have been useful in understanding radiative transfer and were instrumental in uncovering the infrared (IR) radiative losses (thermal offset) that skewed the DHI measurements made using high-quality thermopile pyranometers (Cess, Nemesure, Dutton, DeLuisi, Potter, and Morcrette, 1993).

A good example of a physical model is one developed by Pinker and Ewing (1985) that divides the solar spectrum into 12 intervals and applies a radiative transfer model to a three-layer atmosphere. A primary input for this model is cloud optical depth. Pinker and Laszlo (1992) enhanced the model, and cloud information from the International Satellite Cloud Climatology Project (ISCCP) (Schiffer and Rossow, 1983) was used to develop irradiance data for the Surface Radiation Budget database that was created for a $2.5° \times 2.5°$ grid (Whitlock et al., 1995). The clouds in the ISCCP climatology are separated into low, middle, and high clouds with three different optical thicknesses. Low and middle clouds are also categorized into water and ice clouds, whereas high clouds are always ice clouds. This creates 15 different cloud types for the ISCCP. The ISCCP climatology is used for cloud input for many models (Stoffel et al., 2010).

Empirical models take less computer time to run, are easier to apply, and don't demand as detailed input information as required by the physical models. These empirical models are based on regression relationships between satellite observations and ground-based instrument measurements. The *CI* and regression relationships with other meteorological data are used to estimate the solar irradiance. Accurate clear-sky DNI modeling is important for all models because it is these clear-sky DNI values that are modulated by the cloud cover index. Good atmospheric turbidity values are necessary for accurate clear-sky DNI estimates.

B.3.1   GLOBAL IRRADIANCE (GHI)

Early work by Cano, Monget, Aubuisson, Guillard, Regas, and Wald (1986) is based upon the observation that shortwave (solar) atmospheric transmissivity is linearly related to the earth's planetary albedo (Schmetz, 1989). This is a good first-order approximation. Without any clouds and with a fixed albedo, the irradiance incident on the earth's surface is proportional to the intensity of the reflected irradiance as measured using the counts in the image pixel. The initial task in modeling is to assign the image pixel to a precise ground-based location. As the satellite wobbles in space, this limits the precision to which a pixel can be assigned. The original Perez procedure for extracting irradiance data is the basis for the following general discussion, and much of this material is extracted from Vignola and Perez (2004). A more detailed description can be found in Perez et al. (2002). As with many models, improvements are made over time, and a third-generation Perez model is being validated as of 2012.

There are two distinct steps for this process:

1. Pixel-to-cloud index (*CI*) conversion
2. *CI* to global irradiance conversion

B.3.2   PIXEL-TO-CLOUD INDEX CONVERSION

Individual pixel brightness values in the satellite image are stored as counts. The pixel counts are first normalized by the cosine of the solar zenith angle to account for the solar geometry. This normalized pixel is then gauged against the satellite pixel's dynamic range at the chosen location to estimate a cloud index (*CI*). The lowest pixel count is assumed to be the ground brightness under clear sky conditions.

An additional normalization is necessary to adjust for a secondary atmospheric air mass effect and for the "hot spot" (Zelenka et al., 1999). The hot spot is cause by the sun–satellite angle and incorporates both atmospheric back-scatter brightness intensification and the fact that the ground surface becomes brighter as the sun–satellite angle diminishes due to the reduction of ground shadows seen by the satellite (e.g., Pinty and Verstraete, 1991).

The dynamic range represents the range of values a normalized pixel can assume at a given location from its lowest (darkest pixel, i.e., clearest conditions) to its highest value (brightest pixel, i.e., cloudiest conditions). A record of the normalized pixel counts at a given location is kept to track the changes in the dynamic range. While the upper bound of the range remains relatively constant except for the drift in the satellite's calibration, the lower bound evolves over time as the local ground albedo changes (chiefly from changes in snow, moisture, and vegetation). Obtaining an accurate lower bound that represents the true ground albedo is difficult, especially during periods with long-term cloudiness. This can be especially important during the winter months when snow covers the ground and significantly reduces the dynamic range. Techniques using the IR satellite image are being considered to better distinguish between snow on the ground and clouds.

In the Perez procedure, a sliding time frame is used to determine this lowest bound. The lower bound is determined as the average of the 10 lowest pixels in the sliding time frame. Note that the secondary normalization is applied to the lower bound of the dynamic range. Subtracting the lower bound from the normalized pixel count and dividing this number by the normalized dynamic range defines the cloud index. The *CI* can then vary from 0 (clearest) to 1 (cloudiest).

B.3.3   CLOUD-INDEX-TO-GHI CONVERSION

In the Perez procedure GHI is determined by

$$GHI = (0.02 + (1 - CI))\, GHI_c \tag{B.1}$$

where GHI_c is the clear-sky global irradiance per Kasten (1984). GHI_c is adjustable for broadband turbidity as quantified by the Linke turbidity coefficient (Kasten, 1980) and ground elevation.

$$GHI_c = (0.0000509 \times alt + 0.868) \times I_o \times \cos(sza) \times \exp\{-(0.0000392 \times alt + 0.0387)$$

$$\times am \times (\exp(-alt/8000) + \exp(-alt/1250) \times (TL - 1))\} \times \exp(0.01 \times am^{1.8}) \tag{B.2}$$

where I_o is the extraterrestrial normal incident irradiance, sza is the solar zenith angle, *am* is the elevation-corrected air mass, *TL* is the Linke turbidity coefficient, and *alt* is the ground elevation in meters.

B.3.4 Direct Irradiance (DNI)

DNI is obtained, as GHI, by modifying the clear-sky direct irradiance (DNI_c). An example of the clear-sky model (Ineichen and Perez, 2002) is

$$DNI_c = \min\{0.83 \times I_o \times \exp(-0.09 \times am \times [TL-1]) \times (0.8 + 0.196/\exp[-alt/8000]),$$

$$(GHI_c - DHI_c)/\cos(sza)\}, \tag{B.3}$$

where DHI_c is the minimum clear-sky diffuse irradiance given by

$$DHI_c = GHI_c \times \{0.1 \times [1 - 2 \times \exp(-TL)]\} \times \{1/[0.1 + 0.882/\exp(-alt/8000)]\} \tag{B.4}$$

Unlike GHI, the direct modulating factor is not derived from CI but from global, using a GHI-to-DNI model. One quasi-physical model, DIRINT (Perez, Ineichen, Maxwell, Seals, and Zelenka, 1992), is based on the DISC model developed by Maxwell (1987) that uses a clear-sky model that is a function of the clearness index k_t. In the DIRINT model, a modified clearness index is used along with atmospheric water vapor and the change in clearness index.

DNI is obtained from $DNI = DNI_c \times DIRINT(GHI)/DIRINT(GHI_c)$, where $DIRINT(GHI)$ is the DIRINT model using the input global irradiance GHI.

Other global–beam irradiance models have been used (Ineichen, 2008).

B.3.5 Diffuse Irradiance (DHI)

The DHI is obtained by subtracting the DNI times the cosine of the solar zenith angle from the GHI. When the DNI value is very small, DHI is approximately equal to GHI and the uncertainty in the DHI value is the same as the percent uncertainty in the GHI value. When DNI is large, then the DHI value is small and the uncertainty percent in the DHI value is large compared with the uncertainty percent in the GHI value. When it is neither clear nor totally overcast, the uncertainty in the DHI value can be very large. Therefore, it is important to precisely state the conditions when specifying the uncertainty in the DHI value.

B.4 STATUS OF SATELLITE IRRADIANCE MODELS

Satellite irradiance models provide a good sampling of the location's solar resource. The bias error is very small—a few percent for hourly GHI and random error of about 25% when compared with ground-based measurements. Of course, the different perspectives of ground-based and satellite views account for a good portion of this random error. In addition, the models have been adjusted to reproduce the statistical distributions in the solar resource while minimizing any bias errors.

As with any modeling effort, one has to be mindful of the accuracy of the data with which the model is derived and validated. The absolute accuracy of ground-based global irradiance is ±3%, at best, so a claim of 2% mean bias error has to take into account the measurement error.

TABLE B.1

Uncertainty in the NASA-Modeled Satellite Data for Monthly Averaged Values

Measurement	MBE (%)	RMS (%)
GHI	−0.0	10.3
DHI	7.5	29.3
DNI	−4.1	22.7

Note: The RMS errors are smaller for irradiance values obtained between ±60° latitude and larger for values for locations closer to the poles.

The uncertainty in the NASA-modeled 1-degree gridded data compared with high-quality BSRN data is given in Table B.1 (from Stoffel et al., 2010). While the mean bias error (MBE) appears small overall, it can vary several percent depending on the site examined. For example, the DNI MBE varies from −15.7% above 60 degrees north latitude to 2.4% below 60 degrees. It is difficult to compare ground-based BSRN-site-measured data with satellite-derived data on a 1-degree grid because of the differences in the size of the coverage.

According to Perez, Seals, Ineichen, Stewart, and Menicucci (1987), satellite-based GHI estimates are accurate to 10–12%. According to Renné, Perez, Zelenka, Whitlock, and Di Pasquale (1999) and Zelenka, Perez, Seaks, and Renne (1999), the target-specific comparison with ground-based observations will have a root mean square error (RMSE) of at least 20%; the time-specific area-wide accuracy is 10–12% on an hourly basis.

The RMSE decreases with averaging time, as expected. An hourly RMSE for one site will decrease from 20 to 25% for hourly estimates to 10 to 12% for daily estimates and down to 5 to 10% or less for monthly estimates. Mean bias errors generally range from +5% to −5%, with most studies reporting MBEs in the 2 to 3% range. Considering that the best GHI measurements have an absolute accuracy of 3%, it is necessary to be sure of what is being compared and cognizant of the uncertainties in the comparison measurements. The uncertainty in the GHI estimates increase if there is snow on the ground, the albedo in the image pixel varies considerably, or the terrain creates significant shadowed areas.

DNI and DHI have larger fractional RMSE and MBE than GHI estimates with the DNI, being up to twice the RMSE compared with the GHI estimate uncertainties and the DHI having a slightly larger RMSE than DNI.

With satellite images not centered on the hour, but at other times, the correspondence between ground-based data and data derived from satellite is not always easy to make. For example, when an image is taken at 9:15, it has to be shifted to 9:00 to be compatible with other meteorological data. The satellite-derived data in the National Solar Radiation Data Base is a weighted average of irradiance values that correspond to the hourly averaged meteorological data. This smoothing reduces some of the variability in the data values but gives an overall better representation when the irradiance data are used with other meteorological values for system performance calculations.

Satellite-derived values near sunrise and sunset have high uncertainties. This is caused by two factors: (1) the large incident angles, and (2) the fact that sometimes the satellite images are taken when the sun is below the horizon but there is some irradiance during the hour. As an illustration of the type of problem that can occur, if sunrise is 6:30 and the satellite image is taken at 6:15, there will be no irradiance recorded for the time period when there really is GHI between 6:30 and 7:00. Using an averaging method with the 7:15 value can give some value at 7:00. Of course the irradiance values are relatively small and large uncertainties do not significantly affect the usefulness of the data, but it is important to understand the limitations of the data values used.

B.5 COMMENTS ON MODELING AND MEASUREMENT

As with any modeling effort, it is important to know the estimated uncertainty of the measurements used to develop and validate a model. Specifically, the systematic errors, which are part of the type B uncertainties, are different for an Eppley PSP pyranometer, a LI-COR LI-200 pyranometer, and a Kipp and Zonen CM 21 pyranometer. A model developed using one type of instrument will work best with that type of instrument if the systematic errors are not removed. Therefore, care should be taken when developing a model to use data where known systematic errors of the instruments have been removed. If they are not, then it is likely that the model will agree best with the measurements made with the instruments used to develop the model, but not necessarily the best measurements.

If an instrument is fully characterized, it is possible to remove most of the systematic errors in the measurements. Alternatively it is possible to use very accurate, and usually more expensive, instruments like absolute cavity radiometers to make the measurements. It should be possible to make significant refinement to models based on very accurate measurements. However, if one wants to develop models with data taken with lesser-quality instruments, the systematic errors associated with these less accurate instruments have to be accounted for or else they will skew the modeled result. Alternatively, if one uses measured data without correcting for systematic errors, the modeled results will be skewed. Therefore, it is important to know the instrument's measurement performance characteristics. Important information such as temperature dependence, calibration uncertainty and frequency, and known systematic errors do affect the usefulness of the data and the accuracy of any results.

Deriving irradiance data from satellite images is dependent on the characteristics of the local albedo and also may be affected by local physical and environmental conditions. While models using satellite images are validated for a variety of sites, there is always the possibility that conditions at a given site are such that the satellite-derived data may be skewed. For example, the satellite pixel encompasses two types of surfaces, say, a salt bed and a lake. Any wobble in the pointing of the satellite will pick up more albedo from the salt bed or the lake. This wobble makes it difficult to determine the surface albedo and hence decreases the accuracy of the modeled data. Therefore, it is always useful to have some ground-based measurements to validate the satellite-derived data for the specific location under study.

QUESTIONS

1. What type of satellites are used to estimate solar radiation?
2. Briefly describe the process of obtaining global solar radiation data from satellite images.
3. Compare and contrast the strengths and weaknesses of solar radiation values obtained using satellite images and data from ground based stations.

REFERENCES

ARM. 2002. Atmospheric radiation measurement program. Available at: http://www.arm.gov/

Broesamle, H., H. Mannstein, C. Schillings, and F. Trieb. 2001. Assessment of solar electricity potentials in North Africa based on satellite data and a geographic information system. *Solar Energy* 70:1–12.

BSRN. 2002. Baseline surface radiation network. Available at: http://bsrn.ethz.ch/

Cano, D., J. M. Monget, M. Aubuisson, H. Guillard, N. Regas, and L. Wald. 1986. A method for the determination of global solar radiation from meteorological satellite data. *Solar Energy* 37:31–39.

Cess, R. D., S. Nemesure, E. G. Dutton, J. J. DeLuisi, G. L. Potter, and J. Morcrette. 1993. The impact of clouds on shortwave radiation budget of the surface-atmosphere system: Interfacing measurements and models. *Journal of Climate* 6:308–316.

Ineichen, P. 2008. Comparison and validation of three global-to-beam irradiance models against ground measurements, *Solar Energy* 82: 501–512.

Ineichen, P. and R. Perez. 1999. Derivation of cloud index from geostationary satellites and application to the production of solar irradiance and daylight illuminance data. *Theoretical and Applied Climatology* 64:119–130.

Ineichen, P. and R. Perez. 2002. A new airmass independent formulation for the Linke turbidity coefficient. *Solar Energy* 73(3): 151–157.

Ineichen, P., R. Perez, M. Kmiecik, and D. Renne. 2000. Modeling direct irradiance from GOES visible channel using generalized cloud indices. In *Proceedings of the 80th AMS Annual Meeting*, Long Beach, CA.

Kasten, F. 1980. A simple parameterization of two pyrheliometric formulae for determining the Linke turbidity factor. *Meteorologische Rundschau* 33:124–127.

Kasten, F. 1984. Parametriesirung der Globalstahlung durch Bedeckungsgrad und Trubungsfaktor. *Annalen der Meteorologie Neue Folge* 20:49–50.

NOHRSC. 2002. National Operational Hydrologic Remote Sensing Center. Available at: http://www.nohrsc.nws.gov/

NSRDB. 1995. National Solar Radiation Data Base—final technical report, Volume 2, 1995. NREL/TP-463-5784.

Perez, R., P. Ineichen, E. Maxwell, R. Seals, and A. Zelenka. 1992. Dynamic global-to-direct irradiance conversion models. *ASHRAE Transactions-Research Series*, pp. 354–369.

Perez, R., P. Ineichen, K. Moore, M. Kmiecik, C. Chain, R. George, and F. Vignola. 2002. A new operational satellite-to-irradiance model. *Solar Energy* 73(5):307–317.

Perez, R., R. Seals, P. Ineichen, R. Stewart, D. Menicucci. 1987. A new simplified version of the Perez Diffuse Irradiance Model for tilted surfaces. Description performance validation. *Solar Energy* 39:221–232.

Pinker, R. and J. Ewing. 1985. Modeling surface solar radiation: Model formulation and validation. *Journal of Climate and Applied Meteorology* 24: 389–401.

Pinker, R. and I. Laszlo. 1992. Modeling surface solar irradiance for satellite applications on a global scale. *Journal of Applied Meteorology* 31:194–211.

Pinty, B. and M. M. Verstraete. 1991. Extracting information on surface properties from bidirectional reflectance measurements. *Journal of Geophysical Research* 96:2865–2874.

Schmetz, J. 1989. Towards a surface radiation climatology: Retrieval of downward irradiances from satellites. *Atmospheric Research* 23:287–321.

Stoffel, T., D. Renné, D. Myers, S. Wilcox, M. Sengupta, R. George, and C. Turchi. 2010. CONCENTRATING SOLAR POWER best practices handbook for the collection and use of solar resource data. NREL/TP-550-47465.

SWERA. 2002. Solar and wind resource assessment. Available at: http://www.uneptie.org/energy/act/re/fs/swera.pdf

Vignola, F. and R. Perez. 2004. Solar resource GIS data base for the Pacific Northwest using satellite data—final report. Project ID DE-FC26-00NT41011.

Vignola, F., P. Harlan, R. Perez, and M. Kmiecik. 2007. Analysis of satellite derived beam and global solar radiation data. *Solar Energy* 81:768–772.

Whitlock, C. H., T. P. Charlock, W. F. Staylor, R. T. Pinker, I. Laszlo, A. Ohmura, H. Gilgen, T. Konzelman, R. C. DiPasquale, C. D. Moats, S. R. LeCroy, and N. A. Ritchey. 1995. First global WCRP shortwave surface radiation budget dataset. *Bulletin of the American Meteorological Society* 76:905–922.

Zelenka, A., R. Perez, R. Seals, and D. Renné. 1999. Effective accuracy of satellite-derived irradiance. *Theoretical and Applied Climatology* 62:199–207.

Appendix C: Sun Path Charts

For setting up a solar monitoring station or evaluating the performance of a solar energy system, it is useful to know the position of the sun in the sky. Solar algorithms exist (Appendix D) to calculate the solar position. Sometimes it is useful to get a visual image of the position of the sun, especially when evaluating shading caused by nearby objects. Usage of sun path charts are discussed in Chapter 2 and Chapter 14. In this appendix, both Cartesian and polar plots are given for every 10° of latitude. Sun path charts for a given latitude can also be created at http://solardata. uoregon.edu/SunChartProgram.php. Sundial plots, showing the shadow from the tip of a pole as it moves across the ground, can be created at http://solardata.uoregon. edu/SunDialProgram.html. The plots are all created from the same information, but the perspective is different.

C.1 CARTESIAN SUN PATH CHARTS

Sun path charts drawn using Cartesian coordinates are illustrated first. The solar elevation ($90° - sza$) is the vertical axis, and the solar azimuth is the horizontal axis. True north has an azimuth angle of 0°. Seven sun paths are shown on the chart: December 21, January 21 and November 21, February 20 and October 21, March 20 and September 22, April 20 and August 22, May 21 and July 21, and June 21. The paths on December 21 and June 21 represent the extremes of the sun path and illustrate the paths during the winter and summer solstices. The days of the months of the paths in between were chosen to be about 30 days apart so that the spring and fall paths would be identical.

Sun paths charts were created at latitudes 10° apart and from the North Pole to the South Pole. The time shown on the charts is solar time. This means that the sun is highest in the sky at solar noon. Plots can be made in local standard time, which is useful if the charts are used to locate the azimuth angle of obstructions (Figures C.1 to C.19).

C.2 POLAR SUN PATH CHARTS

The sun path charts in Cartesian and polar coordinates show exactly the same solar zenith and solar azimuth angles. The difference is that the polar sun path charts project the sun path onto polar coordinates. The highest point in the sky has a solar altitude of 90°. Concentric circles show the decreasing solar altitude at 10° increments with zero altitude being the horizon. The spokes (or radii) of the wheel are the solar azimuth angle and vary from 0° to 360° starting from zero degrees at true north. The solar azimuth values are given on the outside of the largest circle.

The polar sun paths should be viewed as if one were standing in the center of the plot and looking at the sky dome. The sun path and the hour lines form a grid that changes shape with latitude. At the poles, the sun paths become circular (Figures C.20 and C.38).

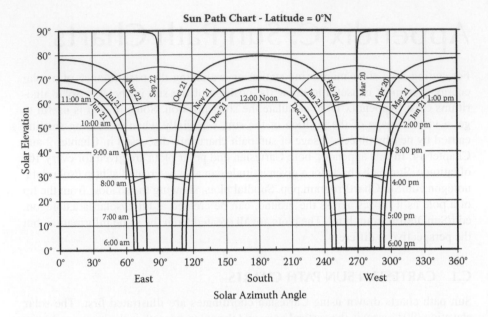

FIGURE C.1 Cartesian sun path chart facing the equator at latitude 0° N.

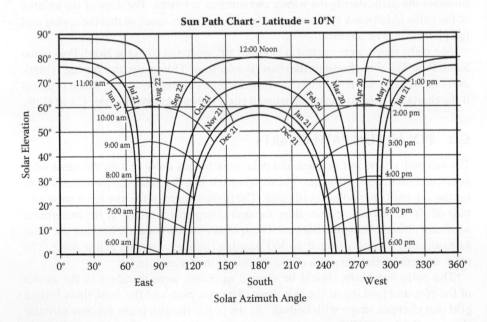

FIGURE C.2 Cartesian sun path chart facing the equator at latitude 10° N.

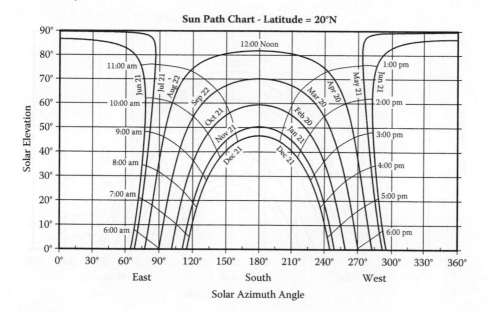

FIGURE C.3 Cartesian sun path chart facing the equator at latitude 20° N.

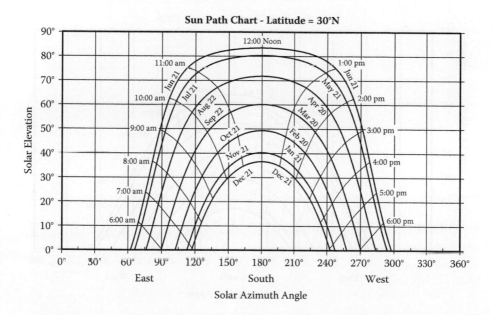

FIGURE C.4 Cartesian sun path chart facing the equator at latitude 30° N.

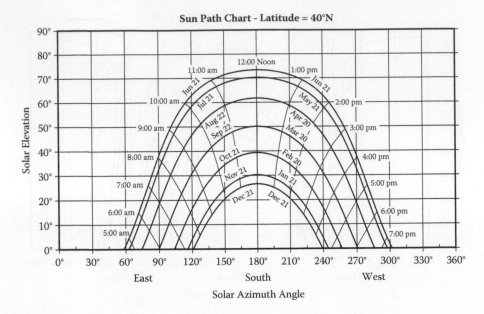

FIGURE C.5 Cartesian sun path chart facing the equator at latitude 40° N.

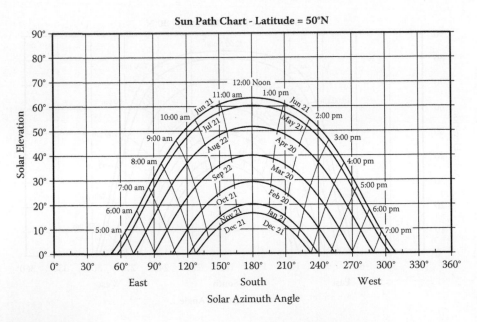

FIGURE C.6 Cartesian sun path chart facing the equator at latitude 50° N.

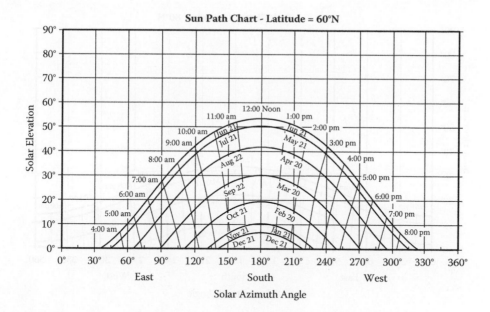

FIGURE C.7   Cartesian sun path chart facing the equator at latitude 60° N.

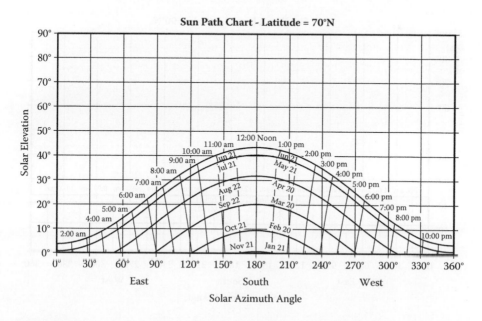

FIGURE C.8   Cartesian sun path chart facing the equator at latitude 70° N.

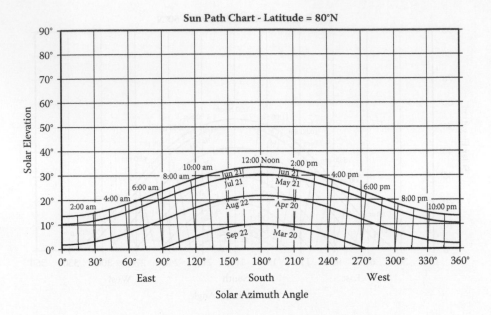

FIGURE C.9   Cartesian sun path chart facing the equator at latitude 80° N.

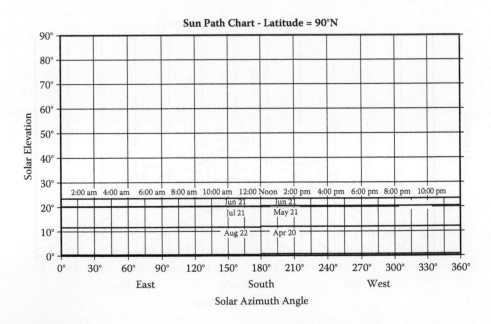

FIGURE C.10   Cartesian sun path chart facing the equator at latitude 90° N.

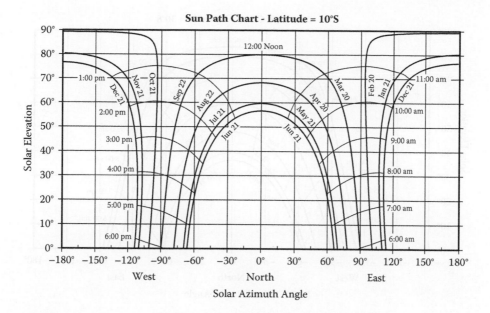

FIGURE C.11   Cartesian sun path chart facing the equator at latitude 10° S.

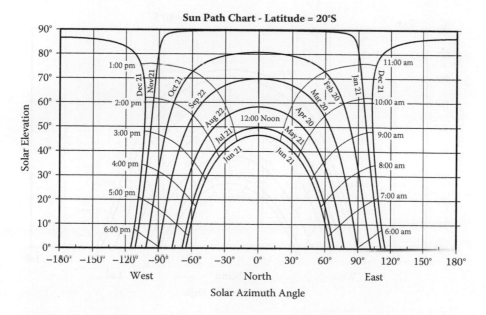

FIGURE C.12   Cartesian sun path chart facing the equator at latitude 20° S.

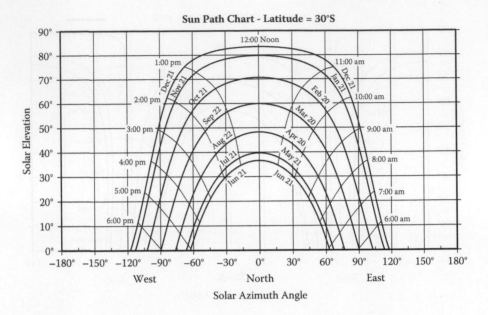

FIGURE C.13 Cartesian sun path chart facing the equator at latitude 30° S.

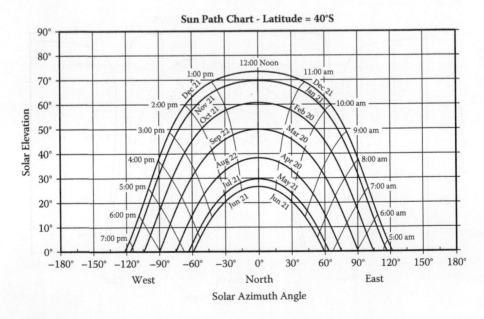

FIGURE C.14 Cartesian sun path chart facing the equator at latitude 40° S.

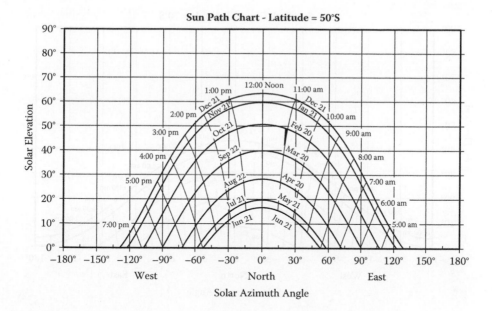

FIGURE C.15   Cartesian sun path chart facing the equator at latitude 50° S.

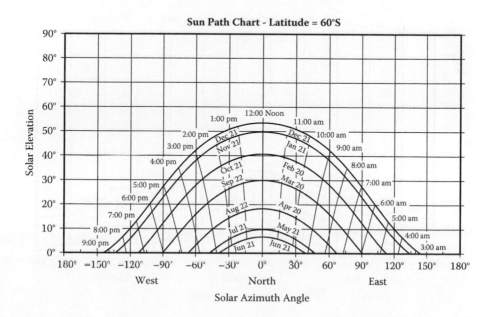

FIGURE C.16   Cartesian sun path chart facing the equator at latitude 60° S.

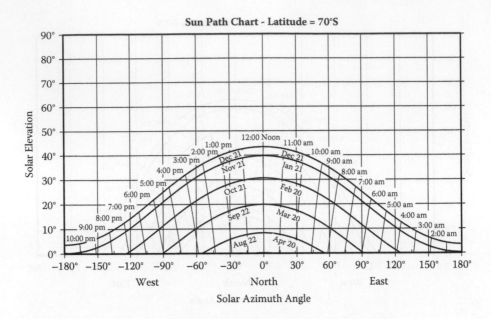

FIGURE C.17   Cartesian sun path chart facing the equator at latitude 70° S.

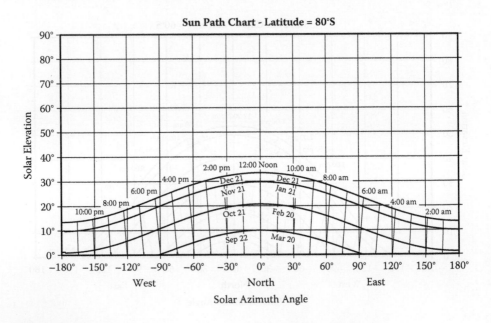

FIGURE C.18   Cartesian sun path chart facing the equator at latitude 80° S.

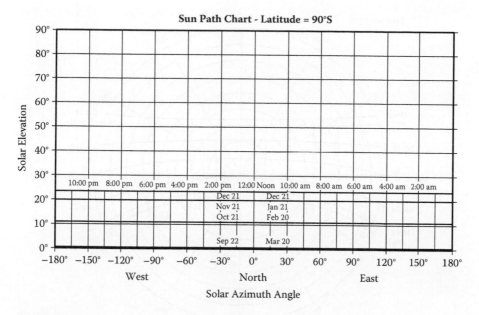

FIGURE C.19 Cartesian sun path chart facing the equator at latitude 90° S.

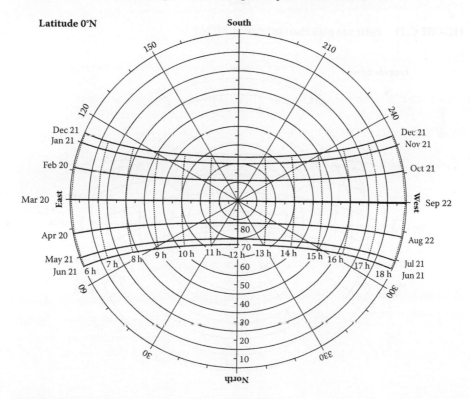

FIGURE C.20 Polar sun path chart at latitude 0° N.

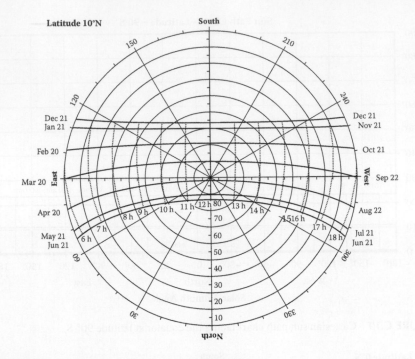

FIGURE C.21   Polar sun path chart at latitude 10° N.

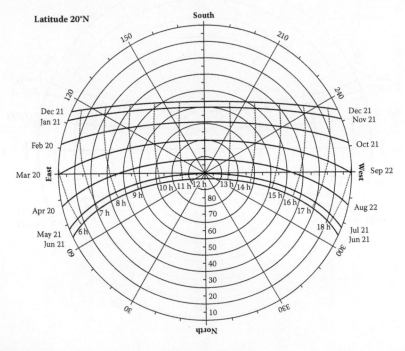

FIGURE C.22   Polar sun path chart at latitude 20° N.

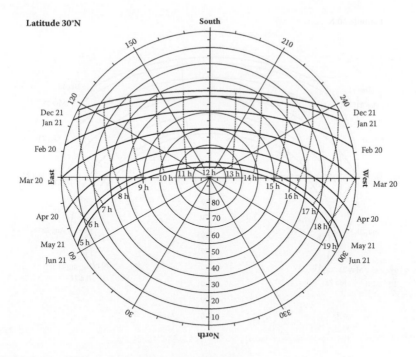

FIGURE C.23 Polar sun path chart at latitude 30° N.

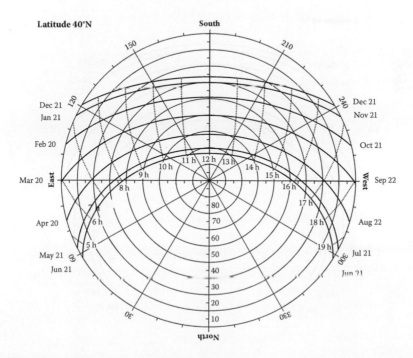

FIGURE C.24 Polar sun path chart at latitude 40° N.

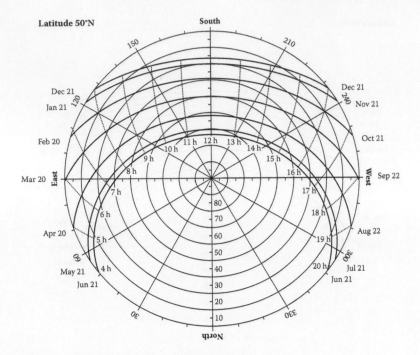

FIGURE C.25   Polar sun path chart at latitude 50° N.

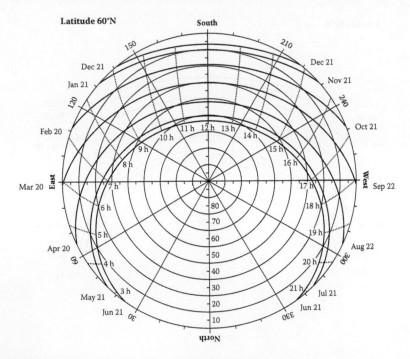

FIGURE C.26   Polar sun path chart at latitude 60° N.

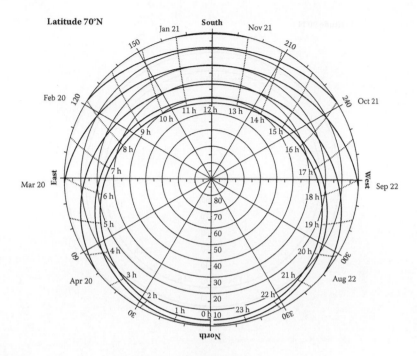

FIGURE C.27   Polar sun path chart at latitude 70° N.

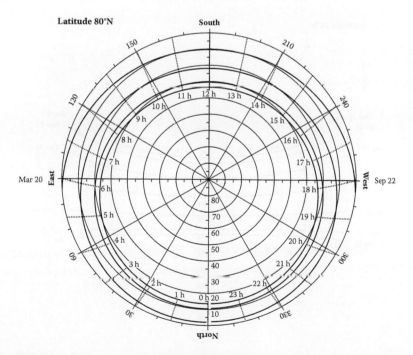

FIGURE C.28   Polar sun path chart at latitude 80° N.

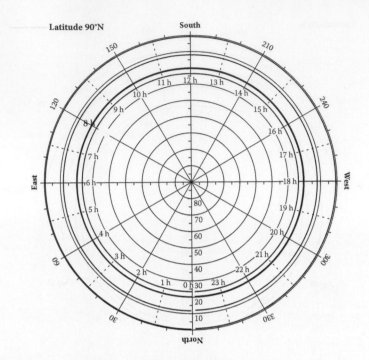

FIGURE C.29   Polar sun path chart at latitude 90° N.

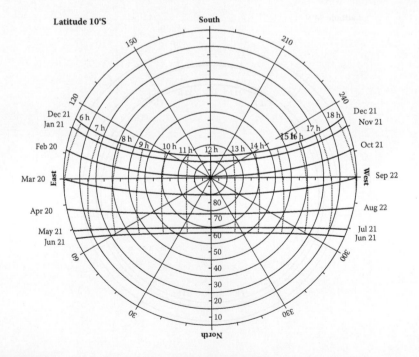

FIGURE C.30   Polar sun path chart at latitude 10° S.

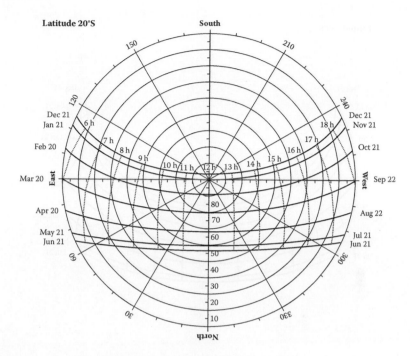

FIGURE C.31 Polar sun path chart at latitude 20° S.

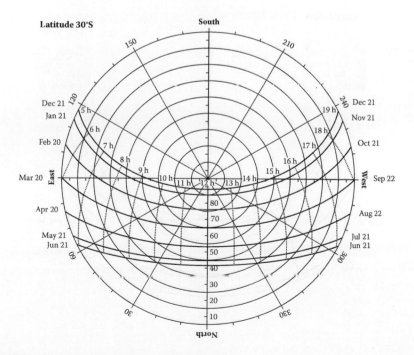

FIGURE C.32 Polar sun path chart at latitude 30° S.

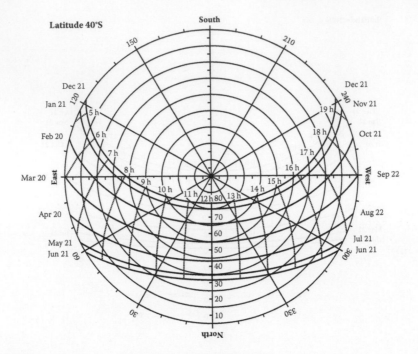

FIGURE C.33   Polar sun path chart at latitude 40° S.

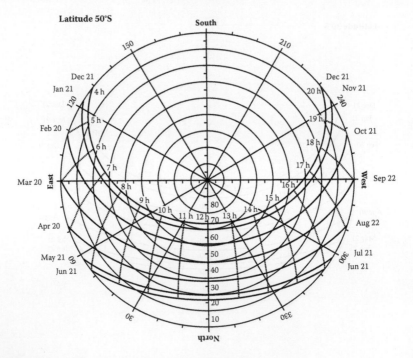

FIGURE C.34   Polar sun path chart at latitude 50° S.

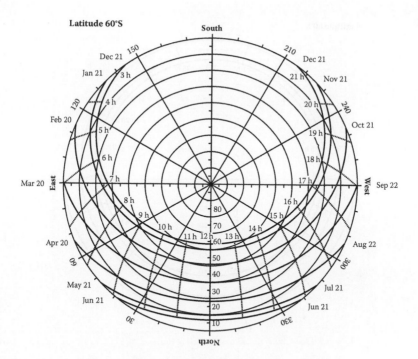

FIGURE C.35 Polar sun path chart at latitude 60° S.

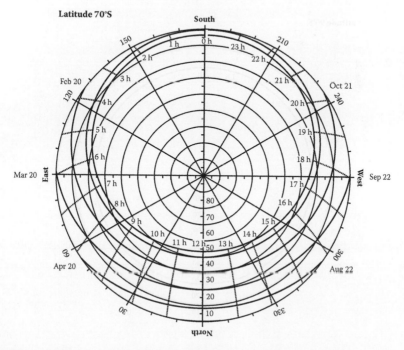

FIGURE C.36 Polar sun path chart at latitude 70° S.

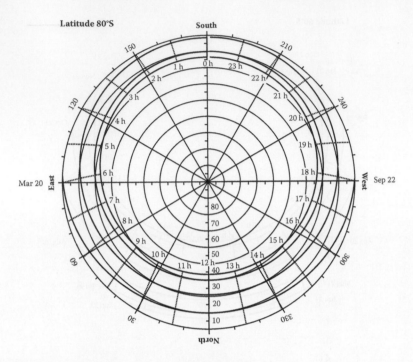

FIGURE C.37 Polar sun path chart at latitude 80° S.

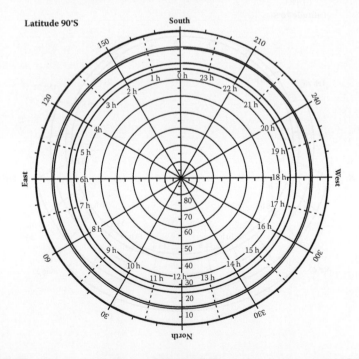

FIGURE C.38 Polar sun path chart at latitude 90° S.

Appendix D: Solar Position Algorithms

Solar position calculations are used for a wide variety of tasks from setting up solar monitoring stations to positioning photovoltaic panels to maximize electricity output. Solar position is used in the analysis of irradiance data and is essential for modeling solar irradiance for arbitrarily tilted surfaces. Many algorithms are available that estimate the solar position to varying degrees of accuracy; these solar algorithms and their accuracy are discussed in this appendix.

Every year the *Astronomical Almanac* (2011) publishes next year's sun position with accuracies approaching 0.1 seconds of arc. To achieve these accuracies, the Nautical Almanac Office of the United States Naval Observatory and Her Majesty's Nautical Almanac Office within the United Kingdom Hydrographic Office use an extremely complex algorithm that includes small effects on the earth's orbit, such as lunar and planetary perturbations, plus optical effects such as stellar aberration, parallax, and refraction.

Approximations to the sun's position relative to a location on the earth's surface have become increasingly accurate. Blanco-Muriel, Alarcón-Padilla, Lópex-Moratalla, and Lara-Coira (2001) give a thorough history of the development of the more accurate solar position algorithms up to the time of their paper. This history is briefly discussed here.

Spencer (1971) provided truncated Fourier series expressions for the equation of time and the solar declination that can be used to approximate the solar position with maximum biases of between 10 and 20 minutes of arc. However, these are accurate enough for positioning fixed flat-plate thermal panels or fixed flat-plate photovoltaic panels but are not useful for tracking the sun for concentrating solar power installations. As an example, Vant-Hull and Hildebrandt (1976) estimated that a tracking accuracy of 3.5 minutes of arc would be needed for a concentration ratio of 1,000. Spencer's (1971) approximations are inadequate for this application.

In the late 1970s, at about the same time and apparently unaware of each other's efforts, Pitman and Vant-Hull (1978) and Walraven (1978, 1979) developed simplifications of the Nautical Almanac's algorithms to produce much-improved approximations for the solar position. These were accurate to approximately one arc minute. Walraven (1978) was noteworthy since it provided a Fortran subroutine to calculate solar position; Walraven (1979) corrected two lines of code in an erratum within a year of the first publication. Archer (1980) and Wilkinson (1981) followed with a comment and a note that suggested a correction for refraction to the algorithm. To our knowledge there is no hyperlink to the corrected code with refraction.

Progress continued with the Fortran program developed by Michalsky (1988a) using the low-precision formulae published in the *Astronomical Almanac* (1984). These formulae were expected to be good to about 0.01° between the years 1950 and 2050. The formulae in the current version of the *Astronomical Almanac* (2011) are identical to those in the 1984 publication of this almanac. An erratum was published that fixed some typesetting issues (Michalsky, 1988b) in the original, and then another fixed typesetting issues in the erratum (Michalsky, 1989). However, the most significant issue was corrected in a letter from Spencer (1989) that changed the code to give the correct assignment to the azimuth in the southern hemisphere. The Fortran program for *sunpos* is given at the end of this appendix to illustrate the parameters needed and the terminology used in calculating the position of the sun. The ftp site ftp://ftp.srrb.noaa.gov/pub/users/joe/asked_for_stuff/asunpos.f also contains the *sunpos* Fortran code with all corrections and has been stable for years. Alternately, Wiscombe created another version (ftp://climate1.gsfc.nasa.gov/wiscombe/Solar_Rad/SunAngles/sunae.f) in standard Fortran that is similar and yields identical results. This same algorithm is used in the National Renewable Energy Laboratory's *solpos* program that calculates solar position and much more (available at http://rredc.nrel.gov/solar/codesandalgorithms/solpos/).

A little more than a decade had passed when Blanco-Muriel et al. (2001) published their PSA algorithm (available as C++ code at http://www.psa.es/sdg/sunpos.htm). The accuracy was improved over Michalsky's (1988a) code by restricting the years approximated to the range 1999–2015, modifying the "low-precision" formulae in the Astronomical Almanac (1984), and including a correction for parallax.

Reda and Andreas (2004) published solar position code based on Meeus's (1998) algorithm. This code is the most accurate to date with a stated accuracy of ±0.0003° for the inclusive years 2000 BC–AD 6000. The C code name *SPA* can be found at http://www.nrel.gov/midc/spa/. There is a link to an online calculator from that page. Cornwall provides a Meeus-based online calculator for the solar position at http://www.esrl.noaa.gov/gmd/grad/solcalc/ that includes links to spreadsheet versions of the solar position calculator for a day or year via the "Calculation Details" link on that page.

An algorithm that provided higher accuracy than Blanco-Muriel et al.'s (2001) algorithm and lower accuracy than Reda and Andreas (2004) but with a shorter run time and less complexity was provide in Grena (2008). The maximum error for this algorithm is stated to be 0.0027° with a run time comparable to somewhat less accurate algorithms (Blanco-Muriel et al., 2001; Michalsky, 1988). This accuracy is for the period 2003 to 2022. The code is included in Grena (2008) as an appendix.

Table D.1 contains a summary of the available codes with maximum errors and years for which the codes are valid. The maximum error is from the information in Grena (2008).

More complex codes run more slowly but deliver greater accuracy, sometimes with a limited window over which the accuracy is guaranteed. However, run time with today's processors is hardly ever an issue.

One point that does not seem to be emphasized in any of these papers is the issue of accuracy given the generally unknown true refraction. Refraction is usually calculated for standard conditions but depends on the true temperature and moisture structure of the atmosphere in the direction of the sun and on atmospheric pressure. Van der Werf (2003) demonstrates arc minutes of differences for a modified standard

TABLE D.1
Summary of Solar Position Accuracies and Years of Applicability

Source	Maximum Error	Years Included
Spencer (1971)	~0.25°	—
Michalsky (1988a)	0.011°	1950–2050
Blanco-Muriel et al. (2001)	0.008°	1999–2015
Reda and Andreas (2004)	0.0003	2000 BC–AD 6000
Grena (2008)	0.0027°	2003–2022

atmosphere with different temperature and humidity profiles. These changes are exacerbated at the lowest sun angles, especially from the horizon to 5° above the horizon.

SUNPOS.F

This Fortran program calculates the solar position given the year, day, time, latitude, and longitude. It outputs solar azimuth, elevation, hour angle, declination, and air mass.

```
c    this program calculates the solar position given the year, day,
c    time, latitude, and longitude
c
c    it outputs solar azimuth, elevation, hour angle, declination, and
c    air mass
c
c
     program sunpos
     real lat,long, mn
     write(6,'(3x,"This program calculates sun position if you
   # specify time and location!")')
     write(6,'(/)')
     write(6,'(3x,"Type the latitude like this sxx.xx where s is a
   # sign!")')
     read(5,'(f8.4)')lat
     write(6,'(f8.4)')lat
     write(6,'(3x,"Type the longitude like this sxxx.xx!")')
     read(5,'(f9.4)')long
     write(6,'(f9.4)')long
     write(6,'(3x,"Type the year like this xxxx.!")')
     read(5,'(f5.0)')year
     write(6,'(f5.0)')year
     write(6,'(3x,"Indicate time zone,e.g., est=+5,mst=+7,etc.(The
   # sign is important)!")')
     read(5,'(i3)')itz
     write(6,'(i3)')itz
     tz=float(itz)
```

```
60      write(6,'(3x,"Type the day of year(Feb 1=32) like this xxx.!")')
        read(5,'(f4.0)')day
        write(6,'(f4.0)')day
50      write(6,'(3x,"Type the local standard time(not daylight savings
#       time) like this xx.xx.xx., e.g., 12.15.47.!")')
        read(5,'(3f3.0)')hour, mn,sec
        write(6,'(3f3.0)')hour, mn,sec
        tm=((tz+hour)*3600+ mn*60+sec)/3600.
        call sunae(year,day,tm,lat,long,az,el,ha,dec,soldst)
        write(6,'(//)')
        write(6,'("The solar azimuth and elevation are")')
        write(6,100)az,el
100     format(3x,f9.4,3x,f8.4,/)
        write(6,'("The solar hour angle and declination are")')
        write(6,100)ha,dec
        am=armass(el)
        write(6,'("The air mass is")')
        write(6,200)am
200     format(3x,f6.2)
        write(6,'("The solar distance is")')
        write(6,250)soldst
250     format(3x,f7.4)
        write(6,'(//)')
        write(6,'("If you want another time same day, type 9!")')
        read(5,'(i1)')i
        if(i.eq.9)go to 50
        write(6,'("If you want another day, type 99!")')
        read(5,'(i2)')i
        if(i.eq.99)go to 60
        end
c----------------------------------------------------------------------+
c                                                                      |
c    subroutine sunae(year,day,hour,lat,long,az,el,ha,dec,soldst)      |
c                                                                      |
c    this subroutine calculates the local azimuth and elevation of the |
c       sun at a specific location and time using an approximation to  |
c       equations used to generate tables in The Astronomical Almanac. |
c       refraction correction is added so sun position is apparent one.|
c                                                                      |
c    The Astronomical Almanac, U.S. Gov't Printing Office, Washington, |
c       D.C. (1985).                                                   |
c                                                                      |
c    input parameters                                                  |
c       year=year, e.g., 1986                                          |
c       day=day of year, e.g., feb 1=32                                |
c       hour=hours plus fraction in UT, e.g., 8:30 am eastern daylight |
c          time is equal to 8.5 + 5(5 hours west of Greenwich) -1(for  |
c          daylight savings time correction)                           |
c       lat=latitude in degrees (north is positive)                    |
c       long=longitude in degrees (east is positive)                   |
c                                                                      |
c    output parameters                                                 |
c       a=sun azimuth angle (measured east from north, 0 to 360 degs)  |
c       e=sun elevation angle (degs)                                   |
c       plus others, but note the units indicated before return       |
c                                                                      |
c----------------------------------------------------------------------+
```

```
c
      subroutine sunae(year,day,hour,lat,long,az,el,ha,dec,soldst)
c   work with real variables and define some constants, including
c   one to change between degs and radians
      implicit real (a-z)
      data twopi,pi,rad/6.2831853,3.1415927,.017453293/
c
c   get the current julian date (actually add 2,400,000 for jd)
      delta=year-1949.
      leap=aint(delta/4.)
      jd=32916.5+delta*365.+leap+day+hour/24.
c   1st no. is mid. 0 jan 1949 minus 2.4e6; leap=leap days since 1949
c   the last yr of century is not leap yr unless divisible by 400
      if(amod(year,100.).eq.0.0.and.amod(year,400.).ne.0.0)jd=jd-1.
c
c   calculate ecliptic coordinates
      time=jd-51545.0
c   51545.0 + 2.4e6 = noon 1 jan 2000
c
c   force mean longitude between 0 and 360 degs
      mnlong=280.460+.9856474*time
      mnlong=mod(mnlong,360.)
      if(mnlong.lt.0.)mnlong=mnlong+360.
c
c   mean anomaly in radians between 0 and 2*pi
      mnanom=357.528+.9856003*time
      mnanom=mod(mnanom,360.)
      if(mnanom.lt.0.)mnanom=mnanom+360.
      mnanom=mnanom*rad
c
c   compute the ecliptic longitude and obliquity of ecliptic in radians
      eclong=mnlong+1.915*sin(mnanom)+.020*sin(2.*mnanom)
      eclong=mod(eclong,360.)
      if(eclong.lt.0.)eclong=eclong+360.
      oblqec=23.439-.0000004*time
      eclong=eclong*rad
      oblqec=oblqec*rad
c
c   calculate right ascension and declination
      num=cos(oblqec)*sin(eclong)
      den=cos(eclong)
      ra=atan(num/den)
c   force ra between 0 and 2*pi
      if(den.lt.0)then
          ra=ra+pi
      elseif(num.lt.0)then
          ra=ra+twopi
      endif
c
c   dec in radians
      dec=asin(sin(oblqec)*sin(eclong))
c
c   calculate Greenwich mean sidereal time in hours
      gmst=6.697375+.0657098242*time+hour
c   hour not changed to sidereal time since 'time' includes
c   the fractional day
      gmst=mod(gmst,24.)
```

```fortran
      if(gmst.lt.0.)gmst=gmst+24.
c
c     calculate local mean sidereal time in radians
      lmst=gmst+long/15.
      lmst=mod(lmst,24.)
      if(lmst.lt.0.)lmst=lmst+24.
      lmst=lmst*15.*rad
c
c     calculate hour angle in radians between -pi and pi
      ha=lmst-ra
      if(ha.lt.-pi)ha=ha+twopi
      if(ha.gt.pi)ha=ha-twopi
c
c     change latitude to radians
      lat=lat*rad
c
c     calculate azimuth and elevation
      el=asin(sin(dec)*sin(lat)+cos(dec)*cos(lat)*cos(ha))
      az=asin(-cos(dec)*sin(ha)/cos(el))
c
c     this puts azimuth between 0 and 2*pi radians
      if(sin(dec)-sin(el)*sin(lat).ge.0.)then
      if(sin(az).lt.0.)az=az+twopi
      else
      az=pi-az
      endif
c
c     calculate refraction correction for US stan. atmosphere
c     need to have el in degs before calculating correction
      el=el/rad
c
      if(el.ge.19.225) then
         refrac=.00452*3.51823/tan(el*rad)
      else if (el.gt.-.766.and.el.lt.19.225) then
         refrac=3.51823*(.1594+.0196*el+.00002*el**2)/
     #   (1.+.505*el+.0845*el**2)
      else if (el.le.-.766) then
         refrac=0.0
      end if
c
c     note that 3.51823=1013.25 mb/288 C
      el=el+refrac
c     elevation in degs
c
c     calculate distance to sun in A.U. & diameter in degs
      soldst=1.00014-.01671*cos(mnanom)-.00014*cos(2.*mnanom)
      soldia=.5332/soldst
c
c     convert az and lat to degs before returning
      az=az/rad
      lat=lat/rad
      ha=ha/rad
      dec=dec/rad
c     mnlong in degs, gmst in hours, jd in days if 2.4e6 added;
c     mnanom,eclong,oblqec,ra,and lmst in radians
      return
      end
```

```
c
c
c    this function calculates air mass using kasten's
c    approximation to bemporad's tables
c
c
     function armass(el)
     z=(90.-el)*3.141592654/180.
     armass=1./(cos(z)+.50572*(6.07995+el)**-1.6364)
     return
     end
```

REFERENCES

Archer, C. B. 1980. Comments on "Calculating the position of the sun." *Solar Energy* 25:91.

Blanco-Muriel, M., D. C. Alarcón-Padilla, T. Lópex-Moratalla, and M. Lara-Coira. 2001. Computing the solar vector. *Solar Energy* 70:431–441.

Grena, R. 2008. An algorithm for the computation of the solar position. *Solar Energy* 82:462–470.

Meeus, J. 1998. *Astronomical algorithms*, 2d ed. Richmond, VA: Willmann-Bell, Inc.

Michalsky, J. J. 1988a. The Astronomical Almanac's algorithm for approximate solar position (1950–2050). *Solar Energy* 40:227–235.

Michalsky, J. J. 1988b. Errata. *Solar Energy* 41:113.

Michalsky, J. J. 1989. Errata. *Solar Energy* 43:323.

Nautical Almanac Office of the United State Naval Observatory and Nautical Almanac Office of Great Britain. 1984. *The astronomical almanac for the year 1985*. Washington, DC: U.S. Government Printing Office and London: Her Majesty's Stationery Office.

Nautical Almanac Office of the United State Naval Observatory and Nautical Almanac Office of Great Britain. 2011. *The astronomical almanac for the year 2012*. Washington: U.S. Government Printing Office and London: Her Majesty's Stationery Office.

Pitman, C. L. and L. L. Vant-Hull. 1978. Errors in locating the sun and their effect on solar intensity predictions. In *Proceedings of the American Section of the International Solar Energy Society,* Denver, CO.

Reda, I. and A. Andreas. 2004. Solar position algorithm for solar radiation applications. *Solar Energy* 76:577–589.

Spencer, J. W. 1971. Fourier series representation of the position of the sun. *Search* 2:172.

Spencer, J. W. 1989. Comments on "The astronomical almanac's algorithm for approximate solar position (1950-2050)." *Solar Energy* 42:353.

Van der Werf, S. 2003. Ray tracing and refraction in the modified US1976 atmosphere. *Applied Optics* 42:354–366.

Vant-Hull, L. L. and A. F. Hildebrandt. 1976. Solar thermal power system based on optical transmission. *Solar Energy* 18:31–39.

Walraven, R. 1978. Calculating the position of the sun. *Solar Energy* 20:393–397.

Walraven. R. 1979. Erratum. *Solar Energy* 22:195.

Wilkinson, B. J. 1981. An improved FORTRAN program for the rapid calculation of the solar position. *Solar Energy* 27:67–68.

Appendix E: Useful Conversion Factors

Key terms and units of measure important to radiometry are presented in Table E.1. Converting units of measure for other selected quantities can be accomplished with the information presented in Table E.2. The multipliers needed to convert from the unit in the first column to the unit in the second column are given in the third column of Table E.2.

E.1 THE MANY MEANINGS OF WM^{-2}

Solar irradiance, the amount of solar radiation received by a surface (i.e., *radiant energy density*) for the minute, hour, day, month, or year is often reported in units of average Wm^{-2} or averaged *radiant power* per unit *area* for a given time period. The energy per unit area per unit time (Wm^{-2}) multiplied by the time interval becomes the *radiant energy density* because energy is *power* multiplied by *time*. For example, one Colorado town received an average of 5.6 kWhm^{-2} of solar radiation per day. This value is also reported as an average of 233 Wm^{-2} over the day.

$$233\frac{W \cdot day}{m^2} = 233\frac{W \cdot 24 \cdot hr}{m^2} = 5592\frac{W \cdot hr}{m^2} = 5.6\frac{kW \cdot hr}{m^2} \quad \text{(E.1)}$$

Another example of quantifying energy and power is to assume the measured solar irradiance (power density) was 1,000 Wm^{-2} over a 5-minute period. The equivalent energy density is 83 Whm^{-2}.

$$1000\frac{W \cdot 5\,min}{m^2} = 1000\frac{W \cdot 5 \cdot min \cdot \dfrac{1hr}{60\,min}}{m^2} = 83\frac{W \cdot hr}{m^2} \quad \text{(E.2)}$$

The general public is interested in kilowatt-hours (kWh) that can be produced by a photovoltaic (PV) system. Consider that the average U.S. household consumes between 900 and 1,000 kilowatt-hours of electricity each month. In the previous example, a town in Colorado received an annual average of 5.6 kilowatt-hours of solar energy per meter squared per day on a surface that is tilted to the south by an angle equal to the site latitude (the angle made with the horizontal is 40° for a site at 40°). In an average 30-day month each square meter of a tilted surface will receive about 168 kilowatt-hours of energy; therefore, 6 square meters would be needed to intercept a little over 1,000 kilowatt-hours of energy. However, if the PV panels receiving this energy are about 16% efficient at converting sunlight into electricity, then 36 square meters of PV panels are required to meet the average monthly electrical load of the house. This is roughly a square area that is 6 meters (~20 feet) on a side.

TABLE E.1
Radiometric Terminology and Units of Measurement

Quantity	SI Unit	Abbreviation	Description
Radiant Energy	Joule	J	Energy (ability to do work)
Radiant Flux	Watt	W	Radiant energy per unit time (radiant power)
Radiant Intensity	Watt per steradian	Wsr^{-1}	Power per unit solid angle
Radiant Power Density	Watt per square meter	Wm^{-2}	Radiant power per unit area
Radiant Energy Density	Watt-hour per square meter	Whm^{-2}	Radiant power in one hour (energy) per unit area
Radiant Emittance	Watt per square meter	Wm^{-2}	Power emitted from a surface
Radiance	Watt per steradian per square meter	$Wsr^{-1}m^{-2}$	Power per unit solid angle per unit of projected source area
Spectral Irradiance	Watt per square meter per nanometer	$Wm^{-2}nm^{-1}$	Power incident on a surface per unit wavelength

TABLE E.2
Useful Conversion Factors

To convert from	to	Multiply by
Watt	Joule(J)/sec	1.0000
Watt	Btu/hour	3.4094
Watt	Horsepower	1.3410×10^{-3}
Joules/sec	Watt	1.0000
Btu/hour	Watt	0.2933
Horsepower	Watt	745.71
Kilowatt-hour (kWh)	Btu	3409.4
Kilowatt-hours/meter^2 (kWhm^{-2})	kilojoule/meter^2 (kJm^{-2})	3600.0
Kilowatt-hours/meter^2 (kWhm^{-2})	Btu/foot^2	316.74
Kilowatt-hours/meter^2 (kWhm^{-2})	Langley (Ly)	86.04
Kilowatt-hours/meter^2 (kWhm^{-2})	Calories/centimeter^2	86.04
Btu	kiloWatt-hour (kWh)	2.9331×10^{-4}
Kilojoules/meter^2	kiloWatt-hour/meter^2 (kWhm^{-2})	2.7777×10^{-4}
Btu/foot^2	kiloWatt-hour/meter^2 (kWhm^{-2})	3.1571×10^{-3}
Langley (Ly)	kilojoule/meter^2 (kJm^{-2})	41.84
Langley (Ly)	kiloWatt-hour/meter^2 (kWhm^{-2})	0.011622
Calories/centimeter^2	kiloWatt-hour/meter^2 (kWhm^{-2})	0.011622
Joules (J)	Btu	0.94706×10^{-3}
Kilojoules/meter^2 (kJm^{-2})	kiloWatt-hour meter^{-2} (KWhm^{-2})	0.27778×10^{-3}
Kilojoules/meter^2 (kJm^{-2})	Langley (Ly)	0.023901
Kilojoules/meter^2 (kJm^{-2})	Btu foot^{-2}	0.087982

TABLE E.2 (CONTINUED)
Useful Conversion Factors

To convert from	to	Multiply by
Btu foot^{-2}	Langley (Ly)	0.27144
Btu foot^{-2}	kilojoules/meter^2 (kJm^{-2})	11.366
Btu/hour-foot^2	Watt meter^{-2}	3.1572
Foot-candles	Lux	10.76
Lux	Foot-candles	0.09294
Watts/meter^2-C	Btu/hour-foot^2-F	0.17611
Meters	Inches	39.37
Meters	Feet	3.2808
Meters	Yards	1.094
Kilometer	Mile	0.6213712
Meter^2	Foot^2	10.764
Kilometer^2	Acre	247.10
Liter (L)	Foot^3	0.035316
Meter^3	Gallon (U.S.)	264.172
Gallon	Foot^3	0.13368
Liter	Gallon	0.26418
Kilogram	Pound (Lb)	2.2046
Kilogram-meter^{-2}	Pounds/square inch (Lb-in^{-2})	0.001422
Celsius	Fahrenheit (°F)	C°*1.8+32
Celsius	Kelvin	C°+273.15
Fahrenheit (°F)	Celsius	°F – 32/1.8
Kelvin	Celsius	K – 273.15
Meters/second	Miles/hr (mph)	2.2369
Meters/second	Kilometer/hr	3.6
Meters/second	Knots	1.9438
Millibars	Pascals	100.0
Millibars	Atmospheres	0.0009869
Millibars	Pounds per square inch (Lb-in^{-2})	0.014504
Degrees	Radians	0.017453293
Radians	Degrees	57.2957795

QUESTIONS

1. Given that an average 5.6 kWhm^{-2} is incident on a surface and solar panels are 16% efficient, what is the area of photovoltaic panels needed to supply an average of 1200 kWh to a house? What is the answer in meters? Feet?
2. What is the difference between energy and power?

Appendix F: Sources for Equipment

Manufacturers of solar instrumentation are listed here. Contact and website information are given for each entry. This list includes manufacturers of cavity radiometers, pyrheliometers, rotating shadowband radiometers, pyranometers, pyrgeometers, spectral radiometers, weather stations, and/or automatic solar trackers. This list is not complete and inclusion in the list does not constitute a recommendation. A review of the literature should be conducted before purchasing any solar equipment.

Analytical Spectral Devices, Inc.
5335 Sterling Drive, Suite A
Boulder, Colorado 80301
Telephone: (303) 444-6522
Telefax: (303) 444-6852
http://www.asdi.com
Spectral irradiance measurements

Brusag
Chapfwiesenstrasse 14, CH-8712
Stäfa, Switzerland
Voice: +41 1 926 74 74
Fax: +41 1 926 73 34
Automatic solar trackers

Campbell Scientific, Inc.
815 West 1800
North Logan, Utah 84321-1784
Telephone: (435) 753-2342 (Info)
Telephone: (435) 750-9681 (Orders)
Telefax: (435) 750-9540
Email: info@campbellsci.com
http://www.campbellsci.com
Data logger systems and weather stations

Casella London Limited
Regent House
Britannia Walk, London N1 7ND UK
Telephone: 01-253-8581
Telex: 26 16 41
Radiometers

Davis Instruments, Corp.
3465 Diablo Avenue
Hayward, California 94545
Telephone: (510) 732-9229
Telefax: (510) 670-0589
http://www.davisnet.com
Weather stations

DAYSTAR
3250 Majestic Ridge
Las Cruces, New Mexico 88011
Telephone: (505) 522-4943
http://www.raydec.com/daystar
Radiometers

Delta-T Devices Ltd.
130 Low Road
Burwell, Cambridge, CB25 0EJ UK
Radiometers, weather stations, and data loggers

U.S. Distributor: Gary L. Woods, Sales Manager
garywoods@dynamax.com
http://www.dynamax.com
Telephone: (800) 896-7108 (toll-free); (281) 564-5100
Fax: (281) 564-5200

EKO Instruments Trading Co., Ltd.
21-8 Hatagaya 1-chome
Shibuyaku, Tokyo 151 Japan
Voice: 81-3-3469-4511
Fax: 81-3-3469-4593
http://www.eko.co.jp/eko/english/03/a.html

USA: Tsukasa Kobashi
95 South Market Street, Suite 300
San Jose, California 95113
Tel: (408) 977-7751
Fax: (408) 977-7741
http://www.eko-usa.com

Dr. Alexander Los
Lulofsstraat 55, 2521 AL
Den Haag, The Netherlands

Voice: +31 (0)70 3050117
Fax: +31 (0)70 3840607
http://www.eko-eu.com
Radiometers, automatic solar trackers, and data loggers

The Eppley Laboratory, Inc.
12 Sheffield Avenue
Newport, Rhode Island 02840
Voice: (401) 847-1020
Fax: (401) 847-1031
http://www.eppleylab.com
Cavity radiometers, pyrheliometers, pyranometers, ultraviolet radiometers, and manual and automatic solar trackers

Hukseflux Thermal Sensors B.V.
Elektronicaweg 25 2628 XG
Delft, The Netherlands
Voice: +31-15-2142669
Fax: +31-15-2574949
http://www.hukseflux.com

U.S.: Robert Dolce
P.O. Box 850
Manorville, New York 11949
Voice: (631) 251-6963
Pyrheliometers, pyranometers, pyrgeometers, albedometers, and net radiometers

Irradiance. Inc.
41 Laurel Drive
Lincoln, Massachusetts 01773 USA
Phone/Fax: (781) 259-1134
http://www.irradiance.com/rsr.html
Rotating shadowband radiometers

Kipp & Zonen, Delft BV
P.O. Box 507
2600 AM Delft, The Netherlands
Mercuriusweg 1
2624 BC Delft, The Netherlands
Voice: 015-561 000
Fax: 015-620 351
Telex: 38137
http://www.kippzonen.com

U.S.: Victor Cassella
125 Wilbur Place
Bohemia, New York 11716
Voice: (631) 589-2065 ext. 22
Fax: (631) 589-2068
Pyrheliometers, pyranometers, pyrgeometers, albedometers, net radiometers, ultra-violet radiometers, automatic solar trackers, PAR sensors, sunshine duration sensors, and data loggers

LI-COR Environmental
4421 Superior Street
Lincoln, Nebraska 68504
Telephone: (402) 467-3576; (800) 447-3576
Telefax: (402) 467-2819
http://licor.com
Pyranometers, PAR sensors (terrestrial and underwater), photometric sensors, and data loggers

Matrix, Inc.
537 S. 31st Street
Mesa, Arizona 85204
Telephone: (480) 832-1380
Radiometers

Medtherm Corporation
P.O. Box 412
Huntsville, Alabama 35804
Telephone: (256) 837-2000
Telefax: (256) 837-2001
http://www.medtherm.com
Cavity radiometers

Middleton Solar
20/155 Hyde Street
Yarraville, Victoria 3013 Australia
Voice: +61-3-9396 1890
Fax: +61-3-9689 2384
http://www.middletonsolar.com
Pyrheliometers, pyranometers, pyrgeometers, albedometers, net radiometers, ultra-violet radiometers, active solar trackers, PAR sensors, sunshine duration sensors, spectrometers, and data loggers

Ocean Optics, Inc.
830 Douglas Avenue
Dunedin, Florida 34698
Telephone: (727) 733-2447
Telefax (727) 733-3962
http://www.oceanoptics.com
Spectroradiometers

European Sales Office:
Geograaf 24
6921 EW DUIVEN, The Netherlands
Telephone: +31 (0) 26 319 0500
Fax +31 (0) 26 319 05 05

PH. Schenk GmbH & Co KG
Jedleseer Strasse 59 A-1210
Wien, Austria
Telephone: (+43/1) 271 51 31-0
Telefax: (+43/1) 271 12 28 12
E-Mail: office@schenk.co.at
http://www.schenk.co.at/schenk
Radiometers

Physikalisch-Meteorologisches Observatorium Davos
World Radiation Center
Dorfstrasse 33, CH-7260
Davos Dorf, Switzerland
Voice: +41 81 417 51 11
Fax: +41 81 417 51 00
http://www.pmodwrc.ch
Cavity radiometers and precision filter radiometers

Solar Light Company
721 Oak Lane
Philadelphia, Pennsylvania 19126-3342
Telephone: (215) 927-4206
http://www.solar.com/
Radiometers

Solar Millennium
Nägelsbachstr. 40
91052 Erlangen, Germany
Voice: +49 (0) 9131 9409-145
Fax +49 (0) 9131 9409-111
http://www.solarmillennium.de
Rotating shadowband radiometers

Yankee Environmental Systems, Inc.
101 Industrial Road
P.O. Box 746
Turners Falls, Massachusetts 01376
Voice: (413) 863-0200
http://www.yesinc.com
Pyranometers, rotating shadowband radiometers, ultraviolet radiometers, sky imagers, and data systems

Appendix G: BORCAL Report

The following pages in Appendix G provide a sample broadband outdoor radiation calibration (BORCAL) report for instruments calibrated at the National Renewable Energy Laboratory (NREL) Solar Radiation Research Laboratory (SRRL). The BORCAL process is based on the summation method for calibration of pyranometers and direct comparison of pyrheliometers with electrically self-calibrating absolute cavity radiometers traceable to the World Radiometric Reference (see Chapter 4). BORCAL reports for many instruments can be found at http://www.nrel.gov/aim/borcal.html. Recently SRRL calibrations have become ISO certified, and this has resulted in some changes in the BORCAL reports.

Broadband Outdoor Radiometer Calibration

BORCAL
2009-02

Customer:
University of Oregon

Calibration Facility:
Solar Radiation Research Laboratory

Longitude: 105.180°W
Latitude: 39.742°N
Elevation: 1828.8 meters AMSL
Time Zone: -7.0

Calibration date
06/22/2009 to 06/26/2009

Report Date
July 2, 2009

NOTICE

Broadband Outdoor Radiometer Calibration Report

BORCAL 2009-02 / University of Oregon

Introduction

This report compiles the calibration results from a Broadband Outdoor Radiometer Calibration (BORCAL). The work was accomplished at the Radiometer Calibration Facility shown on the front of this report. The calibration results reported here are traceable to the World Radiometric Reference and to the National Institute of Standards and Technology.

This report includes these sections:

- Calibration Environment - meteorological conditions and irradiance reference data encountered during the event.

- Control Instruments - a group of instruments included in each BORCAL event that provides a measure of process consistency.

- Results Summary - a table of all instruments included in this report summarizing their calibration results and uncertainty.

- Instrument Details - the calibration certificates and suggested methods of applying results for each instrument.

The BORCAL process is described in "Improved Methods for Broadband Outdoor Radiometer Calibration (BORCAL)," Wilcox, S., Andreas, A., Reda, I., and Myers, D., Proceedings of the ARM Science Team Meeting, St. Petersburg, Florida, April 2002.

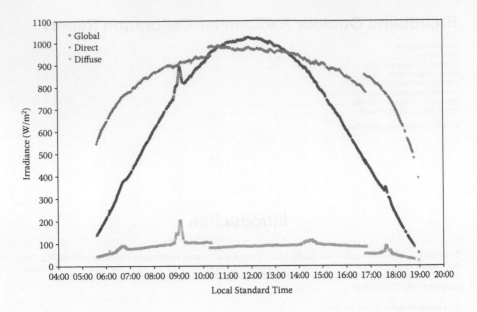

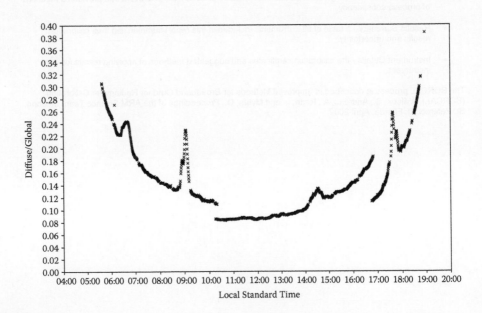

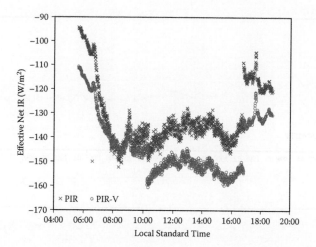

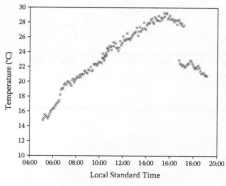

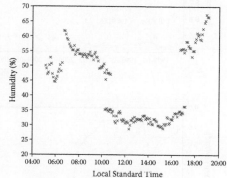

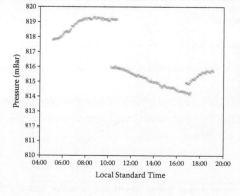

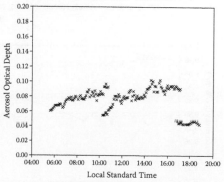

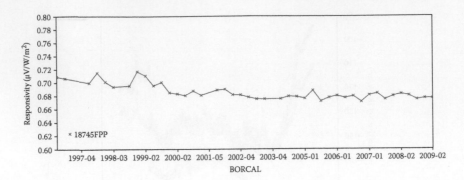

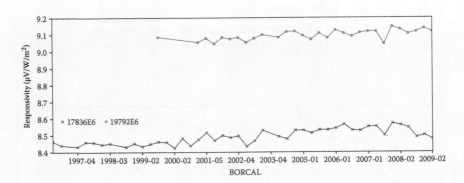

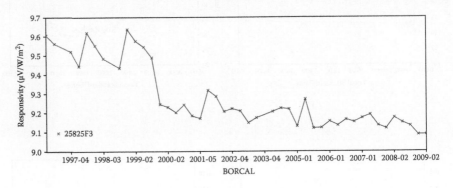

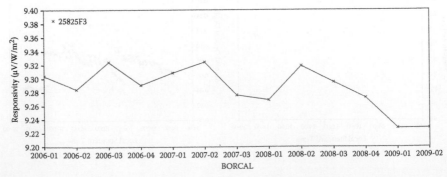

Results Summary

Table 2. Results Summary

Instrument	RS@45 [1] (µV/W/m²)	CF@45 [1] (W/m²/mV)	U95 (%)	RSc@45 [1][2] (µV/W/m²)	CFc@45 [1][2] (W/m²/mV)	U95 corr. [2] (%)	RS net [3] (µV/W/m²)	Page
13365F3	7.3961	135.21	+3.61 / -5.57	n/a	n/a	n/a	n/a	A1-2
17668E6	8.7043	114.89	+1.57 / -1.10	n/a	n/a	n/a	n/a	A1-7
PY22978	14.508	68.928	+1.86 / -1.26	n/a	n/a	n/a	n/a	A1-10

Note: Ancillary Data for BORCAL starts on page A1-15.

[1] CF = 1000 / Rs   [2] Effective Net IR Corrected
[3] Instrument's Effective Net IR Response

Appendix 1
Instrument Details

Calibration Certificates: 2 Pages for each Pyrheliometer/Shaded Pyranometer and 3 Pages for each Unshaded Pyranometer.
Suggested Methods: 1 Page for each Pyrheliometer/Shaded Pyranometer and 2 Pages for each Unshaded Pyranometer.
Ancillary Data for BORCAL: Last Page of a Calibration Certificate. Note: This appears only once, at the end of Appendix 1.

National Renewable Energy Laboratory
Solar Radiation Research Laboratory
Metrology Laboratory

Calibration Certificate

Test Instrument:	Precision Spectral Pyranometer	**Manufacturer:**	Eppley
Model:	PSP	**Serial Number:**	13365F3
Calibration Date:	6/26/2009	**Due Date:**	6/26/2010
Customer:	University of Oregon	**Calibration Site Parameters:**	see Ancillary Data

Environmental Conditions: Outdoors, under natural sunlight (see Ancillary Data)

Data Acquisition Dates: 6/22-23, 6/26

Table 1. Traceability

Measurement Type	Instrument	Calibration Date	Calibration Due Date
Beam Irradiance †	Eppley Absolute Cavity Radiometer Model HF, S/N 31104	09/27/2008	09/27/2009
Diffuse Irradiance †	Eppley Black and White Pyranometer Model 8-48, S/N 32858	03/28/2009	03/28/2010
Diffuse Irradiance †	Eppley Black and White Pyranometer Model 8-48, S/N 32871	03/28/2009	03/28/2010
Data Acquisition ‡	NREL Data Proof Scanner System Model RAP-DAQ, S/N 2005-998	04/01/2009	04/01/2010
Data Acquisition ‡	NREL Data Proof Scanner System Model RAP-DAQ, S/N 2005-999	04/01/2009	04/01/2010

† Traceable to the World Radiometric Reference
‡ Traceable to the National Institute of Standards and Technology

Number of pages of certificate: 4

Calibration Procedure: [1] Myers, D., Stoffel, T., Reda, I., Wilcox, S., and Andreas, A., 2002, "Recent Progress in Reducing the Uncertainty in and Improving Pyranometer Calibrations." Journal of Solar Energy Engineering, vol. 124, pp. 44-50. The American Society of Mechanical Engineers, Transactions of the ASME.
[2] "Improved Methods for Broadband Outdoor Radiometer Calibration (BORCAL)," Wilcox, S., Andreas, A., Reda, I., and Myers, D., Proceedings of the ARM Science Team Meeting, St. Petersburg, Florida, April 2002. Available upon request.

This calibration certificate applies only to the item identified above and shall not be reproduced other than in full, without specific written approval by the calibration facility. Calibration certificates without signatures are not valid.

Calibrated by: Afshin Andreas and Peter Gotseff

Certified by:

Afshin M. Andreas

Title: Scientist II

Date: _____

Quality Assured by:

Thomas Stoffel

Title: Principal Group Manager

Date: _____

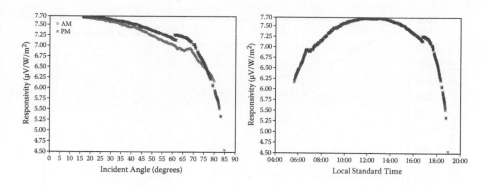

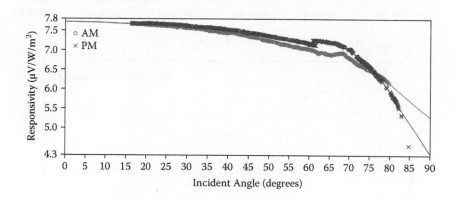

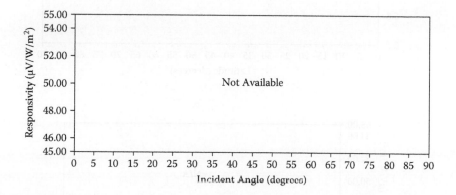

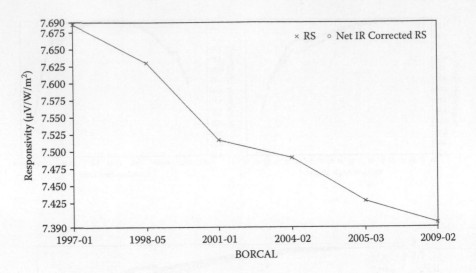

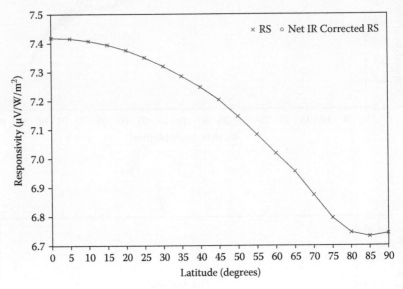

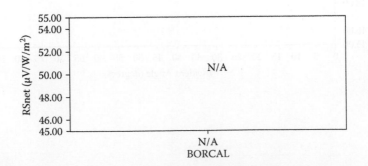

National Renewable Energy Laboratory
Solar Radiation Research Laboratory
Metrology Laboratory

Calibration Certificate

Test Instrument:	Normal Incidence Pyrheliometer	**Manufacturer:**	Eppley
Model:	NIP	**Serial Number:**	17668E6
Calibration Date:	6/26/2009	**Due Date:**	6/26/2010
Customer:	University of Oregon	**Calibration Site Parameters:**	see Ancillary Data

Environmental Conditions: Outdoors, under natural sunlight (see Ancillary Data)

Data Acquisition Dates: 6/22-23, 6/26

Table 1. Traceability

Measurement Type	Instrument	Calibration Date	Calibration Due Date
Beam Irradiance †	Eppley Absolute Cavity Radiometer Model HF, S/N 31104	09/27/2008	09/27/2009
Diffuse Irradiance †	Eppley Black and White Pyranometer Model 8-48, S/N 32858	03/28/2009	03/28/2010
Diffuse Irradiance †	Eppley Black and White Pyranometer Model 8-48, S/N 32871	03/28/2009	03/28/2010
Data Acquisition ‡	NREL Data Proof Scanner System Model RAP-DAQ, S/N 2005-998	04/01/2009	04/01/2010
Data Acquisition ‡	NREL Data Proof Scanner System Model RAP-DAQ, S/N 2005-999	04/01/2009	04/01/2010

† Traceable to the World Radiometric Reference
‡ Traceable to the National Institute of Standards and Technology

Number of pages of certificate: 3

Calibration Procedure: [1] Myers, D., Stoffel, T., Reda, I., Wilcox, S., and Andreas, A., 2002, "Recent Progress in Reducing the Uncertainty in and Improving Pyranometer Calibrations." Journal of Solar Energy Engineering, vol. 124, pp. 44-50. The American Society of Mechanical Engineers, Transactions of the ASME.
[2] "Improved Methods for Broadband Outdoor Radiometer Calibration (BORCAL)." Wilcox, S., Andreas, A., Reda, I., and Myers, D., Proceedings of the ARM Science Team Meeting, St. Petersburg, Florida, April 2002. Available upon request.

This calibration certificate applies only to the item identified above and shall not be reproduced other than in full, without specific written approval by the calibration facility. Calibration certificates without signatures are not valid.

Calibrated by: Afshin Andreas and Peter Gotseff

Certified by:

———————————————————————
Afshin M. Andreas

Title: Scientist II

Date: _____

Quality Assured by:

———————————————————————
Thomas Stoffel

Title: Principal Group Manager

Date: _____

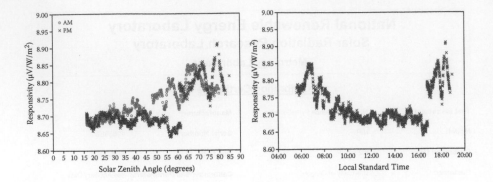

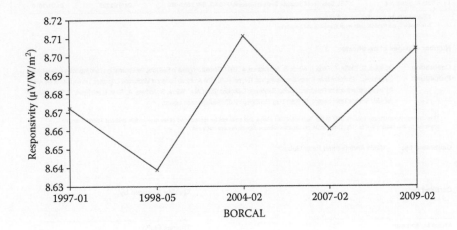

National Renewable Energy Laboratory
Solar Radiation Research Laboratory
Metrology Laboratory

Calibration Certificate

Test Instrument:	Silicon Pyranometer	**Manufacturer:**	Licor
Model:	LI200	**Serial Number:**	PY22978
Calibration Date:	6/26/2009	**Due Date:**	6/26/2010
Customer:	University of Oregon	**Calibration Site Parameters:**	see Ancillary Data

Environmental Conditions: Outdoors, under natural sunlight (see Ancillary Data)

Data Acquisition Dates: 6/22-23, 6/26

Table 1. Traceability

Measurement Type	Instrument	Calibration Date	Calibration Due Date
Beam Irradiance †	Eppley Absolute Cavity Radiometer Model HF, S/N 31104	09/27/2008	09/27/2009
Diffuse Irradiance †	Eppley Black and White Pyranometer Model 8-48, S/N 32858	03/28/2009	03/28/2010
Diffuse Irradiance †	Eppley Black and White Pyranometer Model 8-48, S/N 32871	03/28/2009	03/28/2010
Data Acquisition ‡	NREL Data Proof Scanner System Model RAP-DAQ, S/N 2005-998	04/01/2009	04/01/2010
Data Acquisition ‡	NREL Data Proof Scanner System Model RAP-DAQ, S/N 2005-999	04/01/2009	04/01/2010

† Traceable to the World Radiometric Reference
‡ Traceable to the National Institute of Standards and Technology

Number of pages of certificate: 4

Calibration Procedure: [1] Myers, D., Stoffel, T., Reda, I., Wilcox, S., and Andreas, A., 2002, "Recent Progress in Reducing the Uncertainty in and Improving Pyranometer Calibrations." Journal of Solar Energy Engineering, vol. 124, pp. 44-50. The American Society of Mechanical Engineers, Transactions of the ASME. [2] "Improved Methods for Broadband Outdoor Radiometer Calibration (BORCAL)," Wilcox, S., Andreas, A., Reda, I., and Myers, D., Proceedings of the ARM Science Team Meeting, St. Petersburg, Florida, April 2002. Available upon request.

This calibration certificate applies only to the item identified above and shall not be reproduced other than in full, without specific written approval by the calibration facility. Calibration certificates without signatures are not valid.

Calibrated by: Afshin Andreas and Peter Gotseff

Certified by:	**Quality Assured by:**
_____	_____
Afshin M. Andreas	Thomas Stoffel
Title: Scientist II	Title: Principal Group Manager
Date: _____	Date: _____

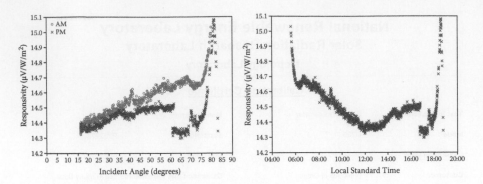

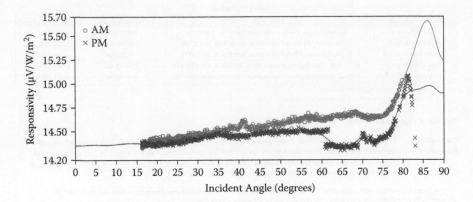

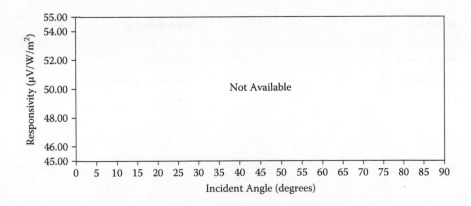

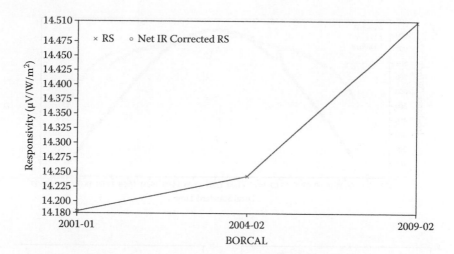

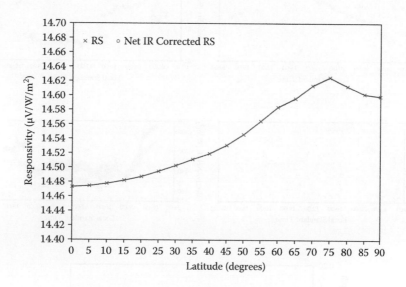

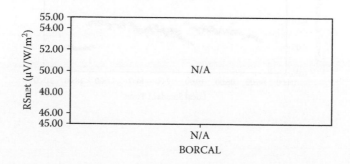

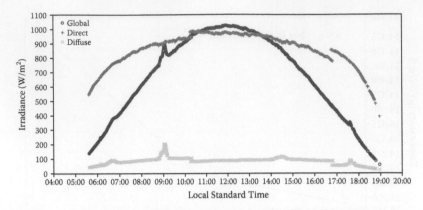

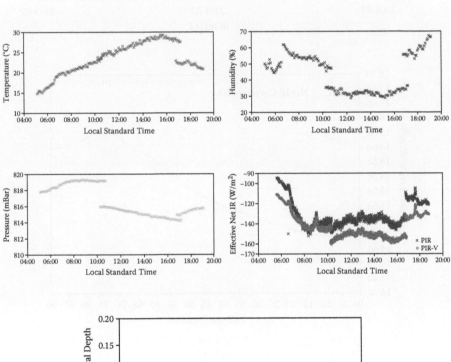

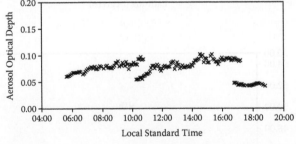

Appendix H: Sunshine Duration

The measurement of sunshine hours or percent sunshine was the typical "measurement" of irradiance until the last few decades. This measurement is still made routinely in many parts of the world. A number of ways to estimate sunshine duration are explained in the CIMO Guide (WMO, 2008).

The measurement has been used to estimate global horizontal irradiance using this equation from Ångström (1924):

$$GHI\big/GHI_0 = a + b \cdot S\big/S_0 \tag{H.1}$$

In this equation GHI_0 is the global irradiance at the top of the atmosphere, S_o is the total possible minutes of sunshine during the day, S is the sum of measured minutes of sunshine, and a and b are constants that are empirically derived for each site. These estimates were used when the use of pyranometers was not widespread, but they are of little use today where even low-cost pyranometers make better measurements than the estimates that are possible using Equation H.1. Today, sunshine duration measurements are most useful to cities and resorts that want to promote tourism.

Qualitatively, the sun is shining brightly whenever the shadow of an object can be discerned. The predominant measurement of sunshine worldwide has been the Campbell–Stokes sunshine recorder described in Section 3.3.2. It consists of a glass ball that focuses direct normal irradiance (DNI) on specially treated paper that is suppose to burn whenever the DNI exceeds 120 Wm^{-2}. This is WMO's (2008) definition of bright sunshine; however, the Campbell–Stokes burns over a range of conditions. For example, in a moist, cold environment, the irradiance will need to be higher before the burn occurs. Figure H.1 is the burn pattern from a Campbell–Stokes sunshine recorder. The handwritten numbers are estimates of the fraction of the hour (measured in tenths) during which there was bright sunshine.

The very best measurement of sunshine duration is made using a reasonably well-calibrated pyrheliometer. There is some discussion concerning the acceptable field of view that a pyrheliometer should have to make this measurement ($5°$ versus $5.7°$), but discrepancies from these minor differences are likely to be insignificant.

The next best estimate is made using two pyranometers of the same model and calibrated at the same time. If one is shaded and a shade-corrected value for diffuse derived, as discussed in Chapter 6, then sunshine minutes are tallied whenever

$$\frac{(GHI - DHI)}{\cos(sza)} \geq 120 \ Wm^{-2}$$

Akin to this idea, but inferior to it, is the Foster-Foskett sunshine recorder (Foster and Foskett, 1953) that was first used at airports by the U.S. National Weather Service

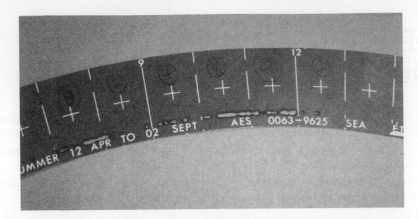

FIGURE H.1   (See color insert.) Northern hemisphere summertime Campbell–Stokes measurement of sunshine duration. Fractional hours with bright sunshine are estimated by an analyst and hand recorded.

(NWS) in 1953. The NWS appears to have decommissioned these instruments in 2009 (see, e.g., http://www.startribune.com/local/62760757.html, or search for "foster foskett switch"). This instrument used two matched selenium photocells with one shaded and one unshaded. The shade ring was adjusted four times a year; therefore, it was much broader than the typical fixed shade band described in chapter 6. It records a minute of sunshine when the difference exceeds a threshold setting that is manually adjustable on the instrument. Michalsky (1992) compares this instrument's measurement of sunshine minutes to a nearby measurement using a calibrated pyrheliometer. Although there was a reasonable correlation between the two instruments, the two measurements of sunshine differed significantly compared with two collocated pyrheliometers.

The burn method, employed in the Campbell–Stokes sunshine recorders, is inferior to the first two methods that use a pyrheliometer or two pyranometers, but, as stated earlier, it is still widely used outside the United States. Other methods less often used methods are described in WMO (2008).

New instruments have been developed that are probably superior to all methods except the pyrheliometer measurement. Two of these newer model sunshine recorders are the MS-093 from EKO Instruments Co., Ltd. and the CSD 3 from Kipp and Zonen.

REFERENCES

Ångström, A. 1924. Solar and terrestrial radiation. *Quarterly Journal of the Royal Meteorological Society* 50:121–126.

Foster, N. B. and L. W. Foskett. 1953. A photoelectric sunshine recorder. *Bulletin of the American Meteorological Society* 34:212–215.

Michalsky, J. J. 1992. Comparison of a National Weather Service Foster sunshine recorder and the World Meteorological Organization for sunshine duration. *Solar Energy* 48:133–141.

World Meteorological Organization (WMO). 2008. Measurement of radiation and sunshine. In *Guide to meteorological instruments and observing practices,* 7th ed., WMO-No. 8. Available at: http://www.wmo.int/pages/prog/www/IMOP/IMOP-home.html

Appendix I: Failure Modes

The following is a list of possible problems that have been compiled from past experiences, mainly from the Atmospheric Radiation Measurement (ARM) research facilities and the experiences from developing and operating a solar measurement station at our respective institutions.

Some key points before addressing possible equipment failures:

- *Proper maintenance* is the final step in *quality control* of measurements. After a measurement is recorded, only *data quality assessment* is possible.
- A situation with *no data* is often better than one that produces *bad data*. Unless bad data is identified and flagged, it can destroy confidence in and value of the set data.

The first section will cover observed problems, and possible causes are noted. Next the corrective maintenance actions will be listed. Because all ARM and UO solar monitoring network sites use Campbell data loggers for solar and infrared measurements, comments on data loggers will be specific for Campbell data loggers, but the gist applies to all data loggers.

I.1 OBSERVED PROBLEMS

I.1.1 No Signal

Possible Sources of Problem
1. An open circuit can be the result of a damaged signal cable, loose signal cable conductor connections at the data logger input terminal, improperly seated connector at the radiometer body, or broken thermopile winding (typically due to lightning strike).
2. The data logger system program may also have been changed or lost.
3. The data logger internal battery voltage may be below nominal limits.

I.1.2 Unstable Signal

Possible Sources of Problem
1. Loss of proper electrical grounding and shielding of the low-level direct current (DC) signal cable may cause erratic voltage readings.
2. The radiometers have a time-constant on the order of 1 sec to changes in irradiance (voltage) due to clouds. Electrical noise will generally have a higher frequency than this natural variation.
3. Data logger system has failed.
   a. The data logger system program may have been changed or lost
   b. The data logger internal battery voltage may be below nominal limits

4. Other possible causes include
   a. Moisture inside the radiometer case or signal cable connector
   b. An improperly seated connector at the radiometer body
   c. Loose signal cable connections at the data logger input terminal
   d. Cold solder joint inside the signal cable connector

I.1.3 SIGNAL GREATER THAN PHYSICAL LIMITS

Possible Sources of Problem

1. The GHI pyranometer can, with unique reflections from properly positioned towering cumulus clouds, exceed the solar constant for periods of less than tens of minutes.
2. Moisture inside the pyranometer dome or the pyrheliometer window can form small lenses that could increase the radiation on the radiometer detector.
3. Half-melted frost or snow on the dome can reflect sunlight onto the sensor receiver.
4. More often, high signals are the result of an open circuit causing the data logger to over range (6999), a ground loop introducing a spurious voltage, or an incorrect radiometer calibration factor or logger program.

I.1.4 SIGNAL LESS THAN PHYSICAL LIMITS

1. Signal has been shorted by damaged cable.
2. Corrosion at signal cable connections has caused increased resistance.
3. Wrong calibration factor or application associated with instrument.
4. Moisture inside the pyranometer dome or behind the pyrheliometer window.

I.1.5 GHI, DNI, AND DHI NOT INTERNALLY CONSISTENT

This is most easily determined from Equation 2.1. In addition to a bad radiometer calibration factors (Cf), possible causes are

1. DNI low
   a. Incorrect tracker alignment
   b. Dirty water, ice on NIP window
   c. Nearby obstructions (trees, poles, structures) that shade or sometimes reflect sunlight onto the detector
2. DNI high
   a. Suspect electrical problem first (see the section on signal greater than physical limit)
   b. Not many physical reasons for NIP to have prolonged *increased* output—can have temporary increase when sun is next to the edge of a cloud
   c. Contaminated NIP window, especially due to frost or dew, typically reduces reading, but may cause increased signal with specific solar geometry
3. GHI low
   a. Dirty/iced dome
   b. Possibly bad electrical connections

    c. Nearby obstructions, trees, poles, or structures shade the pyranometer

    d. Alignment problem—pyranometer tilted to north

4. GHI high

    a. If persistent, suspect electrical problem first

    b. Could be leveling problem, tilted toward sun

    c. Water droplets (dew) on dome can focus sun's rays

    d. Ice or frost on dome opposite the sun's location can reflect sun's rays onto detector

    e. Nearby reflective surfaces, such as buildings and poles

    f. Artificial lights (at night)

5. DHI low

    a. Dirty/iced dome

    b. Possibly bad electrical connections

    c. Correct PSP for thermal-offset problem—Model 8-48, B/W, will not exhibit this behavior

    d. Pyranometer not level, tilted to north

6. DHI high

    a. Suspect misaligned solar tracker (dome must be in full shade of ball)

    b. PSP and PIR not coplanar

    c. Pyranometer not level, tilted to south

    d. Possible ground loop or other electrical problems (see the section on no signal or unstable signal)

    e. Nearby reflective surfaces, such as buildings and poles

    f. Artificial lights (at night)

7. Upwelling GHI low

    a. Dirty/iced dome

    b. Possibly bad electrical connections

    c. Field of view of local terrain biased by dark surfaces—water pooling?

    d. Unlevel mounting platform

8. Upwelling GHI high

    a. Fresh snow cover can reflect up to 98% of the GHI

    b. Frost on dome can reflect more radiation onto pyranometer detector

    c. Possible ground loop or other electrical problems

    d. Field of view of local terrain biased by bright surfaces—water pooling?

    e. Unlevel mounting platform

I.1.6 Shortwave Elements (DNI, GHI, and DHI) Internally Consistent but WRONG (K-space)

1. Tracker failure (if pyrheliometer and shaded pyranometers are mounted on the same tracker, then DNI = 0 and DHI = GHI, where GHI is indicative of clear sky irradiance levels)

2. Tracker alignment problem?

3. Uniform soiling of optics

I.1.7 No Data Collected by Site Data System

1. Suspect communications failure
2. Modem failure
3. Loss of electrical power to data logging system

I.1.8 No Data Collected by Card Storage Module

1. Card not initialized at last site visit (card was full)
2. Card battery failure
3. Failed connection between logger and card storage module
4. Loss of electrical power to data logger system for more than 10 days

I.1.9 Asymmetric Diurnal Profiles (symmetry with solar noon, not clock noon)

1. Incorrect data logger time (clock drift? improper switch to daylight time?)
2. Incorrect time zone
3. Incorrect longitude
4. Assignment of hours (0–23 vs 1–24)
5. Instrument alignment (pyranometer not level?)

I.1.10 Bad Data Time Sequence

1. Data buffer overrun due to corrupt memory?

I.2 CORRECTIVE MAINTENANCE ACTION OPTIONS

Based on past experience and an understanding of the irradiance measurement processes, the following actions are suggested for the previous failure modes. The following information should complement the existing instructions for routine maintenance.

I.3 CORRECTIVE MAINTENANCE ACTIONS BY-THE-NUMBER:

I.3.1 Clean Optics

Note: Excessive dew, frost, snow, or dust on the PSP or PIR can indicate ventilator failure.

1. Contamination of optical surface
   a. Wash dome or window with distilled water, wipe dry
   b. Wash dome or window with alcohol, wipe dry
   c. Warm iced dome or window with palm of hand if necessary—do not scrape ice from dome or window
   d. Moisture inside the dome or window?
   e. Check PSP desiccant canister seal is tight
   f. Check PSP desiccant canister window is not cracked

    g. Remove PSP sunshade and check dome collar screws are tight

    h. Check NIP window screws are tight

    i. Change desiccant if granule colors are changed showing moisture saturation.

I.3.2 CHECK/ADJUST ORIENTATION

1. Align Solar Tracker (north/south and base is level)
2. Adjust shading balls to shade PSP and PIR domes
3. PSP/PIR is level using circular spirit level
4. Signal cables not caught on tracker or mounting fixture

I.3.3 EXAMINE INSTANTANEOUS DATA

1. For Campbell Scientific Data Loggers that use Loggernet program in monitor mode to view all data channels
    a. Compare with expected readings
    b. Open channel value = "6999"
    c. Note lightning can burn out delicate thermopile windings

I.3.4 CONFIRM VENTILATOR IS FUNCTIONING

1. Confirm electrical power available
2. Sun shade must be level with black detector or base of PIR dome
    a. The sun shade can shadow or reflect radiation onto detector if not properly positioned at or slightly below the base of the outer dome

I.3.5 CHECK SIGNAL CABLE AND CONNECTIONS

1. Confirm cable is not cut, stretched, or kinked
2. Check electrical connection at data logger wire panel and at radiometer connector
    a. Hot and cold cycling loosens some wire connections

I.3.6 CHECK DATA ACQUISITION SYSTEM GROUND CONNECTIONS

1. Check electrical connection at data logger, enclosure, and ground rod for corrosion or loose fittings

I.3.7 CONFIRM SERIAL NUMBERS WITH INSTRUMENT LOCATION

1. Visually inspect radiometer serial numbers at each mounting location and compare with data logger program locations (Monitor mode) or current inventory listing

I.3.8 CONFIRM CALIBRATION FACTORS WITH INSTRUMENT SERIAL NUMBER

1. Compare calibration sticker information with data logger program locations (Monitor mode) or current inventory listing

I.3.9   Data Logger Power

1. The red LED indicator on power supply should be illuminated
2. Check logger battery voltage is at least 9.2 Vdc using Monitor mode
3. Check the "on–off" switch is in the "on" position
   a. Note confusing label on CR10X-1M wiring panel—switch is pointing away from "on" label when in operating position

I.3.10   Data Logger Clock

1. Confirm CR10X-1M logger clock is within 3 sec of GMT

Index

Printed and bound by CPI Group (UK) Ltd, Croydon, CR0 4YY

18/10/2024

01776262-0016